T0137227

Advances in Industrial Control

Series Editors

Michael J. Grimble, Industrial Control Centre, University of Strathclyde, Glasgow, UK

Antonella Ferrara, Department of Electrical, Computer and Biomedical Engineering, University of Pavia, Pavia, Italy

Editorial Board

Graham Goodwin, School of Electrical Engineering and Computing, University of Newcastle, Callaghan, NSW, Australia

Thomas J. Harris, Department of Chemical Engineering, Queen's University, Kingston, ON, Canada

Tong Heng Lee, Department of Electrical and Computer Engineering, National University of Singapore, Singapore, Singapore

Om P. Malik, Schulich School of Engineering, University of Calgary, Calgary, AB, Canada

Kim-Fung Man, City University Hong Kong, Kowloon, Hong Kong

Gustaf Olsson, Department of Industrial Electrical Engineering and Automation, Lund Institute of Technology, Lund, Sweden

Asok Ray, Department of Mechanical Engineering, Pennsylvania State University, University Park, PA, USA

Sebastian Engell, Lehrstuhl für Systemdynamik und Prozessführung, Technische Universität Dortmund, Dortmund, Germany

Ikuo Yamamoto, Graduate School of Engineering, University of Nagasaki, Nagasaki, Japan

Advances in Industrial Control is a series of monographs and contributed titles focusing on the applications of advanced and novel control methods within applied settings. This series has worldwide distribution to engineers, researchers and libraries.

The series promotes the exchange of information between academia and industry, to which end the books all demonstrate some theoretical aspect of an advanced or new control method and show how it can be applied either in a pilot plant or in some real industrial situation. The books are distinguished by the combination of the type of theory used and the type of application exemplified. Note that "industrial" here has a very broad interpretation; it applies not merely to the processes employed in industrial plants but to systems such as avionics and automotive brakes and drivetrain. This series complements the theoretical and more mathematical approach of Communications and Control Engineering.

Indexed by SCOPUS and Engineering Index.

Proposals for this series, composed of a proposal form downloaded from this page, a draft Contents, at least two sample chapters and an author cv (with a synopsis of the whole project, if possible) can be submitted to either of the:

Series Editors

Professor **Michael J. Grimble**
Department of Electronic and Electrical Engineering, Royal College Building, 204 George Street, Glasgow G1 1XW, United Kingdom
e-mail: m.j.grimble@strath.ac.uk

Professor **Antonella Ferrara**
Department of Electrical, Computer and Biomedical Engineering, University of Pavia, Via Ferrata 1, 27100 Pavia, Italy
e-mail: antonella.ferrara@unipv.it

or the

In-house Editor

Mr. **Oliver Jackson**
Springer London, 4 Crinan Street, London, N1 9XW, United Kingdom
e-mail: oliver.jackson@springer.com

Proposals are peer-reviewed.

Publishing Ethics

Researchers should conduct their research from research proposal to publication in line with best practices and codes of conduct of relevant professional bodies and/or national and international regulatory bodies. For more details on individual ethics matters please see: https://www.springer.com/gp/authors-editors/journal-author/journal-author-helpdesk/publishing-ethics/14214

More information about this series at http://www.springer.com/series/1412

Jianglin Lan · Ronald J. Patton

Robust Integration of Model-Based Fault Estimation and Fault-Tolerant Control

Jianglin Lan
Aeronautical and Automotive Engineering
Loughborough University
Loughborough, UK

Ronald J. Patton
School of Engineering
and Computer Science
University of Hull
Hull, UK

ISSN 1430-9491 ISSN 2193-1577 (electronic)
Advances in Industrial Control
ISBN 978-3-030-58762-8 ISBN 978-3-030-58760-4 (eBook)
https://doi.org/10.1007/978-3-030-58760-4

Mathematics Subject Classification: 15A39, 15B48, 26A18, 37B25, 93-02, 93B52, 93C05, 93C35, 93D09

© The Editor(s) (if applicable) and The Author(s), under exclusive license to Springer Nature Switzerland AG 2021
This work is subject to copyright. All rights are solely and exclusively licensed by the Publisher, whether the whole or part of the material is concerned, specifically the rights of translation, reprinting, reuse of illustrations, recitation, broadcasting, reproduction on microfilms or in any other physical way, and transmission or information storage and retrieval, electronic adaptation, computer software, or by similar or dissimilar methodology now known or hereafter developed.
The use of general descriptive names, registered names, trademarks, service marks, etc. in this publication does not imply, even in the absence of a specific statement, that such names are exempt from the relevant protective laws and regulations and therefore free for general use.
The publisher, the authors and the editors are safe to assume that the advice and information in this book are believed to be true and accurate at the date of publication. Neither the publisher nor the authors or the editors give a warranty, expressed or implied, with respect to the material contained herein or for any errors or omissions that may have been made. The publisher remains neutral with regard to jurisdictional claims in published maps and institutional affiliations.

This Springer imprint is published by the registered company Springer Nature Switzerland AG
The registered company address is: Gewerbestrasse 11, 6330 Cham, Switzerland

To my parents, Songtian and Pingdi, and my wife, Xianxian.

Jianglin Lan

Preface

In real operations, engineering systems may be in face of system faults that lead to performance degradation, instability, or even trigger a chain of failing subsystems and cause major catastrophes in large-scale interconnected systems. Hence, there are strong demands for enhancing control system reliability and safety in the presence of system faults. To maintain robustly acceptable system performance, fault diagnosis function is embedded to detect, isolate and estimate the fault signals and a fault-tolerant control (FTC) policy is used to compensate for the fault effects. The inevitably existing system uncertainties result in the so-called *bi-directional robustness interactions* between the fault diagnosis and FTC functions. This gives rise to an important academic and industrial subject—robust integration of fault diagnosis and FTC. The aim is to guarantee the design effectiveness and closed-loop system stability when assembling the fault diagnosis and FTC functions together.

Albeit the tremendous development in theory and application of fault diagnosis and FTC up to date, the robust integration remains a challenging yet open question. This book presents a first systematic study of this subject. It covers the definition of basic concepts, development of integration strategies, and demonstration of industrial applications. The study lays a basis on model-based fault estimation (FE) and FTC. As a powerful alternative to the traditional fault detection and isolation (FDI) approach, the FE approach can obtain direct reconstruction of the fault shapes which can then be used for fault compensation. This makes it attractive to use FE-based FTC, rather than FDI-based FTC, to reach the true integration.

The book content is composed of three parts. The first part (Chaps. 1–2) starts with introduction of basic concepts of FE and FTC, followed by an extensive insight into the importance and challenges of robust integration of fault diagnosis and FTC system.

The second part (Chaps. 3–7) outlines the following five effective robust integration strategies for linear systems: sequential strategy, iterative strategy, simultaneous strategy, robust decoupling strategy and adaptive decoupling strategy. Tutorial examples are provided in each chapter to illustrate efficacy of the strategies presented. Although the theories are built on model-based FE and FTC, the ideas behind them can be applied to other approaches for fault diagnosis and FTC. Moreover, the strategies are applicable to a broad range of control problems, considering that the FE-based FTC naturally reverts to conventional observer-based control in the absence of faults.

The third part (Chaps. 8–10) includes extension of the proposed strategies to nonlinear and large-scale systems, and their applications to the important industrial areas of renewable energy, robotics and network systems.

Finally, Chap. 11 provides a summary of this book with perspectives on potential future research directions.

This book is intended to serve as a useful resource to researchers who work in the areas of fault diagnosis and FTC systems, both at universities and in the industry. It can also serve as supplementary material for a graduate or postgraduate levels course on fault diagnosis and FTC. We hope that the content of this book will attract more attention to this subject and inspire further developments in the integration strategies of complex systems.

The authors would like to express their thanks to Dr. Tim Scott and Dr. Ming Hou at University of Hull, and Prof. Christopher Edwards at University of Exeter, for valuable discussions on part of the materials.

Loughborough, UK Jianglin Lan
Hull, UK Ronald J. Patton
September 2019

Contents

Part II Strategies for Robust Integration of FE and FTC

Acronyms

Abbreviations

AFTC	Active fault-tolerant control
ALSO	Augmented Luenberger state observer
ASO	Augmented state observer
ASUIO	Augmented state unknown input observer
BMI	Bilinear matrix inequality
FDI	Fault detection and isolation
FE	Fault estimation
FTC	Fault-tolerant control
LMI	Linear matrix inequality
PFTC	Passive fault-tolerant control
SMC	Sliding mode control
SMO	Sliding mode observer
s.p.d.	Symmetric positive definite
s.t.	Subject to
UAV	Unmanned aerial vehicle
UIO	Unknown input observer

Symbols

$\mathbb{C}$	The set of all complex numbers
$\mathbb{C}_+$	The set of all positive complex numbers
$\mathbb{R}$	The set of all real numbers
$\mathbb{R}^n$	The set of n-dimensional real vectors
$\mathrm{trace}(X)$	Sum of the elements on the main diagonal of matrix X
X^{-1}	The inverse of matrix X
$\mathrm{X}^{\dagger}$	The pseudo inverse of matrix X

$X^{\top}$	The transpose of matrix X
$\star$	It induces symmetric in a block matrix
$\mathrm{He}(X)$	The sum of a matrix X and its transpose, $\mathrm{He}(X) = X + X^{\mathrm{T}}$
$\mathrm{rank}(X)$	The rank of matrix X
$\mathrm{Re}(\lambda)$	The real component of the eigenvalue λ
$\lambda_{\min}(X)$, $\lambda_{\max}(X)$	The minimum, maximum eigenvalue of matrix X
$\succ$ $(\succeq)$	Positive definite (semi-definite)
$A \Longleftrightarrow B$	A is equivalent to (if and only if, iff) B
$\lvert\cdot\rvert$	The absolute value of a scalar
$\lvert\lvert\cdot\rvert\rvert_p$	The p-norm in the Euclidean space. $\lvert\lvert\cdot\rvert\rvert$ and $\lvert\lvert\cdot\rvert\rvert_\infty$ represent the 2-norm and $\infty-$norm, respectively
I_p	A $p \times p$ identity matrix
$\kappa_{m\times n}$	A $m \times n$ matrix with elements all equal to a constant κ
G_{zd}	The transfer function from signal d to signal z
$\mathrm{sat}(\cdot)$	The saturation function
$\mathrm{sign}(\omega)$	The signum function of the variable ω. If $\omega \neq 0$, $\mathrm{sign}(\omega) = \frac{\omega}{\lvert\lvert\omega\rvert\rvert}$; If $\omega = 0$, $\mathrm{sign}(\omega) = 0$
$\mathrm{diag}(X_1, X_2)$	A diagonal matrix $\begin{bmatrix} X_1 & 0 \\ 0 & X_2 \end{bmatrix}$

Part I
Introduction

Chapter 1
Introduction

1.1 Background

Engineering systems may have system faults in real operations, which leads to performance degradation, system instability or even trigger a chain of failing subsystems and cause major catastrophes in large-scale interconnected systems. This gives rise to strong requirements on enhancing control system reliability and safety in the presence of system faults. It is crucial to not only determine the onset and development of faults before they become serious, but also adaptively compensate the fault effects within the closed-loop system or replace faulty components by fault-free alternatives (hardware redundancy). The procedure of accounting for faults acting within a control system to render the closed-loop system insensitive to the faults is known as "fault-tolerant control (FTC)", of which the fault estimation and compensation control is one approach.

In 1985, Eterno et al. (1985) developed a reconfigurable flight control system, in which the terminology "failure tolerant control" was first used to define the meaning of control system tolerance to failures or faults. The word "failure" is used when a fault leads to the situation that the system function concerned fails to operate (Isermann 2006). FTC began to develop in the early 1990s and for the last 30 years a significant number of methods for diagnosing and accommodating faults have been established. The results have been summarized in many survey papers, e.g. Amin and Hasan (2019), Fritz and Zhang (2018), Hwang et al. (2009), Yang et al. (2019), Yu and Jiang (2015), Zhang and Jiang (2008). The technical details are provided in many books and monographs, e.g. Alwi et al. (2011), Blanke et al. (2006), Chen and Patton (1999), Clark et al. (1989), Ding (2014), Escobet et al. (2019), Isermann (2006), Jain et al. (2018), Noura et al. (2009), Patan (2019), Richter (2011), Shen et al. (2017), Witczak (2014), Yang et al. (2010), Zhang et al. (2013). Until now, FTC has been applied to many industrial areas including, but are not limited to, UAVs (Meskin and Khorasani 2011), aircrafts (Ducard 2009; Edwards et al. 2010), vehicles (Stetter 2020), trains (Chen and Jiang 2019), power systems (Boem et al.

© The Author(s), under exclusive license to Springer Nature Switzerland AG 2021

J. Lan and R. J. Patton, *Robust Integration of Model-Based Fault Estimation and Fault-Tolerant Control*, Advances in Industrial Control,
https://doi.org/10.1007/978-3-030-58760-4_1

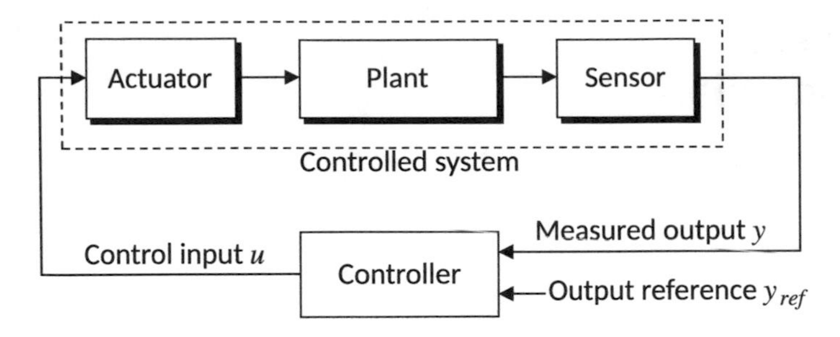

Fig. 1.1 A generic control system

2019; Simani and Farsoni 2018), multi-agent systems (Potiron et al. 2013) and water networks (Quevedo et al. 2010).

1.2 Basic Concepts in FTC System

The structure of a generic closed-loop automatic control system is outlined in Fig. 1.1. It consists of two main parts: the controlled system and the controller. The former includes actuator(s) to actuate the plant based on the applied control input signal $u(t)$, and sensor(s) to measure the plant output $y(t)$. The latter generates the appropriate $u(t)$ based on the $y(t)$ to make $y(t)$ follow the reference signal y_{ref}. The controller can be designed to achieve good output tracking using well-established control approaches, e.g. PID, robust control, adaptive control, etc. However, even the controller is well-tuned, its performance may not be good as expected in real implementation due to the existing of possible faults.

Definition 1.1 (*Fault* (van Schrick 1997)) A fault is defined as an unpermitted deviation of at least one characteristic property or parameter of the system from the acceptable condition.

It is important to determine how a fault should be detected, isolated, estimated and compensated, because there are different types of faults, as outlined in Fig. 1.2, acting at different system locations (Blanke et al. 2003; Chen and Patton 1999; Isermann 2006).

Definition 1.2 (*Actuator fault*) An actuator fault ($f_a(t)$) is defined as a variation of the control input $u(t)$ applied to the controlled plant.

Remark 1.1 The following two types of actuator faults are usually discussed: (1) Loss of actuator effectiveness fault. This may be caused by breakage, burn out of wiring or stuck at a position. In the presence of total loss of effectiveness, an actuator can no longer produce any actuation regardless of the applied input. Hence, it is out of the scope of this book. A partial loss of actuator effectiveness means that

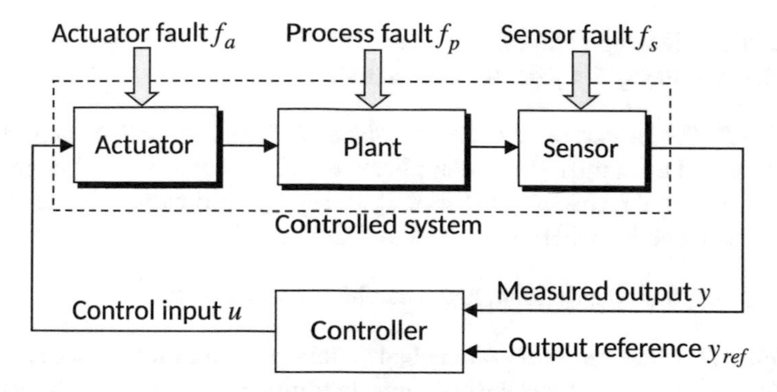

Fig. 1.2 A control system with actuator, process and sensor faults

the actuator becomes less effective, e.g. has degradation in the actuator gain due to a clogged or rusty valve. In such a case, the faults may be compensated by adjusting the actuator action, which will be discussed in Chaps. 5 and 8. (2) Offset actuator fault. It corresponds to a deviation of the actuator action from its nominal situation, due to some parameter changes or unknown disturbances, e.g. oscillatory or drift faults in flight control systems (Goupil 2010). Systems subject to offset actuator faults will be discussed in Chaps. 2–10.

Definition 1.3 (*Sensor fault*) A sensor fault ($f_s(t)$) is defined as a variation of the true measurement taken from the system output ($y(t)$).

Remark 1.2 Sensor faults may be caused by poor calibration, bias, scaling error or sensor dynamic change. Systems subject to bias sensor faults will be discussed in Chaps. 5, 7 and 10.

Definition 1.4 (*Process fault*) A process fault ($f_p(t)$), or called component fault, is a variation from the system structure or parameters used during system modelling.

Remark 1.3 Process faults represent a wide class of faults, e.g. change of mass, damping constant, aerodynamic coefficients, etc. They directly affect the physical system parameters and subsequently the system input and output properties. Systems subject to process faults will be discussed in Chap. 8.

Considering the ways in which they are modelled, faults can be classified as additive or multiplicative faults:

Definition 1.5 (*Additive fault*) An additive fault is defined as a fault that affects the system signal by adding an extra signal to it.

Remark 1.4 Offset actuator and sensor faults can be considered as additive faults. Additive faults are studied extensively in the literature and will be discussed throughout this book.

Definition 1.6 (*Multiplicative fault*) A multiplicative fault is defined as a fault that affects the system signal by multiplying an extra signal.

Remark 1.5 Partial loss of actuator effectiveness fault is a form of multiplicative fault. Compared to additive fault, multiplicative fault has been studied in only a few references, due to the challenges in its diagnosing and estimating. Systems subject to multiplicative faults will be discussed in Chap. 5.

Considering their distribution, faults can be classified as follows:

Definition 1.7 (*Matched fault*) A matched fault is inside the range space spanned by the control input, i.e. satisfying the matching condition $\mathrm{rank}[B\ F] = \mathrm{rank}(B)$, where B and F are the distribution matrices of the control input and fault, respectively.

Remark 1.6 If a fault is matched, then it can be directly compensated through control actions. Therefore, the matching condition is one of the fundamental assumptions and prerequisites for realizing active fault compensation. However, it will be shown in Chap. 6 that unmatched faults may also be compensated through appropriate control design.

Definition 1.8 (*Unmatched fault*) A unmatched fault is outside the range space spanned by the control input, i.e. $\mathrm{rank}[B\ F] \neq \mathrm{rank}(B)$.

In this book, faults are also divided into *differentiable* and *non-differentiable* faults, based on whether they are differentiable or not with respect to time. Some examples for non-differentiable faults are (1) random jumps due to environmental changes or system component failures (Willsky and Jones 1974), and (2) random faults widely existing in networked control systems as a result of the randomly occurring phenomena (Dong et al. 2013).

The above different fault classifications are not independent but have some overlaps, e.g. a certain fault can be viewed as more than one types of faults. Throughout this book, different classifications will be discussed under specific scenarios.

Definition 1.9 (*Fault-tolerant control*) Fault-tolerant control (FTC) is a control strategy and design to ensure that a closed-loop system can continue acceptable operation in the presence of either single or multiple fault actions. When prescribed stability and closed-loop performance metrics are maintained despite the presence of faults, the system is said to be "fault-tolerant" and the control scheme that ensures the fault tolerance is the fault-tolerant controller.

For an FTC system, another important concept is fault diagnosis.

Definition 1.10 (*Fault diagnosis*) Fault diagnosis is defined as a procedure to obtain fault information (location, time occurrence and/or magnitude) used for fault compensation design and the scheduled system maintenance.

1.3 FTC Design

The existing FTC methods can be classified according to whether they are "passive" or "active", using fixed or reconfigurable control strategies (Eterno et al. 1985). The general system schemes of *active* and *passive* FTC (AFTC and PFTC) methods are outlined in Fig. 1.3, where a distinction is made between the "execution" and "supervision" levels. The essential differences and requirements of AFTC and PFTC are also illustrated in the figure.

The PFTC is based solely on the use of robust control in which potential faults are treated as uncertain signals (uncertainties or external disturbances) acting on the system dynamics. This can be related to the concept of reliable control (Veillette et al. 1992). PFTC does not require either online fault information from the fault diagnosis function or control reconfiguration. Several PFTC methods have been developed based on robust control theories, e.g. multi-objective optimization, quantitative feedback theory, H_∞ optimization, absolute stability theory, nonlinear regulation theory, etc. More details for this can be found in the survey papers Benosman (2009), Yu and Jiang (2015). Since a PFTC system uses a controller designed offline based on certain a priori knowledge of the faults, it is able to handle very limited fault scenarios.

The AFTC provides a system with fault-tolerant capability by including two conceptual steps:

(1) Equip the system with a diagnosis mechanism to diagnose the faults and select the required remedial action to maintain acceptable post-fault closed-loop performance. In the absence of faults, a "Baseline controller" is used to ensure good stability and tracking performances. (Supervision level)

(2) Make use of the supervision level information and adapt or restructure the controller to achieve the required remedial activity. (Execution level)

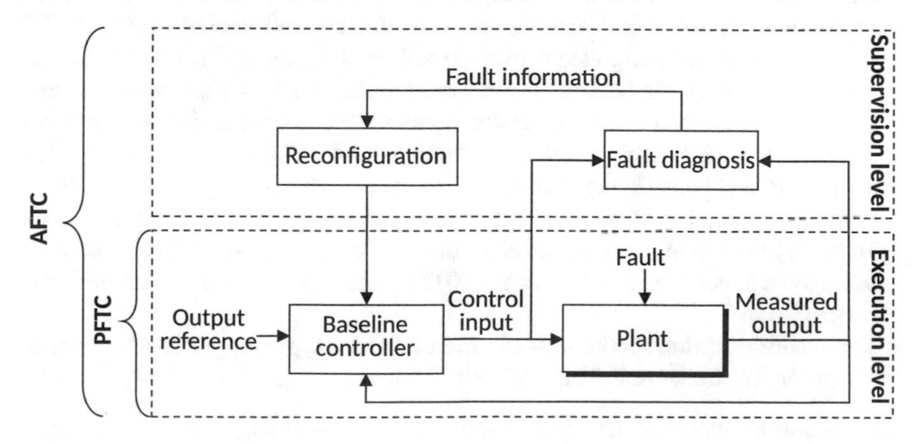

Fig. 1.3 General FTC system architectures

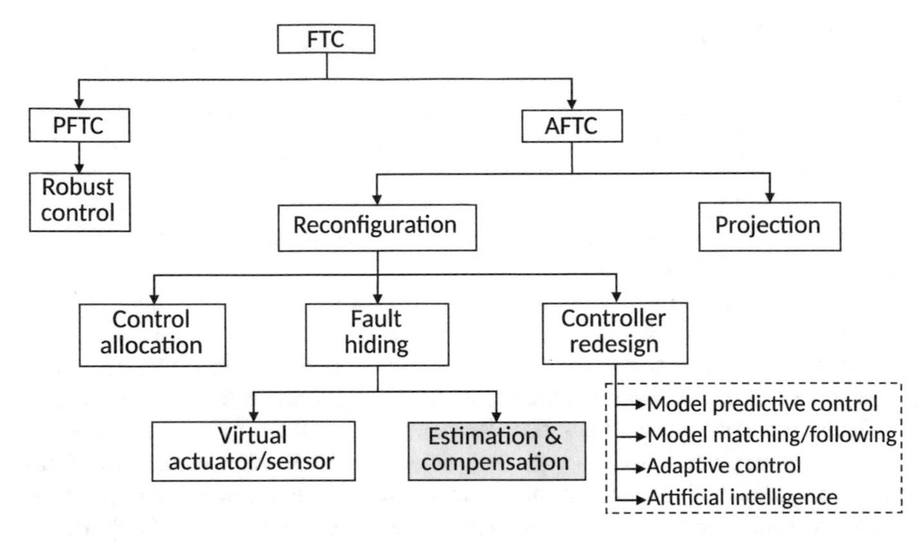

Fig. 1.4 A classification of FTC methods

Compared with PFTC, AFTC is applicable for a broader range of areas and thus has been the major concern of the FTC community, which is also the main focus of this book.

A classification of the existing PFTC and AFTC methods is given in Fig. 1.4.

As shown in the figure, there are two AFTC methods: projection and reconfiguration methods.

- The **projection method** diagnoses the fault occurrence through a fault diagnosis unit and compensates the fault effect using a switching mechanism to select an appropriate control action from the pre-computed controller set. A typical AFTC system architecture using the projection method is shown in Fig. 1.3. The potential fault modes of the controlled plant are known in advance and an associated controller is designed to achieve desired system performance under each fault situation. In the spirit of this method, the multiple model FTC has been developed in the literature (Maybeck and Stevens 1991; Rauch 1995; Zhang and Jiang 2001). A design example will be provided in Sect. 2.3 to illustrate its basic principle. To avoid the need of knowing the plant model, data-driven projection-based FTC has also been developed in Jain et al. (2018) from a behavioural system theoretic perspective.
- The **reconfiguration method** mainly includes three types of approaches: control allocation, controller redesign, and fault hiding.

The control allocation approach re-allocates the required control actions from the faulty actuators to the healthy redundant ones, according to the fault diagnosis results (Buffington et al. 1999). It is an approach for actuator redundancy management and useful for over-actuated control systems, such as flight systems. However, the

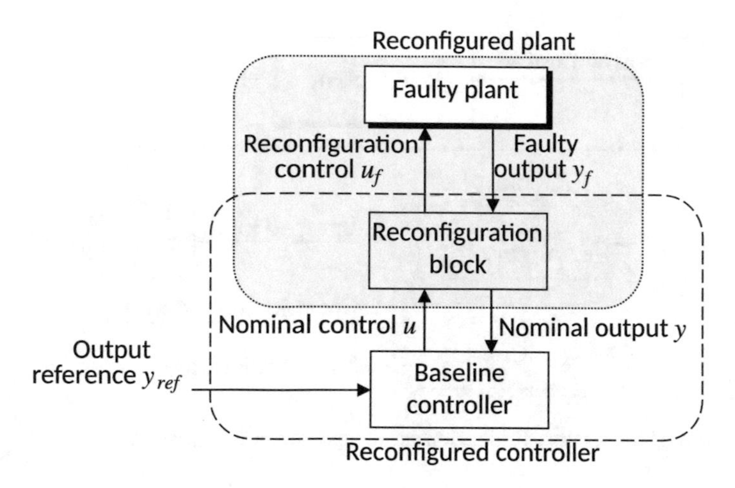

Fig. 1.5 General framework of fault hiding FTC systems

requirement of physical redundancy makes this approach expensive and somehow limited in application.

The controller redesign approach involves the calculation of new controller parameters following control impairment using, e.g. model predictive control (Maciejowski 1999), model matching or following (Staroswiecki 2005), adaptive control (Shen et al. 2017), and artificial intelligence (Patan 2019; Perk et al. 2012), with or without fault diagnosis functions. The use of artificial intelligence offers an opportunity to perform fault diagnosis and FTC for complex industrial plants.

The fault hiding approach has the general framework depicted in Fig. 1.5 and it aims to "hide" the fault from the baseline controller by adding an extra reconfiguration block between the faulty plant and the baseline controller. The reconfiguration block generates the reconfiguration control input u_f to be applied to the faulty plant, by using the measured faulty output y_f and the nominal control input u.

Currently, two mainstream methods have been proposed to achieve fault hiding: virtual actuator/sensor method, and estimation and compensation method.

The virtual actuator/sensor method for FTC design has been treated extensively in the monograph Steffen (2005). It includes three separate design steps: (1) Design a baseline controller for the healthy plant; (2) Design a fault diagnosis block to diagnose the faults and determine the system dynamics of the faulty plant; (3) Design a reconfiguration control signal such that the faulty plant behaves like the original healthy one.

Similarly, the estimation and compensation method uses online fault compensation based on diagnosis of the unanticipated faults and its general framework is sketched in Fig. 1.6. However, compared to the virtual actuator/sensor method, the estimation and compensation method has the following new features: (1) The FTC controller consists of a baseline control component and a fault compensator, which are designed together; (2) The fault estimation (FE) function is embedded with the

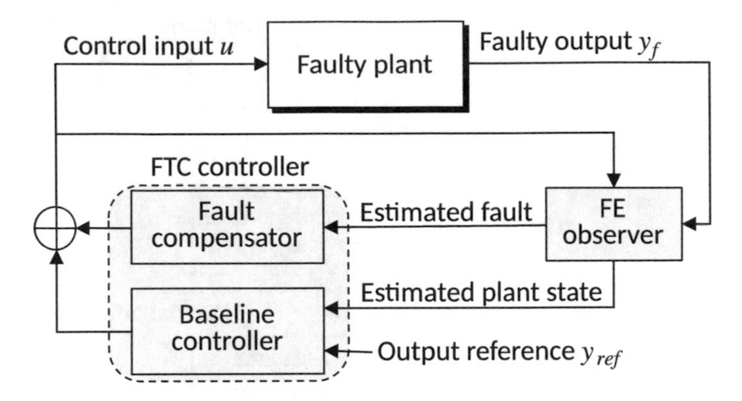

Fig. 1.6 General framework of estimation and compensation FTC systems

controller, automatically estimating the actual fault signals and forwarding the estimates to the FTC controller. The FTC controller can be designed using classical control methodologies such as adaptive control, robust control and sliding mode control (SMC).

Compared with the other AFTC methods, the estimation and compensation method is considered to be a more realistic and robust AFTC strategy. It also offers an opportunity to avoid the control and diagnosis uncertainties and time delays brought by the multiple-step designs in the virtual actuator/sensor method. Considering the background above, **this book focuses on the estimation and compensation method for AFTC systems design**.

1.4 FE Design

In the past three decades, fault detection and isolation (FDI), also called fault detection and diagnosis (FDD), and fault estimation (FE) have been developed as mainstream approaches to realize fault diagnosis.

The current FDI methods mainly make use of the basic principle of residual generation and can be divided into two categories: mode-based and data-driven methods. The former depends on the mathematical system model and uses techniques such as state observer, parity space and parameter estimation. The latter is based on historical data of the systems and uses techniques such as statistics, neural networks and expert systems. A comprehensive review of FDI methods is referred to Gao et al. (2015a, b).

It is attractive to directly reconstruct the fault signal once it occurs through FE. The FDI approach for diagnosing faults involves standard procedures of fault detection and fault isolation, but FE is used to estimate the actual fault signals based on system observer methodologies. Since FDI does not provide fault estimation, some investigators use FDI followed by FE to estimate the faults. It is also important to note that the FE procedure actually includes both detection and isolation in some sense, because the accurate estimation of the faults implies detection and isolation. Hence, FE can totally replace FDI in some situations. For example, **in this book FE totally replaces the FDI functions in the context of fault estimation combined with fault compensation within FTC**.

Several approaches to FE designs have been proposed and the majority treat the faults as auxiliary state. This idea lies on the assumption that the faults are differentiable and it enables casting the FE as a state estimation problem. The FE can then realized by using state observers, e.g. adaptive observer (Kabore and Wang 2001), augmented state observer (ASO) (Gao and Ding 2007), unknown input observer (UIO) (Gao et al. 2016), system-inversion-based Luenberger observer (Kulcsár and Verhaegen 2011) and moving horizon estimator (Feng and Patton 2014). The combination of ASO and zonotopic techniques can also estimate the upper and lower bounds of the fault (Zhang et al. 2020). To avoid the differentiability assumption, techniques such as sliding mode observer (SMO) can be adopted (Edwards et al. 2000). In this book, several new FE observers will be presented, based on further extension and development of ASO, UIO and SMO.

1.5 Robust Integration of FE and FTC

In most situations, the models of engineering systems used in control design are imperfect and have uncertainties. It has long been known that system uncertainty has negative effects on the control performance. Moreover, since both uncertainty and faults can lead to system dynamic changes, it is usually difficult or impossible to distinguish between their effects. Therefore, the two key challenges in AFTC system design are

- *How to extract required fault information from system dynamic changes in the presence of uncertainty?*
- *How to design a closed-loop AFTC system with admissible fault-tolerant performance and good robustness to uncertainty?*

The first challenge is fundamental to an AFTC system where accurate fault information is a prerequisite. Since fault diagnosis depends on the mathematical system model explicitly, within a closed-loop system the fault diagnosis performance is affected by the control system uncertainty (Patton 1997). Initiated by Nett et al. (1988), many works have been published on the integration of control and FDI, where their designs are combined as a joint robustness problem to achieve good robust control and acceptable FDI properties. More details are referred to the review paper Ding (2009).

When combining the functions of fault diagnosis with FTC into an AFTC system, the diagnosis uncertainties (false alarm, time delay, diagnosis error, etc.) also affect the closed-loop system performance. Therefore, there exist mutual interactions between the fault diagnosis and FTC system functions. If these functions are designed separately without considering the coupling effects, they may not fit with each other when assembled together. This will then lead to an FTC system with degraded performance and robustness. Therefore, a necessary consideration is to synthesize the fault diagnosis and FTC functions from a holistic perspective to achieve a robust closed-loop FTC system. This is the problem of robust integration of fault diagnosis and FTC defined in this book as follows:

Definition 1.11 **Robust integration of fault diagnosis and FTC** is a system synthesis process for co-design of the fault diagnosis and FTC functions, by taking into account their mutual interactions, to establish a robust closed-loop FTC system with admissible performance.

Detailed mathematical analysis will be provided in Chap. 2 to discuss the necessity, importance, and challenges of robust integration in the context of FDI and control, FDI and FTC, and, more importantly, FE and FTC. It is worth noting that the focus of this book is on model-based FE and FTC. This scheme is developed as an effective alternative to achieve robust integration of fault diagnosis and FTC. However, it does not exclude the existence of other possible alternatives.

1.6 Book Organization

This book presents novel strategies to FE and FTC and their integration. The fundamental contribution is that the presence of uncertainty in state and fault estimation along with system uncertainty leads to a new concept called **robust integration of FE and FTC**. The book describes several robust integration strategies and their applications in the fields of renewable energy, robotics and network systems. Throughout the book, simulation results are performed on MATLAB and SIMULINK, combined with YALMIP (Löfberg 2004), LMI control toolbox (Gahinet et al. 1995) and MOSEK (Mosek 2018). A graphical view of the book organization is provided in Fig. 1.7. The book consists of 11 chapters which are grouped into 3 parts: introduction, robust integration strategies, and extension and application. Descriptions of them are given in Sects. 1.6.1, 1.6.2 and 1.6.3, respectively. The book content is finally briefly summarized in Chap. 11 along with some future perspectives. The arrows in Fig. 1.7 suggest different paths to read the content.

1.6.1 Part I: Introduction

This part provides necessary background of the book, including the following two chapters:

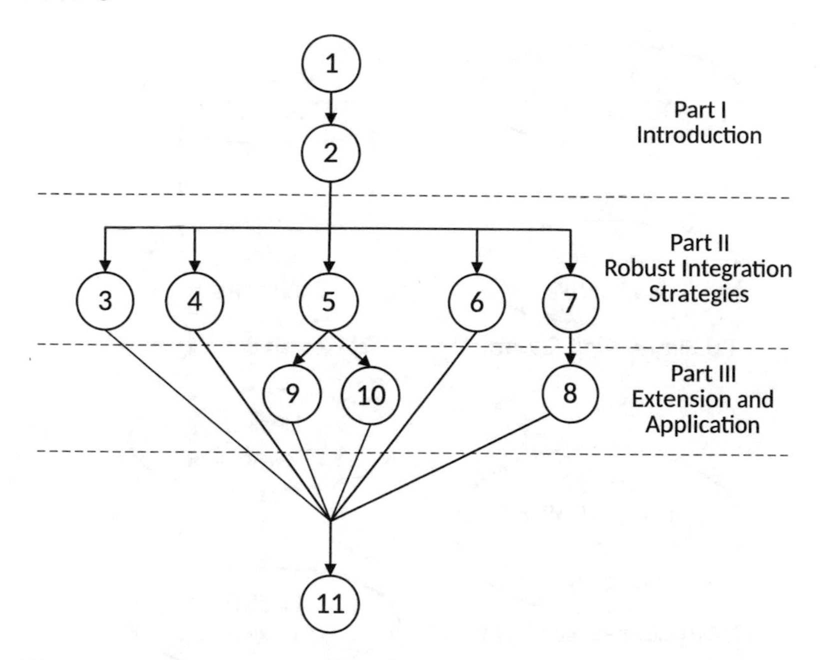

Fig. 1.7 Graphical view of the book organization

Chapter 1 introduces the related definitions, architectures and classification of fault diagnosis and FTC systems. Some lemmas used frequently throughout the book, together with notes on the Separation Principle and the unknown input observer, are outlined.

Chapter 2 discusses the role of robust integration in fault diagnosis and FTC systems. As a result of imperfect system modelling there inevitably exist uncertainties in the mathematical system models that used for control designs. The system uncertainty affects the fault diagnosis accuracy and leads to the existence of *unidirectional robustness interaction* between the control system and fault diagnosis function. The use of fault diagnosis results as feedback to actively adjust the controller to accommodate the fault effects builds an AFTC system. Since the diagnosis and control are in a closed-loop, the control system uncertainty affects the diagnosis and in turn the diagnosis uncertainty affects the control system, which leads to the existence of *bidirectional robustness interactions*. This happens in FTC systems using either FDI or FE. The existence of unidirectional or bidirectional robustness interactions gives rise to the necessity of integrating the designs of fault diagnosis and control, FDI and FTC, or FE and FTC, aiming to obtain good robustness in fault diagnosis and FTC and their functional combination. A motivating example is provided to illustrate the necessity of robust integration of FE and FTC for stable closed-loop FTC system.

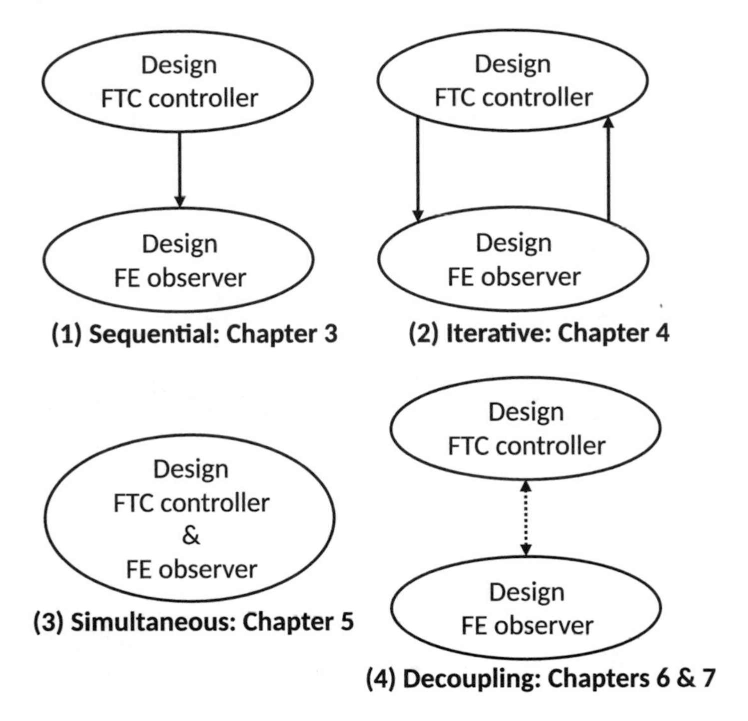

Fig. 1.8 Strategies for robust integration of FE and FTC

1.6.2 *Part II: Strategies for Robust Integration of FE and FTC*

This part describes five strategies (see Fig. 1.8) to achieve robust integration of FE and FTC. For simplicity and clarity, development of the strategies is based on linear systems.

Chapter 3 presents a sequential strategy for robust integration of FE and FTC. An augmented Luenberger state observer (ALSO) is used to estimate the state and faults, and a state feedback FTC controller is used to compensate the faults and stabilize the system. Under this scheme, the FTC controller is designed in the first step and then used in the second step to determine the FE observer. The key difference between this sequential strategy and the separated strategy is that effects of system uncertainty on the FE observer are taken into account by the former (in the second step) but ignored by the latter. The sequential strategy can reduce the design complexity and achieve robustly acceptable, though suboptimal, FTC closed-loop system performance. However, it is worth noting that the effects of FE uncertainty on the FTC control system cannot be handled within this scheme.

Chapter 4 presents an iterative integration strategy, using an augmented state unknown input observer (ASUIO) and a state feedback FTC controller. The FTC controller and FE observer gains are designed in an iterative loop. It starts with an initial FTC controller design, which is then used to determine the FE observer gains. After that, by implementing the FE observer gains the FTC controller is refined to improve the overall system robustness. The iteration continues until the robust performance index reaches its prescribed accuracy. Different from the sequential strategy in Chap. 3, *bidirectional robustness interactions* between the FE observer and FTC controller are taken into account in this strategy through the iteration manner. The iterative procedure has finite convergence, and, like the sequential strategy, can only obtain suboptimal solutions.

Chapter 5 presents a simultaneous integration strategy. The basic idea is to formulate the robust integration of FE and FTC as an observer-based robust control problem and solve it using a single-step linear matrix inequality (LMI) formulation. A class of uncertain linear systems with both actuator and sensor faults are considered, while only actuator faults are considered in Chaps. 3 and 4. Both the cases of state and output feedback controls are studied, using reduced-/full-order ASUIOs, respectively, and adaptive sliding mode FTC controllers. The multiplicative fault case is also discussed, which is not covered by Chaps. 3 and 4.

Chapter 6 presents a robust decoupling strategy for integration of FE and FTC to avoid the bilinear matrix inequality (BMI) issue encountered in Chap. 5 and to have more design freedom. It uses an ALSO FE observer and a state feedback FTC controller that are similar to those used in Chap. 3. The keyword "robust" implies that the *bidirectional robustness interactions* are attenuated using robust control theory, which then approximately recovers the Separation Principle in the spirit of Small Gain Theorem. This is why the dashed lines are used in the conceptual diagram (4) in Fig. 1.7. The keyword "decoupling" means that designs of the FE observer and FTC controller are carried out in two phases, rather than simultaneously as in Chap. 5. In the first phase, the controller is designed to minimize the effects of external disturbance and system uncertainty on the closed-loop state. The estimation error effect is also minimized by maximizing a weighting matrix on the state. In the second phase, the observer is designed to minimize the effects of external disturbance, fault modelling error, and system uncertainty, based on the obtained controller gain and weighting matrix. To improve attenuation of the coupling, an iterative algorithm is developed to minimize the coupling effects simultaneously. The iterative algorithm also realizes a balance of the robustness against external disturbance and the coupling.

Chapter 7 presents an adaptive decoupling strategy for robust integration of FE and FTC. The decoupling recovers the Separation Principle for the designs of FE and FTC functions using adaptation. The word "adaptation" is used to emphasize the way that the *bidirectional robustness interactions* are handled within this scheme: (1) The FE observer actively estimates the perturbation (including uncertainty and/or disturbance) together with the system state and faults; (2) The effects of estimation errors and perturbation are adaptively estimated and compensated together with the faults using the FTC controller action. This is why the dashed lines are used in the conceptual diagram (4) in Fig. 1.7. Moreover, the adaptive compensation makes the proposed decoupling an "active" strategy, in contrast to the "passive" strategy in Chap. 6. The decoupling manner offers great design freedom for the FTC controller and FE observer. Moreover, the estimation and compensation of perturbation contributes to a more robust FE-based FTC system. The adaptive decoupling strategy can handle both differentiable and non-differentiable faults, and both matched and unmatched faults. This is otherwise difficult, or impossible, by using the strategies in Chaps. 3–6.

1.6.3 Part III: Extension and Application

This part describes the extension of the proposed robust integration strategies to Lipschitz nonlinear systems and large-scale interconnected systems. It also describes application of the strategies in the areas of renewable energy, robotics and network systems.

Chapter 8 presents a fault-tolerant wind turbine pitch control example as a simple guide for FE-based FTC systems design. A new FE observer is also developed for reconstructing the real shapes of multiplicative (component) faults, rather than their fictitious replacements as in Chap. 5.

Chapter 9 presents extension of the simultaneous integration strategy in Chap. 5 for Lipschitz nonlinear systems. It also demonstrates the capability of applying the proposed strategy to a Lipschitz nonlinear 3-DOF helicopter system with actuator faults and input saturation constraints.

Chapter 10 presents extension of the strategy in Chap. 5 for large-scale interconnected systems subject to uncertain nonlinear interconnections. Decentralized FE and FTC strategies are developed for both the actuator fault and sensor fault cases and are applied to a 3-machine power system.

Chapter 11 summarizes the content in this book and highlights the pros and cons of each robust integration strategy. Potential future research directions are also discussed.

1.7 Lemmas and Notes Useful Throughout the Book

This section lists together some lemmas that are useful throughout the book. It also provides some notes on the classical Separation Principle for observer-based control design, and the unknown input observer.

1.7.1 List of Useful Lemmas

Lemma 1.1 (Bounded Real Lemma (Anderson and Vongpanitlerd 1973)) *A linear system*

$$\dot{x} = Ax + Dd$$
$$z = Cx$$

is stable with H_∞ performance $\|G_{zd}\|_\infty < \gamma$ if and only if there exists a s.p.d. matrix P and a positive scalar γ such that

$$\begin{bmatrix} \text{He}(PA) & PD & C^\top \\ \star & -\gamma I & 0 \\ \star & \star & -\gamma I \end{bmatrix} \prec 0.$$

Based on Lemma 1.1, the following corollary is given (Boyd et al. 1994):

Corollary 1.1 *Consider an uncertain linear system*

$$\dot{x} = (A + \Delta A)x + Dd$$
$$z = Cx,$$

where the uncertainty ΔA has the norm-bounded (energy bounded) form $\Delta A = \mathcal{M}\mathcal{F}(t)\mathcal{N}$, with known matrices $\mathcal{M}$ and $\mathcal{N}$ of appropriate dimensions, and an unknown matrix $\mathcal{F}(t)$ satisfying $\mathcal{F}^\top(t)\mathcal{F}(t) \leq I$.

This system is stable with the H_∞ performance $\|G_{zd}\|_\infty < \gamma$ if there exists a s.p.d. matrix $\mathcal{P}$ and a scalar $\gamma > 0$ such that

$$\begin{bmatrix} \text{He}(AP) & D & \mathcal{M} & P\mathcal{N}^\top & PC^\top \\ \star & -\gamma I & 0 & 0 & 0 \\ \star & \star & -I & 0 & 0 \\ \star & \star & \star & -I & 0 \\ \star & \star & \star & \star & -\gamma I \end{bmatrix} \prec 0.$$

Lemma 1.2 (Schur Complement (Boyd et al. 1994)) *For any symmetric matrix $\mathcal{S}$ of the form*

$$\mathcal{S} = \begin{bmatrix} \mathcal{S}_{11} & \mathcal{S}_{12} \\ \star & \mathcal{S}_{22} \end{bmatrix},$$

if $\mathcal{S}_{11}$ and $\mathcal{S}_{22}$ are invertible, then the following properties hold:
(1) $\mathcal{S} \prec 0 \iff \mathcal{S}_{11} \prec 0,\ \mathcal{S}_{22} - \mathcal{S}_{21}\mathcal{S}_{11}^{-1}\mathcal{S}_{12} \prec 0$;
(2) $\mathcal{S} \prec 0 \iff \mathcal{S}_{22} \prec 0,\ \mathcal{S}_{11} - \mathcal{S}_{12}\mathcal{S}_{22}^{-1}\mathcal{S}_{21} \prec 0$.

Lemma 1.3 (Pole placement lemma (Chilali and Gahinet 1996)) *The system $\dot{x} = Ax$ is $\mathcal{D}$-stable, if there exists a s.p.d. matrix P such that*

$$\alpha \otimes P + \text{He}\,[\beta \otimes (PA)] \prec 0 \tag{1.1}$$

where α and β are given matrices, and $\otimes$ represents the Kronecker product.

If $\mathcal{D}$ is a strip region: $a < \text{Re}(\lambda) < b$, where a and b are negative constants, then (1.1) is represented as

$$\begin{bmatrix} \text{He}(PA) - 2bP & 0 \\ \star & -\text{He}(PA) + 2aP \end{bmatrix} \prec 0.$$

Lemma 1.4 (Young inequality (Boyd et al. 1994)) *Given matrices X and Y of appropriate dimensions, for any matrix $S \succ 0$, it holds that*

$$X^\top Y + Y^\top X \preceq X^\top SX + Y^\top S^{-1}Y.$$

Lemma 1.5 (Barbalat's Lemma (Slotine et al. 1991)) *If the differentiable function $f(t)$ has a finite limit as $t \to \infty$, and if $\dot{f}$ is uniformly continuous, then $\dot{f}(t) \to 0$ as $t \to \infty$.*

Remark 1.7 According to Barbalat's Lemma, an immediate results widely used for system stability analysis is given below: For a scalar function $V(x, t)$, if

- $V(x, t)$ is lower bounded,
- $\dot{V}(x, t)$ is negative semi-definite,
- $\dot{V}(x, t)$ is uniformly continuous in time,

then $\dot{V}(x, t) \to 0$ as $t \to \infty$.

1.7.2 Notes on Separation Principle

The two-step (separated) design strategy for observer-based control is proposed by Luenberger (1971): "*The first phase is design of the control law assuming that the state vector is available. This may be based on optimization or other design techniques and typically results in a control law without dynamics. The second phase is the design of a system that produces an approximation to the state vector. This system, which in a deterministic setting is called an observer, or Luenberger observer ... The observer is a dynamic system whose characteristics are somewhat free to be determined by the designer, and it is through its introduction that dynamics enter the overall two-phase design procedure when the entire state is not available*". This Separation Principle is illustrated below using a simple example and shown to be unsatisfied in the presence of system uncertainties.

Consider a linear system

$$\begin{aligned} \dot{x} &= Ax + Bu \\ y &= Cx \end{aligned} \tag{1.2}$$

where $x(t) \in \mathbb{R}^n$, $u(t) \in \mathbb{R}^m$, and $y(t) \in \mathbb{R}^p$ are the state, control input and system output vectors, respectively. A, B and C are known constant matrices of compatible dimensions. The system is controllable and observable.

If the system state vector x is unavailable for feedback control design, then they can be estimated using the Luenberger observer

$$\begin{aligned}\dot{\hat{x}} &= A\hat{x} + Bu + L(y - \hat{y}) \\ \hat{y} &= C\hat{x}\end{aligned} \tag{1.3}$$

where $\hat{x} \in \mathbb{R}^n$ and $\hat{y} \in \mathbb{R}^p$ are the estimates of x and y, respectively. L is a design matrix.

Define the state estimation error as $e = x - \hat{x}$, then it follows from (1.2) and (1.3) that the error system is

$$\dot{e} = (A - LC)e. \tag{1.4}$$

A controller to stabilize x using feedback of the state estimate $\hat{x}$ is designed as

$$u = -K\hat{x} \tag{1.5}$$

where K is the designed controller gain.

Substituting the controller (1.5) into (1.2) gives the closed-loop control system

$$\dot{x} = (A - BK)x + BKe. \tag{1.6}$$

The composite closed-loop system composed of (1.4) and (1.6) is obtained as

$$\begin{bmatrix}\dot{x} \\ \dot{e}\end{bmatrix} = \begin{bmatrix}A - BK & BK \\ 0 & A - LC\end{bmatrix}\begin{bmatrix}x \\ e\end{bmatrix}. \tag{1.7}$$

It can be seen from (1.7) that the system matrix

$$A_c = \begin{bmatrix}A - BK & BK \\ 0 & A - LC\end{bmatrix}$$

is block triangular. Hence, eigenvalues of the composite system are the union of those of the control system and the estimation error system, i.e.

$$\lambda(A_c) = \lambda(A - BK) \cup \lambda(A - LC).$$

This implies that the state observer (1.3) does not affect eigenvalues of the original state feedback control system (1.6); nor are the eigenvalues of the observer affected by the connection. Therefore, the design of the state feedback controller (1.5) and the observer (1.3) can be carried out independently. This is called the *Separation Principle* (Chen 1995).

As analyzed below, the Separation Principle no longer holds when there are system uncertainties in the matrices A, B and C. Assume that

$$A_0 = A + \sigma_A, \;\; B_0 = B + \sigma_B, \;\; C_0 = C + \sigma_C,$$

where σ_A, σ_B, and σ_C are parameter uncertainties. By replacing A, B, and C in (1.7) by A_0, B_0 and C_0, respectively, the composite closed-loop system in the presence of uncertainties is represented by

$$\begin{bmatrix} \dot{x} \\ \dot{e} \end{bmatrix} = \begin{bmatrix} A + \sigma_A - (B + \sigma_B)K & (B + \sigma_B)K \\ \sigma_A + \sigma_B K - L\sigma_C & A - LC - \sigma_B K \end{bmatrix} \begin{bmatrix} x \\ e \end{bmatrix}. \tag{1.8}$$

It is obvious that the system matrix

$$A_\sigma = \begin{bmatrix} (A + \sigma_A) - (B + \sigma_B)K & (B + \sigma_B)K \\ \sigma_A + \sigma_B K - L\sigma_C & A - LC - \sigma_B K \end{bmatrix}$$

is no longer block triangular. This means that the observer and control system affect each other and cannot be designed independently. Therefore, the existence of uncertainties breaks down the Separation Principle.

Example 1.1 Consider a linear system in the form of (1.2) with

$$A = \begin{bmatrix} 0 & 1 \\ 1 & 1 \end{bmatrix}, \quad B = \begin{bmatrix} 0 \\ 1 \end{bmatrix}, \quad C = [1\ 0], \quad K = [16\ 9], \quad L = \begin{bmatrix} 18 \\ 89 \end{bmatrix}.$$

According to (1.7), it is obtained that

$$\begin{aligned} \lambda(A_c) &= (-3, -5, -10, -7), \\ \lambda(A - BK) &= (-3, -5), \\ \lambda(A - LC) &= (-10, -7). \end{aligned} \tag{1.9}$$

Hence, $\lambda(A_c) = \lambda(A - BK) \cup \lambda(A - LC)$ and the Separation Principle holds.

Suppose the system suffers from the following uncertainties: $\sigma_A = 0.02A$, $\sigma_B = 0.01B$, and $\sigma_C = 0.05C$, then according to (1.8), it is derived that

$$\begin{aligned} \lambda(A_\sigma) &= (-12.1779, -5.3602 \pm 3.5698\mathrm{i}, -2.2617), \\ \lambda\left(A + \sigma_A - (B + \sigma_B)K\right) &= (-3.1193, -4.9507), \\ \lambda(A - LC - \sigma_B K) &= (-9.6572, -7.4328). \end{aligned} \tag{1.10}$$

It can be seen from (1.10) that the Separation Principle does not hold in the uncertain case. In the real design procedure, if the system uncertainties (σ_A, σ_B and σ_C) are ignored, then the Separation Principle is satisfied and the observer and controller are designed separately with desired eigenvalues in (1.9). However, when the observer and controller are implemented on the real uncertain system, the real eigenvalues of the composite system are those in (1.10), which are quite different from the desired ones. Therefore, in the presence of uncertainties, the separately designed observer and controller cannot guarantee the desired estimation and control performances and may destabilize the closed-loop system.

1.7.3 Notes on Unknown Input Observer

In this book, several new FE observers will be presented based on further extension and development of the techniques used for designing an unknown input observer (UIO). Therefore, this section aims to provide necessary background of UIO based on the results in Chen and Patton (1999).

Consider a linear system

$$\begin{aligned}\dot{x} &= Ax + Bu + Dd \\ y &= Cx\end{aligned} \tag{1.11}$$

where $x \in \mathbb{R}^n$, $u \in \mathbb{R}^m$, $y \in \mathbb{R}^p$, and $d \in \mathbb{R}^l$ are the state, known control input, measured output, and unknown input, respectively. A, B, D and C are known constant matrices of compatible dimensions. The pair (A, C) is observable.

Definition 1.12 (*Unknown input observer* (Chen and Patton 1999)) An observer is defined as an unknown input observer for the system (1.11), if it can achieve asymptotic state estimation despite the presence of unknown input in the system.

A full-order UIO is in the form of

$$\begin{aligned}\dot{z} &= Nz + Gu + Ly \\ \hat{x} &= z + Hy\end{aligned} \tag{1.12}$$

where z is the observer state and $\hat{x}$ is the estimate of x. N, G, L and H are constant matrices to be designed.

Define the estimation error as $e = x - \hat{x}$, then the error system is given by

$$\begin{aligned}\dot{e} = &(\Xi A - L_1 C)e + [N - (\Xi A - L_1 C)]z \\ &+ [L_2 - (\Xi A - L_1 C)H]y \\ &+ (G - \Xi B)u + \Xi Dd\end{aligned} \tag{1.13}$$

where $\Xi = I - HC$ and $L = L_1 + L_2$.

By applying the following relations:

$$\Xi D = 0 \tag{1.14}$$

$$N = \Xi A - L_1 C \tag{1.15}$$

$$L_2 = (\Xi A - L_1 C)H \tag{1.16}$$

$$G = \Xi B \tag{1.17}$$

then the error system (1.13) becomes

$$\dot{e} = Ne. \tag{1.18}$$

If the matrix N is designed to be Hurwitz stable, then $\lim_{t\to\infty} e(t) = 0$ and asymptotic state estimation is obtained.

It is seen from (1.18) that the unknown input d is completely removed from the error dynamics, thanks to the use of the relation (1.14). The necessary and sufficient conditions for the feasibility of relations (1.14)–(1.17), and furthermore for (1.12) to be a UIO for system (1.11) are as follows: (1) The matching condition is satisfied, i.e. $\text{rank}(CD) = \text{rank}(D)$; (2) The pair (C, A_1) is detectable, where $A_1 = A - D(CD)^{\dagger}CA$.

Proof details of the above statement are referred to Sect. 3.2 of Chen and Patton (1999). Readers can also find there, details of using UIO for robust fault detection. Based on extension and further development of the UIO theory, several observers will be developed in this book for FE.

In fact, the UIO (1.12) can be reformulated in a form of Luenberger observer but with an additional derivative term of the output tracking error. To show this, we first add $(-HC\hat{x})$ to both sides of the second equation in (1.12) and get

$$\Xi\hat{x} = z + H(y - \hat{y}) \tag{1.19}$$

where $\hat{y} = C\hat{x}$.

By using (1.12)–(1.19) and $L = L_1 + L_2$, it can then be derived that

$$\begin{aligned}
\Xi\dot{\hat{x}} &= \dot{z} + H(\dot{y} - \dot{\hat{y}}) \\
&= Nz + Gu + Ly + H(\dot{y} - \dot{\hat{y}}) \\
&= (\Xi A - L_1 C)(\hat{x} - Hy) + \Xi Bu + Ly + H(\dot{y} - \dot{\hat{y}}) \\
&= \Xi A\hat{x} - L_1\hat{y} - (\Xi A - L_1 C)Hy + \Xi Bu + Ly + H(\dot{y} - \dot{\hat{y}}) \\
&= \Xi A\hat{x} + \Xi Bu - L_1\hat{y} - L_2 y + Ly + H(\dot{y} - \dot{\hat{y}}) \\
&= \Xi A\hat{x} + \Xi Bu + L_1(y - \hat{y}) + H(\dot{y} - \dot{\hat{y}}).
\end{aligned} \tag{1.20}$$

If the left pseudo-inverse $\Xi^{\dagger}$ of matrix Ξ exists, then (1.20) can be rewritten as

$$\dot{\hat{x}} = A\hat{x} + Bu + \bar{L}(y - \hat{y}) + \bar{H}(\dot{y} - \dot{\hat{y}}) \tag{1.21}$$

with $\bar{L} = \Xi^{\dagger}L_1$ and $\bar{H} = \Xi^{\dagger}H$. It is seen that (1.21) is in the form of the Luenberger observer (1.3) but with an extra term $\bar{H}(\dot{y} - \dot{\hat{y}})$. Hence, the UIO can be viewed as a Proportional-Derivative Luenberger observer, but having disturbance decoupling capability and more design freedom because more matrices can be designed.

References

Alwi H, Edwards C, Tan CP (2011) Fault detection and fault-tolerant control using sliding modes. Springer Science & Business Media, Berlin

Amin AA, Hasan KM (2019) A review of fault tolerant control systems: advancements and applications. Measurement 143:58–68. Elsevier

Anderson B, Vongpanitlerd S (1973) Network analysis and synthesis: a modern systems theory approach. Prentice-Hall, Englewood Cliffs

Benosman M (2009) A survey of some recent results on nonlinear fault tolerant control. Math Probl Eng 2010

Blanke M, Kinnaert M, Lunze J, Staroswiecki M, Schröder J (2003) Diagnosis and fault-tolerant control. Springer, Berlin

Blanke M, Schröder J, Kinnaert M, Lunze J, Staroswiecki M (2006) Diagnosis and fault-tolerant control. Springer Science & Business Media, Berlin

Boem F, Gallo AJ, Raimondo DM, Parisini T (2019) Distributed fault-tolerant control of large-scale systems: an active fault diagnosis approach. IEEE Trans Control Netw Syst 7(1):288–301. IEEE

Boyd SP, El Ghaoui L, Feron E, Balakrishnan V (1994) Linear matrix inequalities in system and control theory, vol 15. SIAM, Philadelphia

Buffington J, Chandler P, Pachter M (1999) On-line system identification for aircraft with distributed control effectors. Int J Robust Nonlinear Control 9(14):1033–1049

Chen CT (1995) Linear system theory and design. Oxford University Press Inc, Oxford

Chen H, Jiang B (2019) A review of fault detection and diagnosis for the traction system in high-speed trains. IEEE Trans Intell Transp Syst 21(2):450–465. IEEE

Chen J, Patton RJ (1999) Robust model-based fault diagnosis for dynamic systems. Kluwer Academic Publishers, London

Chilali M, Gahinet P (1996) H_∞ design with pole placement constraints: an LMI approach. IEEE Trans Autom Control 41(3):358–367

Clark RN, Frank PM, Patton RJ (1989) Fault diagnosis in dynamic systems: theory and applications. Prentice Hall, Upper Saddle River

Ding SX (2009) Integrated design of feedback controllers and fault detectors. Annu Rev Control 33(2):124–135

Ding SX (2014) Data-driven design of fault diagnosis and fault-tolerant control systems. Springer, Berlin

Dong H, Wang Z, Gao H (2013) Filtering, control and fault detection with randomly occurring incomplete information. Wiley, Hoboken

Ducard GJ (2009) Fault-tolerant flight control and guidance systems: practical methods for small unmanned aerial vehicles. Springer Science & Business Media, Berlin

Edwards C, Spurgeon SK, Patton RJ (2000) Sliding mode observers for fault detection and isolation. Automatica 36(4):541–553

Edwards C, Lombaerts T, Smaili H (2010) Fault tolerant flight control: a benchmark challenge. Springer Science & Business Media, Berlin

Escobet T, Bregon A, Pulido B, Puig V (2019) Fault diagnosis of dynamic systems. Springer, Berlin

Eterno J, Weiss J, Looze D, Willsky A (1985) Design issues for fault tolerant-restructurable aircraft control. In: Proceedings of the 24th IEEE conference on decision and control, 24, pp 900–905

Feng X, Patton R (2014) Active fault tolerant control of a wind turbine via fuzzy MPC and moving horizon estimation. IFAC Proc Vol 47(3):3633–3638

Fritz R, Zhang P (2018) Overview of fault-tolerant control methods for discrete event systems. IFAC-PapersOnLine 51(24):88–95

Gahinet P, Nemirovski A, Laub AJ, Chilali M (1995) LMI control toolbox for use with MATLAB. The MathWorks Inc, Natick

Gao Z, Ding SX (2007) Actuator fault robust estimation and fault-tolerant control for a class of nonlinear descriptor systems. Automatica 43(5):912–920

Gao Z, Cecati C, Ding SX (2015a) A survey of fault diagnosis and fault-tolerant techniques-Part II: fault diagnosis with knowledge-based and hybrid/active approaches. IEEE Trans Ind Electron 62(6):3768–3774

Gao Z, Cecati C, Ding SX (2015b) A survey of fault diagnosis and fault-tolerant techniques-Part I: fault diagnosis with model-based and signal-based approaches. IEEE Trans Ind Electron 62(6):3757–3767

Gao Z, Liu X, Chen MZ (2016) Unknown input observer-based robust fault estimation for systems corrupted by partially decoupled disturbances. IEEE Trans Ind Electron 63(4):2537–2547
Goupil P (2010) Oscillatory failure case detection in the A380 electrical flight control system by analytical redundancy. Control Eng Pract 18(9):1110–1119
Hwang I, Kim S, Kim Y, Seah CE (2009) A survey of fault detection, isolation, and reconfiguration methods. IEEE Trans Control Syst Technol 18(3):636–653
Isermann R (2006) Fault-diagnosis systems: an introduction from fault detection to fault tolerance. Springer Science & Business Media, Berlin
Jain T, Yamé JJ, Sauter D (2018) Active fault-tolerant control systems. Springer
Kabore R, Wang H (2001) Design of fault diagnosis filters and fault-tolerant control for a class of nonlinear systems. IEEE Trans Autom Control 46(11):1805–1810
Kulcsár B, Verhaegen M (2011) Robust inversion based fault estimation for discrete-time LPV systems. IEEE Trans Autom Control 57(6):1581–1586
Löfberg J (2004) YALMIP: a toolbox for modeling and optimization in MATLAB. In: Proceedings of the CACSD, Taipei, Taiwan, vol 3
Luenberger D (1971) An introduction to observers. IEEE Trans Autom Control 16(6):596–602
Maciejowski J (1999) Modelling and predictive control: enabling technologies for reconfiguration. Annu Rev Control 23:13–23
Maybeck PS, Stevens RD (1991) Reconfigurable flight control via multiple model adaptive control methods. IEEE Trans Aerosp Electron Syst 27(3):470–480
Meskin N, Khorasani K (2011) Fault detection and isolation: multi-vehicle unmanned systems. Springer Science & Business Media, Berlin
Mosek A (2018) The MOSEK optimization software, version 8.1. http://www.mosek.com
Nett C, Jacobson C, Miller A (1988) An integrated approach to controls and diagnostics: the 4-parameter controller. In: Proceedings of the American control conference, pp 824–835
Noura H, Theilliol D, Ponsart JC, Chamseddine A (2009) Fault-tolerant control systems: design and practical applications. Springer Science & Business Media, Berlin
Patan K (2019) Robust and fault-tolerant control: neural-network-based solutions, vol 197. Springer, Berlin
Patton RJ (1997) Fault-tolerant control systems: the 1997 situation. In: Proceedings of IFAC symposium on fault detection supervision and safety for technical processes, vol 3, pp 1033–1054
Perk S, Shao Q, Teymour F, Cinar A (2012) An adaptive fault-tolerant control framework with agent-based systems. Int J Robust Nonlinear Control 22(1):43–67
Potiron K, Seghrouchni AEF, Taillibert P (2013) From fault classification to fault tolerance for multi-agent systems, vol 67. Springer, Berlin
Quevedo J, Puig V, Cembrano G, Blanch J, Aguilar J, Saporta D, Benito G, Hedo M, Molina A (2010) Validation and reconstruction of flow meter data in the Barcelona water distribution network. Control Eng Pract 18(6):640–651
Rauch HE (1995) Autonomous control reconfiguration. IEEE Control Syst 15(6):37–48
Richter JH (2011) Reconfigurable control of nonlinear dynamical systems: a fault-hiding approach, vol 408. Springer, Berlin
van Schrick D (1997) Remarks on terminology in the field of supervision, fault detection and diagnosis. In: Proceedings of the 3rd IFAC symposium on fault detection, supervision and safety for technical processes, pp 959–964
Shen Q, Jiang B, Shi P (2017) Fault diagnosis and fault-tolerant control based on adaptive control approach, vol 91. Springer, Berlin
Simani S, Farsoni S (2018) Fault diagnosis and sustainable control of wind turbines: robust data-driven and model-based strategies. Butterworth-Heinemann, Oxford
Slotine JJE, Li W et al (1991) Applied nonlinear control. Prentice Hall, Englewood Cliffs
Staroswiecki M (2005) Fault tolerant control: the pseudo-inverse method revisited. IFAC Proc Vol 38(1):418–423
Steffen T (2005) Control reconfiguration of dynamical systems: linear approaches and structural tests, vol 320. Springer Science & Business Media, Berlin

Stetter R (2020) Fault-tolerant design and control of automated vehicles and processes. Springer, Berlin

Veillette RJ, Medanic J, Perkins WR (1992) Design of reliable control systems. IEEE Trans Autom Control 37(3):290–304

Willsky AS, Jones HL (1974) A generalized likelihood ratio approach to state estimation in linear systems subjects to abrupt changes. In: Proceedings of IEEE conference on decision and control including the 13th symposium on adaptive processes, vol 13. IEEE, pp 846–853

Witczak M (2014) Fault diagnosis and fault-tolerant control strategies for non-linear systems, vol 266. Springer, Berlin

Yang H, Jiang B, Cocquempot V (2010) Fault tolerant control and hybrid systems. In: Fault tolerant control design for hybrid systems. Springer, Berlin, pp 1–9

Yang H, Han QL, Ge X, Ding L, Xu Y, Jiang B, Zhou D (2019) Fault-tolerant cooperative control of multiagent systems: a survey of trends and methodologies. IEEE Trans Ind Inform 16(1):4–17. IEEE

Yu X, Jiang J (2015) A survey of fault-tolerant controllers based on safety-related issues. Annu Rev Control 39:46–57

Zhang K, Jiang B, Shi P (2013) Observer-based fault estimation and accomodation for dynamic systems. Springer, Berlin

Zhang W, Wang Z, Raïssi T, Wang Y, Shen Y (2020) A state augmentation approach to interval fault estimation for descriptor systems. Eur J Control 51:19–29

Zhang Y, Jiang J (2001) Integrated active fault-tolerant control using IMM approach. IEEE Trans Aerosp Electron Syst 37(4):1221–1235

Zhang Y, Jiang J (2008) Bibliographical review on reconfigurable fault-tolerant control systems. Annu Rev Control 32(2):229–252

Chapter 2
Robust Integration in Fault Diagnosis and FTC

2.1 Introduction

In the presence of system uncertainties, it is necessary to synthesize the functions of fault diagnosis and FTC from a holistic perspective to guarantee robust closed-loop FTC system performance. To this end, the concept of robust integration of fault diagnosis and FTC is defined in Definition 1.11 in Chap. 1. The purpose of this chapter is to provide a detailed mathematical analysis of the necessity, importance, and challenges of robust integration in the context of FDI and control, FDI and FTC, and FE and FTC. The reason of choosing FE-based FTC rather than FDI-based FTC to achieve robust integration is also given.

2.2 Robust Integration of FDI and Control

A general scheme of closed-loop FDI systems is outlined in Fig. 2.1, where an FDI block is used to detect and isolate the fault using the control input signal and output measurement. It has been known for some time that the FDI performance within closed-loop systems is affected by external disturbance acting on the plant and system uncertainty. This is illustrated below through a simple example of model-based FDI with a residual generator.

Consider a generic linear controlled system modelled by

$$\begin{aligned} \dot{x} &= Ax + Bu + Ff + \Delta(x, u, d) \\ y &= Cx \end{aligned} \tag{2.1}$$

where x, u and y are the state, control input and measured output, respectively. f is the actuator fault. $\Delta(x, u, d)$ is the lumped uncertainty given by $\Delta(x, u, d) =$

© The Author(s), under exclusive license to Springer Nature Switzerland AG 2021

J. Lan and R. J. Patton, *Robust Integration of Model-Based Fault Estimation and Fault-Tolerant Control*, Advances in Industrial Control,
https://doi.org/10.1007/978-3-030-58760-4_2

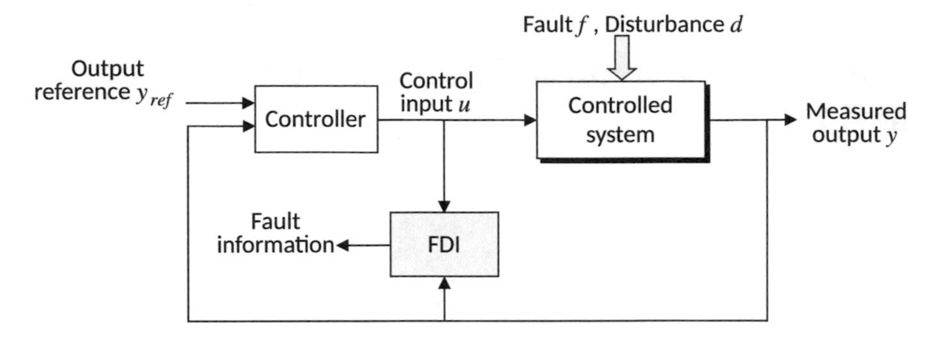

Fig. 2.1 A general scheme of closed-loop FDI systems

$\Delta Ax + \Delta Bu + Dd$, with ΔA and ΔB characterizing, respectively, the parametric uncertainties of A and B, and Dd characterizing the external disturbance d with a distribution matrix D. The constant matrices A, B, F and C are known and the pair (A, C) is observable.

The following Luenberger observer is used as a residual generator (Patton 1997):

$$\begin{aligned} \dot{\hat{x}} &= A\hat{x} + Bu + L(y - \hat{y}) \\ \hat{y} &= C\hat{x} \\ r &= y - \hat{y} \end{aligned} \tag{2.2}$$

where $\hat{x}$ and $\hat{y}$ are the estimates of x and y, respectively. r is the residual signal reflecting the discrepancy between the real and estimated outputs. L is a design matrix such that $(A - LC)$ is Hurwitz stable.

Define the state estimation error as $e = x - \hat{x}$. By using (2.1) and (2.2), the estimation error system is obtained as

$$\begin{aligned} \dot{e} &= (A - LC)\, e + Ff + \Delta(x, u, d) \\ r &= Ce. \end{aligned} \tag{2.3}$$

According to (2.3), the time response of r is characterized by

$$\begin{aligned} r(t) &= Ce^{(A-LC)t} e(0) \\ &\quad + C \int_0^t e^{(A-LC)(t-\tau)} \left[Ff(\tau) + \Delta(x(\tau), u(\tau), d(\tau))\right] \mathrm{d}\tau. \end{aligned} \tag{2.4}$$

The principle of residual generation for fault diagnosis is to make r sensitive to the fault f. However, it can be seen from (2.4) that r is sensitive to both the fault and the lumped uncertainty $\Delta(x, u, d)$. *Since FDI relies on the residual signal to detect and isolate the fault, the subsequently designed FDI function is also affected by the uncertainty* $\Delta(x, u, d)$. Since the state-of-the-art FDI methods explicitly use either

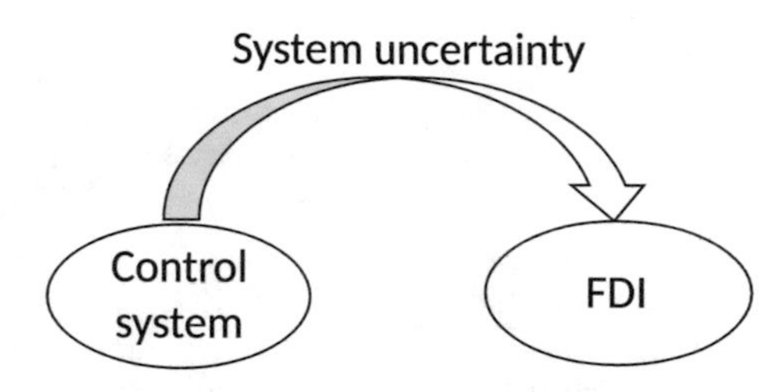

Fig. 2.2 Unidirectional robustness interaction between FDI and control

the mathematical system model or process data, their performances are inevitably affected by the system uncertainty. Therefore, it can be concluded that, in the presence of system uncertainty, there always exists a *unidirectional robustness interaction* between the control system and FDI unit, as shown in Fig. 2.2.

To achieve robust FDI, the tool of H_-/H_∞ optimization was first developed in Hou and Patton (1996) for robust residual generation, with the purpose of minimizing the disturbance effect while maximizing the fault sensitivity. The LMI formulation to this tool has been used for linear time-invariant system (Wang et al. 2007) and linear time-varying system (Li and Zhou 2009). The matrix factorization to H_-/H_∞ optimization has been proposed in Jaimoukha et al. (2006), Zhang and Ding (2008). Combined H_-/H_∞ optimization with linear parameter varying (Chen et al. 2016; Grenaille et al. 2008) or T-S fuzzy modellings (Chadli et al. 2013) have also been used to design FDI for uncertain nonlinear systems. However, the above works only focus on the robust design of FDI.

In 1998, Nett et al. (1988) first defined the concept of "integrated design" for combining the control and FDI designs into a joint robustness problem to achieve robust control and robust FDI. They also proposed a four-parameter controller for integrated FDI and control design. Following this direction, many works have been published for linear and nonlinear systems, see for example, the review paper Ding (2009) and some recent papers Davoodi et al. (2011), Du et al. (2016), Weng et al. (2008), Zhai et al. (2016), Zhong and Yang (2015). Nevertheless, all these works focus on the integration of FDI and control without considering FTC design.

2.3 Robust Integration of FDI and FTC

In order to actively use the FDI results in system feedback control to accommodate the faults, FDI-based FTC systems have been developed following the scheme in Fig. 2.3. Within this scheme, the plant is assumed to have N potential fault scenarios that are known a priori, corresponding to the N different operating modes. For each operating mode an associated controller is pre-computed to achieve desired system performance. Meanwhile, a bank of estimators are used for FDI purpose to identify the current operating mode with the help of the "Supervisor". Once a certain mode is

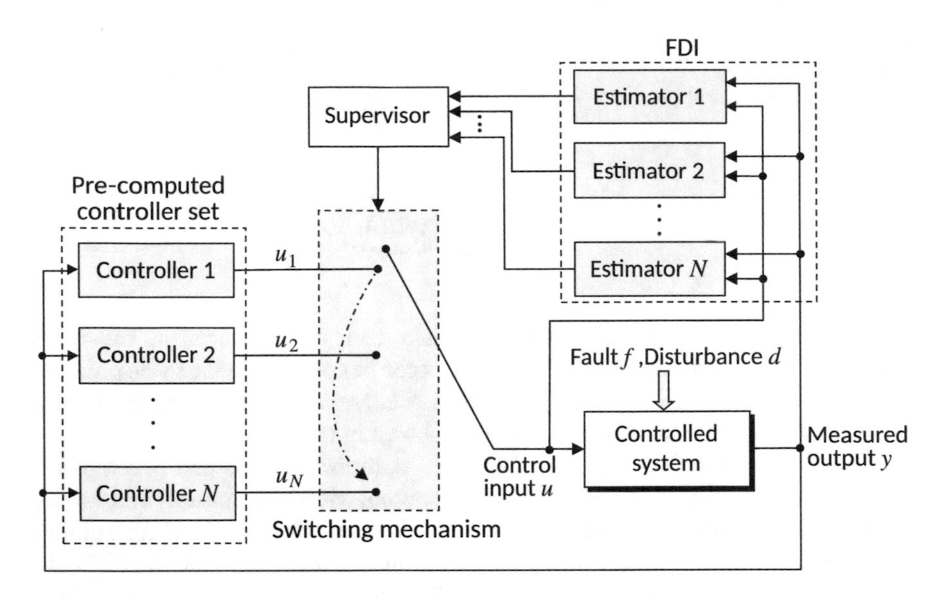

Fig. 2.3 A general scheme of FDI-based FTC systems

identified by the Supervisor, a switching signal is sent to the "Switching mechanism" to select the corresponding controller from the pre-computed controller set $\{u_i\}_{i=1}^{N}$.

Without loss of generality, in the following a simple example is used to delineate the basic idea and design challenges of FDI-based FTC systems. Representative FTC controllers and FDI estimators are adapted from Yang et al. (2009) and Cieslak et al. (2015).

Consider a class of linear controlled systems represented by

$$\begin{aligned} \dot{x} &= Ax + Bu + \Delta(x, u, d) \\ y &= Cx \end{aligned} \tag{2.5}$$

where x, u and y are the state, control input and measured output, respectively. $\Delta(x, u, d)$ denotes the lumped uncertainty as defined in (2.1). The constant matrices A, B and C are known and of compatible dimensions.

The system (2.5) has N operating modes, one is fault-free and the other $N - 1$ are faulty. The system dynamics in the ith operating mode are represented by

$$\begin{aligned} \dot{x} &= A_i x + B_i u + \Delta_i(x, u, d) \\ y &= Cx \end{aligned} \tag{2.6}$$

where A_i, B_i and $\Delta_i(x, u, d)$ are the system matrices and lumped uncertainty in the ith faulty mode. Assume that the pairs (A_i, B_i) and (A_i, C) are controllable and observable, respectively.

An output feedback controller to stabilize the system (2.6) is designed as

$$u_i = -K_i y \tag{2.7}$$

where K_i is the control gain designed such that $(A_i - B_i K_i C)$ is Hurwitz stable.

Substituting (2.7) into (2.6) gives the closed-loop system

$$\dot{x} = (A_i - B_i K_i C)x + \Delta_i(x, u, d). \tag{2.8}$$

To identify the ith operating mode, the following estimator is used:

$$\begin{aligned} \dot{z}_i &= A_i z_i + B_i u + L_i(y - Cz_i) \\ r_i &= y - Cz_i \end{aligned} \tag{2.9}$$

where z_i is the estimate of x and r_i is a residual signal reflecting the discrepancy between the real measured output y and the estimated ith operating mode output Cz_i. The observer gain L_i is designed such that $(A_i - L_i C)$ is Hurwitz stable.

Define the state estimation error as $e_i = x - z_i$. By using (2.6) and (2.9), the estimation error system is obtained as

$$\begin{aligned} \dot{e}_i &= (A_i - L_i C)e_i + \Delta_i(x, u, d) \\ r_i &= Ce_i. \end{aligned} \tag{2.10}$$

(1) Ideal case: $\Delta(x, u, d) = 0$ **and** $\Delta_i(x, u, d) = 0$

In this case, if the plant is currently working in the ith operating mode, then $r_i = 0$ and all other r_j, $j \neq i$, $j = 1, 2 \ldots, N$, are non-zero. Hence, the Supervisor can identify the current operating mode of the plant by collecting and comparing all the residual signals r_i, $i = 1, 2, \ldots, N$, from the estimators.

After the ith operating mode being identified, the control signal u is set as $u = u_i$. Substituting (2.7) into (2.5) yields the closed-loop control system

$$\dot{x} = (A - BK_i C)x. \tag{2.11}$$

Since the system is working at the ith operating mode, $A = A_i$ and $B = B_i$. Hence, (2.11) is equivalent to

$$\dot{x} = (A_i - B_i K_i C)x. \tag{2.12}$$

Since $(A_i - B_i K_i C)$ is stable, so is the closed-loop system (2.12).

(2) Uncertain case: $\Delta(x, u, d) \neq 0$ **and** $\Delta_i(x, u, d) \neq 0$

In this case, it follows from (2.10) that the time response of r_i is

$$r_i(t) = Ce^{(A_i - L_i C)t} e_i(0) + C \int_0^t e^{(A_i - L_i C)(t-\tau)} \Delta_i(x(\tau), u(\tau), d(\tau)) \mathrm{d}\tau. \tag{2.13}$$

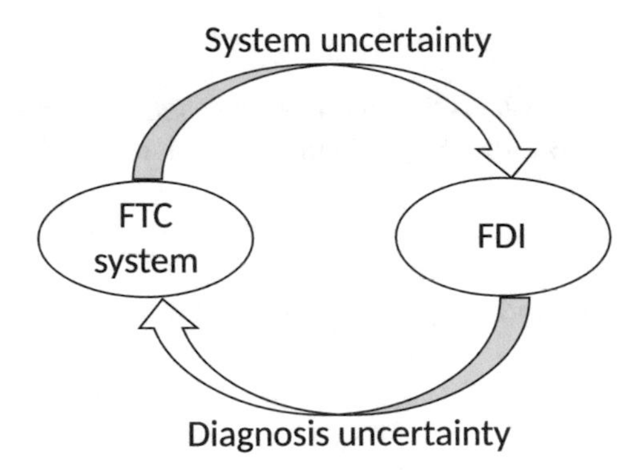

Fig. 2.4 Bidirectional robustness interactions between FDI and FTC

Obviously, *all the residual signals* r_i, $i = 1, 2, \ldots, N$, *are affected by the system uncertainty* $\Delta_i(x, u, d)$. *So* $r_i \neq 0$, *even when the plant is working at the* i*th operating mode. This gives rise to the challenge of identifying the actual operating mode.*

For example, if the actual operating mode is i, but it is wrongly identified as j by the Supervisor. Then the controller u is set as $u = u_j$ and the resulting closed-loop control system is

$$\dot{x} = (A - BK_jC)x + \Delta(x, u, d). \tag{2.14}$$

Since $A = A_i$ and $B = B_i$, the closed-loop system (2.14) is equivalent to

$$\begin{aligned}\dot{x} &= (A_i - B_iK_jC)x + \Delta(x, u, d) \\ &= (A_j - B_jK_jC)x + (\Delta A_{ij} - \Delta B_{ij}K_jC)x + \Delta(x, u, d)\end{aligned} \tag{2.15}$$

where $\Delta A_{ij} = A_i - A_j$ and $\Delta B_{ij} = B_i - B_j$.

Although K_j is designed such that $(A_j - B_jK_jC)$ is stable, *stability of the FTC system* (2.15) *is affected by both the uncertainty* $\Delta(x, u, d)$ *and the identification (FDI) uncertainty* $(\Delta A_{ij} - \Delta B_{ij}K_jC)x$.

According to the analysis above, when applying the FDI approach to an FTC system design (see Fig. 2.3), the identification of (fault) operating mode directly affects the control actions applied to the plant. In the presence of system uncertainty, the FDI is affected by the uncertainty, referred as the *unidirectional robustness interaction* described in Sect. 2.2. Moreover, the closed-loop FDI-based FTC system performance is affected by not only the system uncertainty, but also the diagnosis uncertainties. This results in *bidirectional robustness interactions* between the FDI and FTC functions, as shown in Fig. 2.4.

Within the closed-loop FDI-based FTC system the diagnosis and control functions may not fit with each other, if they are designed separately without taking into account the *bidirectional robustness interactions*. Therefore, it is necessary to consider the mutual interactions by integrating the designs of FDI and FTC into a simultaneous procedure.

The integration of FDI and FTC discussed here is a hard challenge, because the reconfiguration and FDI roles have a bidirectional uncertainty which is more complex when compared with the integration of FDI and control. The complexity arises from the joint multi-objectives of robust closed-loop stability, robust residual performance (requiring optimal fault detection thresholds), and robust fault tolerance with stable reconfiguration, generally operating in the presence of various time delays and uncertainties. Integrated designs of FDI and FTC have been considered in several works, see for example Cieslak et al. (2015), Mhaskar et al. (2006), Yang et al. (2009). However, the robust integration defined in this book as a single procedure to obtain the FDI and FTC parameters simultaneously (see Definition 1.11) has not been achieved. **It is very difficult or impossible to achieve the true integration using the FDI approach**, because it involves (1) discrete-event structure with complex decision, (2) variable and unknown time delay and (3) a control reconfiguration that is very complex. Therefore, the FDI approach to FTC can be one of the most difficult problems of adaptive control and in general is not suitable for achieving true robust integration of fault diagnosis and FTC as well as practical application.

2.4 Robust Integration of FE and FTC

2.4.1 Theoretic Analysis

The FE-based FTC design follows the scheme in Fig. 2.5, where the baseline controller u_b is used to maintain nominal system performance while the fault compensator u_f is automatically activated once a fault is estimated by the FE observer.

Most existing FE-based FTC systems follow this scheme by assuming satisfaction of the Separation Principle (see Sect. 1.7.2), see, for Example, the papers Gao and Ding (2007), Jiang et al. (2006), de Loza et al. (2015) and the monograph Zhang

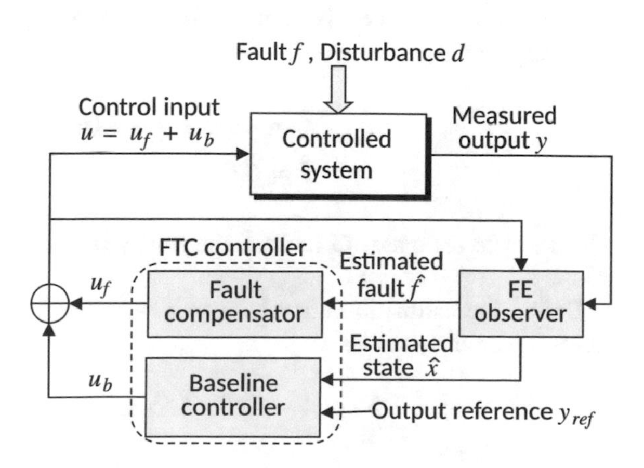

Fig. 2.5 A general scheme of FE-based FTC systems

et al. (2012). The challenge of true integration is that the Separation Principle no longer holds when in addition to modelling uncertainty there is also uncertainty associated with the FE role used in compensating the faults. Errors in the state and fault estimation will cause uncertainty in the FTC function.

For simplicity but without loss of generality, the basic idea and design challenges of FE-based FTC are illustrated below using a Luenberger-type FE observer together with a state feedback FTC controller.

Consider a class of linear controlled systems represented by

$$\begin{aligned} \dot{x} &= (A + \Delta A)x + (B + \Delta B)u + Ff + Dd \\ y &= Cx \end{aligned} \tag{2.16}$$

where x, u, y, f and d are the state, control input, measured output, actuator fault and external disturbance, respectively. The constant matrices A, B, F, D and C are known. The matrices ΔA and ΔB denotes the unknown parametric uncertainties. This system satisfies the following assumptions: (1) The pairs (A, B) and (A, C) are controllable and observable, respectively; (2) The actuator fault f is matched (see Definition 1.7), and both f and its first-order time derivative $\dot{f}$ are bounded.

By considering f as auxiliary state, the system (2.16) can be augmented as

$$\begin{aligned} \dot{\bar{x}} &= (\bar{A} + \Delta\bar{A})\bar{x} + (\bar{B} + \Delta\bar{B})u + \bar{D}\bar{d} \\ y &= \bar{C}\bar{x} \end{aligned} \tag{2.17}$$

with

$$\bar{x} = \begin{bmatrix} x \\ f \end{bmatrix}, \ \bar{d} = \begin{bmatrix} d \\ \dot{f} \end{bmatrix}, \ \bar{A} = \begin{bmatrix} A & F \\ 0 & 0 \end{bmatrix}, \ \Delta\bar{A} = \begin{bmatrix} \Delta A & 0 \\ 0 & 0 \end{bmatrix},$$
$$\bar{B} = \begin{bmatrix} B \\ 0 \end{bmatrix}, \ \Delta\bar{B} = \begin{bmatrix} \Delta B \\ 0 \end{bmatrix}, \ \bar{D} = \begin{bmatrix} D & 0 \\ 0 & I \end{bmatrix}, \ \bar{C} = [C \ 0].$$

The augmented state $\bar{x}$ is estimated by the following augmented Luenberger state observer (ALSO):

$$\begin{aligned} \dot{\hat{\bar{x}}} &= \bar{A}\hat{\bar{x}} + \bar{B}u + L(y - \hat{y}) \\ \hat{y} &= \bar{C}\hat{\bar{x}} \end{aligned} \tag{2.18}$$

where $\hat{\bar{x}}$ is the estimate of $\bar{x}$ and L is the gain designed such that $(\bar{A} - L\bar{C})$ is Hurwitz stable.

Define the estimation error as $e = \bar{x} - \hat{\bar{x}}$, then subtracting (2.18) from (2.17) yields the error dynamics

$$\dot{e} = (\bar{A} - L\bar{C})e + \Delta\bar{A}\bar{x} + \Delta\bar{B}u + \bar{D}\bar{d}. \tag{2.19}$$

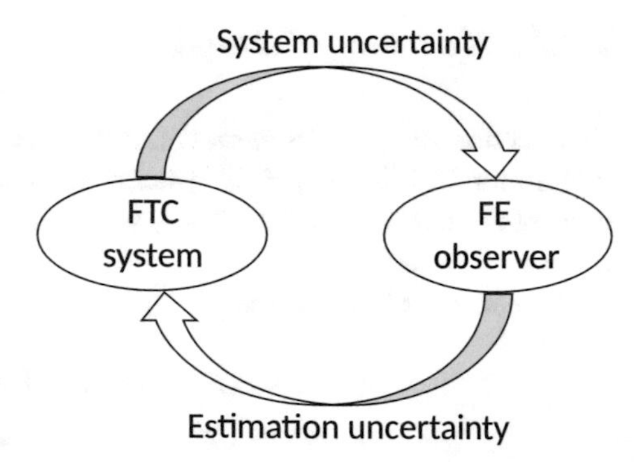

Fig. 2.6 Bidirectional robustness interactions between FE and FTC functions

It is seen from (2.19) *that the state estimation and FE are affected by the system uncertainties* ΔAx *and* ΔBu, *disturbance* d, *and fault modelling error* $\dot{f}$.

In the literature, the following FE-based state feedback FTC controller is widely adopted to regulate the state of (2.16) to zero:

$$u = u_b + u_f \tag{2.20}$$

where the baseline controller is designed as $u_b = K\hat{x}$ with a constant gain K such that $(A + BK)$ is Hurwitz stable. Since the actuator fault is matched, the fault compensator u_f is optimally designed as $u_f = -B^{\dagger}F\hat{f}$ such that $Bu_f + Ff = F(f - \hat{f})$. Hence, f is completely compensated if it is accurately estimated.

Substituting controller (2.20) into (2.16) gives the closed-loop FTC system

$$\dot{x} = (A + BK)x + B_e e + \Delta Ax + \Delta Bu + Dd \tag{2.21}$$

where $B_e = [-BK \;\; F]$.

It can be seen from (2.21) *that the closed-loop FTC system is perturbed by the system uncertainties* ΔAx *and* ΔBu, *disturbance* d *and estimation error* e.

According to (2.19) and (2.21), there exist *bidirectional robustness interactions* between the FE observer and FTC control system, as conceptually demonstrated in Fig. 2.6. **Since uncertainties and estimation errors inevitably exist in real systems, the *bidirectional robustness interactions* appear at all observer-based FE and FTC systems.** Without taking into account of these interactions, the well-designed FE observer and FTC controller may not fit well when assembled together, resulting in a closed-loop system with degraded performance and robustness, and even instability.

2.4.2 Motivating Example

This section presents a numerical simulation to further demonstrate the need of robust integration in FE-based FTC system. The example uses the FE observer and FTC controller described in Sect. 2.4.1, which are designed separately by ignoring their mutual interactions.

Consider the linear system

$$\begin{aligned} \dot{x} &= (A + \Delta A)x + Bu + Ff + Dd \\ y &= Cx \end{aligned} \tag{2.22}$$

with

$$A = \begin{bmatrix} 24 & 12 \\ -4.4379 & 2.2189 \end{bmatrix}, \ B = \begin{bmatrix} 1 \\ 0 \end{bmatrix}, \ D = \begin{bmatrix} 1 \\ 1 \end{bmatrix}, \ \Delta A = \delta A, \ F = B, \ C = [1\ 0],$$

where δ is a scalar function satisfying $|\delta| \leq 0.1$.

Within the separated synthesis scheme, the uncertainty term ΔAx is ignored in the observer design. By using the ALSO FE observer described in Sect. 2.4.1, the estimation error system is obtained as

$$\begin{aligned} \dot{e} &= (\bar{A} - L\bar{C})e + \bar{D}\bar{d} \\ z_e &= C_e e \end{aligned} \tag{2.23}$$

where z_e is the performance output with a constant weight C_e. Based on the Bounded Real Lemma (see Sect. 1.7.1), by solving the following optimization problem:

$$\begin{aligned} &\min\ \gamma_e \\ &\text{s.t.} \begin{bmatrix} \text{He}(Q\bar{A} - Y\bar{C}) & Q\bar{D} & C_e^\top \\ \star & -\gamma_e I & 0 \\ \star & \star & -\gamma_e I \end{bmatrix} \prec 0, \\ &Q = Q^\top \succ 0,\ \gamma_e > 0, \end{aligned} \tag{2.24}$$

then the observer gain is obtained as $L = Q^{-1}Y$. This gain ensures that the error system (2.23) is robustly stable against the disturbance $\bar{d}$ with $\|G_{z_e\bar{d}}\|_\infty < \gamma_e$.

Solving (2.24) with $C_e = I_2$ gives

$$\gamma_e = 4.4671,\ L = [1180\ 54150\ \ -516500]^\top. \tag{2.25}$$

In the controller design, the estimation error e is ignored by assuming that the real state x is available, so $\hat{x}$ is replaced by x. Hence, the closed-loop FTC system is obtained as

$$\begin{aligned} \dot{x} &= (A + BK)x + \Delta A x + Dd \\ z_c &= C_x x \end{aligned} \tag{2.26}$$

where z_c is the performance output with a weight C_x.

Notice that the uncertainty matrix ΔA satisfies $\Delta A = \mathcal{M}\mathcal{F}\mathcal{N}$, with $\mathcal{M} = \mathcal{F} = I_2$ and $\mathcal{N} = 0.1\,A$. This condition on the uncertainty matrix has been widely used in the area of robust control. Based on the Bounded Real Lemma, by solving the following optimization problem:

$$\min \ \gamma_c \tag{2.27}$$

$$\text{s.t.} \quad \begin{bmatrix} \mathrm{He}(AP + BX) & D & \mathcal{M} & P\mathcal{N}^\top & PC_x^\top \\ \star & -\gamma_c I & 0 & 0 & 0 \\ \star & \star & -I & 0 & 0 \\ \star & \star & \star & -I & 0 \\ \star & \star & \star & \star & -\gamma_c I \end{bmatrix} \prec 0,$$

$$P = P^\top \succ 0, \ \gamma_c > 0,$$

then the control gain is obtained as $K = XP^{-1}$. This gain ensures that the closed-loop system (2.26) is robustly stable against the uncertainty $\Delta A x$ and disturbance d with $\|G_{z_c d}\|_\infty < \gamma_c$.

Solving (2.27) with $C_x = I_2$ gives

$$\gamma_c = 1.0321, \ \ K = [-251.6 \ 1322]. \tag{2.28}$$

The system is simulated under initial conditions $x(0) = [0.1 \ 0]^\top$ and $\hat{x}(0) = 0$, and the following fault and disturbance:

$$f(t) = \begin{cases} 0, & 0\text{ s} \le t \le 10\text{ s} \\ 0.1, & 10\text{ s} < t \le 20\text{ s} \\ 0, & 20\text{ s} < t \le 30\text{ s} \end{cases}, \ \ d(t) = 0.001\sin(t), \ 0\text{ s} \le t \le 30\text{ s}.$$

To verify the nominal control performance, the system is simulated by using the baseline controller $u = Kx$. The state response shown in Fig. 2.7 indicates that the baseline control achieves acceptable closed-loop performance even in the presence of uncertainty $\Delta A = 0.1A$, disturbance d and fault f.

To verify the closed-loop system performance when assembling FE and FTC, two cases are simulated: (1) the system has no uncertainty, i.e. $\Delta A = 0$, and (2) the system has uncertainty $\Delta A = 0.018A$.

Results of the first case are depicted in Figs. 2.8 and 2.9. It is seen from Fig. 2.8 that the FE observer obtains accurate estimation of the state and fault, though there are small errors at the transients. It is shown in Fig. 2.9 that the two state variables are both controlled to be zero with small errors at the transients. These results confirm that the FE-based FTC design achieves good estimation and stabilization performance in the absence of uncertainty. However, when the uncertainty occurs (even small

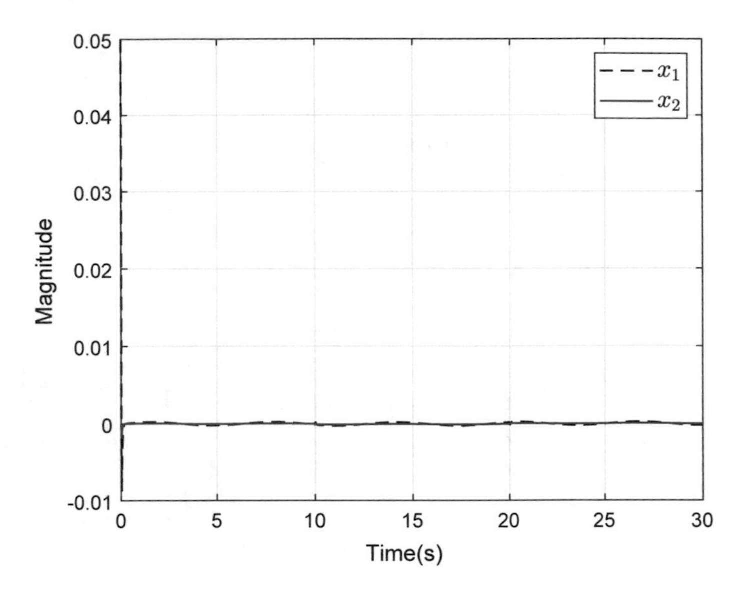

Fig. 2.7 State response under baseline control $u = Kx$ with $\Delta A = 0.1A$

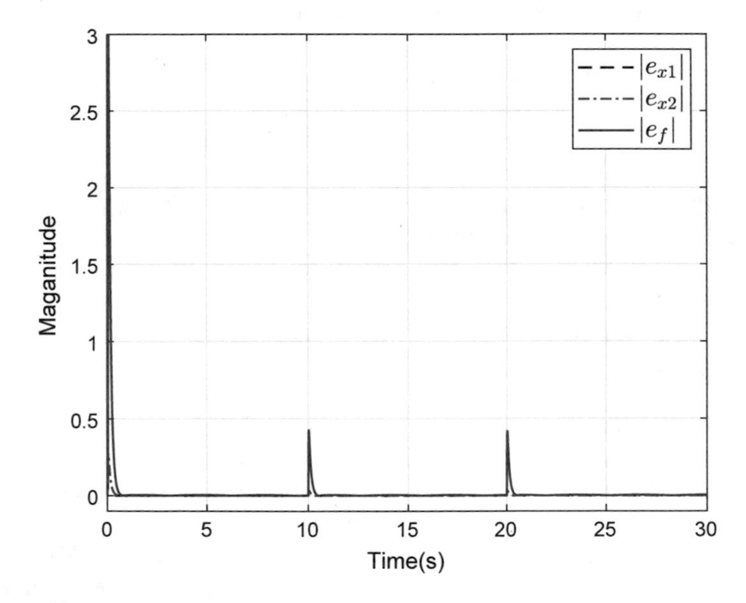

Fig. 2.8 Estimation performance under FE-based FTC with $\Delta A = 0$

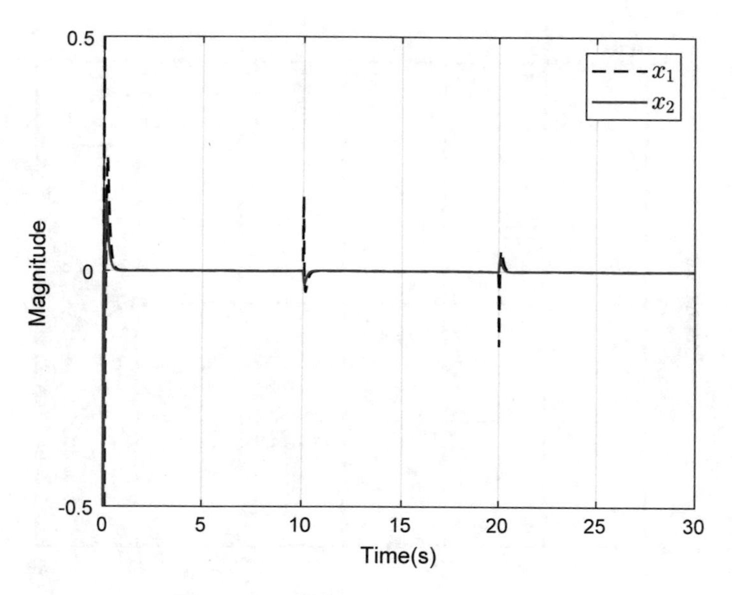

Fig. 2.9 State response under FE-based FTC with $\Delta A = 0$

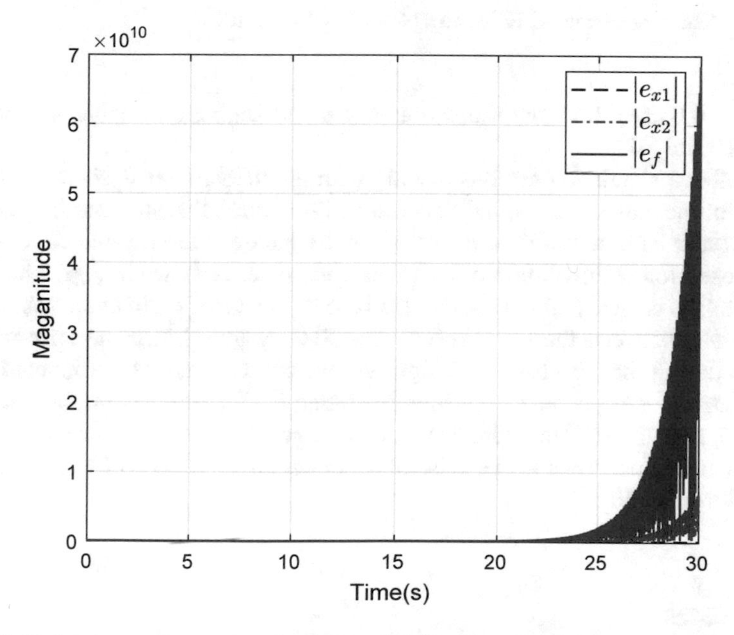

Fig. 2.10 Estimation performance under FE-based FTC with $\Delta A = 0.018A$

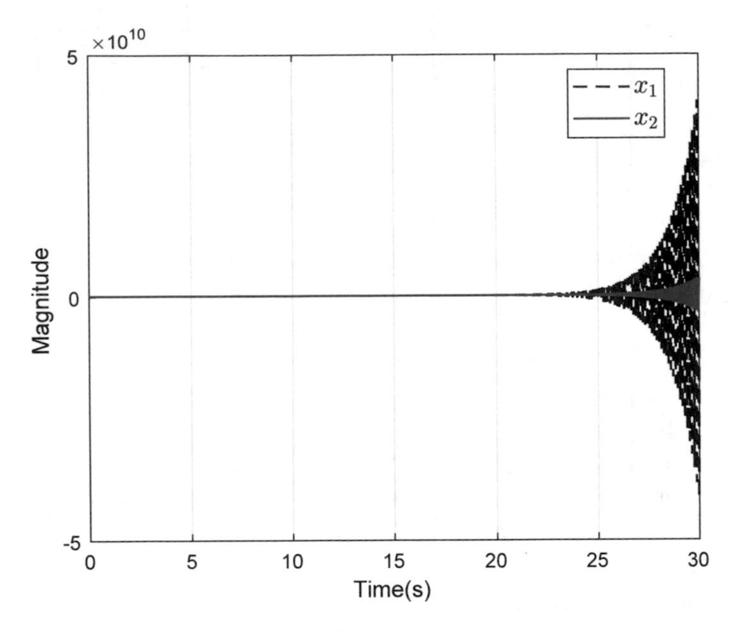

Fig. 2.11 State response under FE-based FTC with $\Delta A = 0.018A$

as $\Delta A = 0.018A$), both the estimation errors and state are unstable, as shown in Figs. 2.10 and 2.11.

The above simulation results coincide with the theoretic analysis in Sect. 2.4.1. When no uncertainty exists, the Separation Principle holds. In such case, the FE observer and FTC controller can be designed separately and assembled to achieve good closed-loop estimation and control performance, as shown in Figs. 2.8 and 2.9. However, the presence of uncertainty leads to existence of *bidirectional robustness interactions* between the FE observer and FTC system. Since the uncertainty is ignored during the FE observer design while estimation errors are ignored in the FTC controller design, the closed-loop FE-based FTC system is unstable, as shown in Figs. 2.10 and 2.11. Therefore, the simulation results clearly confirm the theoretic analysis and demonstrate the need in robust integration of FE and FTC to guarantee closed-loop stability.

References

Chadli M, Abdo A, Ding SX (2013) H_-/H_∞ fault detection filter design for discrete-time Takagi-Sugeno fuzzy system. Automatica 49(7):1996–2005

Chen L, Patton R, Goupil P (2016) Robust fault estimation using an LPV reference model: ADDSAFE benchmark case study. Control Eng Pract 49:194–203

Cieslak J, Efimov D, Henry D (2015) Transient management of a supervisory fault-tolerant control scheme based on dwell-time conditions. Int J Adapt Control Signal Process 29(1):123–142

Davoodi MR, Talebi H, Momeni HR (2011) A novel simultaneous fault detection and control approach based on dynamic observer. In: Proceedings of the IFAC world congress. IEEE, pp 12036–12041

Ding SX (2009) Integrated design of feedback controllers and fault detectors. Annu Rev Control 33(2):124–135

Du Y, Budman H, Duever TA (2016) Integration of fault diagnosis and control based on a trade-off between fault detectability and closed loop performance. J Process Control 38:42–53

Gao Z, Ding SX (2007) Actuator fault robust estimation and fault-tolerant control for a class of nonlinear descriptor systems. Automatica 43(5):912–920

Grenaille S, Henry D, Zolghadri A (2008) A method for designing fault diagnosis filters for LPV polytopic systems. J Control Sci Eng 2008:1

Hou M, Patton R (1996) An LMI approach to H_-/H_∞ fault detection observers. In: Proceedings of the UKACC international conference on control, vol 1. IET, pp 305–310

Jaimoukha IM, Li Z, Papakos V (2006) A matrix factorization solution to the H_-/H_∞ fault detection problem. Automatica 42(11):1907–1912

Jiang B, Staroswiecki M, Cocquempot V (2006) Fault accommodation for nonlinear dynamic systems. IEEE Trans Autom Control 51(9):1578–1583

Li X, Zhou K (2009) A time domain approach to robust fault detection of linear time-varying systems. Automatica 45(1):94–102

de Loza AF, Cieslak J, Henry D, Zolghadri A, Fridman LM (2015) Output tracking of systems subjected to perturbations and a class of actuator faults based on HOSM observation and identification. Automatica 59:200–205

Mhaskar P, Gani A, El-Farra NH, McFall C, Christofides PD, Davis JF (2006) Integrated fault-detection and fault-tolerant control of process systems. AIChE J 52(6):2129–2148

Nett C, Jacobson C, Miller A (1988) An integrated approach to controls and diagnostics: the 4-parameter controller. In: Proceedings of the American control conference, pp 824–835

Patton RJ (1997) Fault-tolerant control systems: the 1997 situation. In: Proceedings of IFAC symposium on fault detection supervision and safety for technical processes, vol 3, pp 1033–1054

Wang JL, Yang GH, Liu J (2007) An LMI approach to H_- index and mixed H_-/H_∞ fault detection observer design. Automatica 43(9):1656–1665

Weng Z, Patton RJ, Cui P (2008) Integrated design of robust controller and fault estimator for linear parameter varying systems. In: Proceedings of the IFAC world congress, pp 4535–4539

Yang H, Jiang B, Staroswiecki M (2009) Supervisory fault tolerant control for a class of uncertain nonlinear systems. Automatica 45(10):2319–2324

Zhai D, Lu AY, Li JH, Zhang QL (2016) Simultaneous fault detection and control for switched linear systems with mode-dependent average dwell-time. Appl Math Comput 273:767–792

Zhang K, Jiang B, Shi P (2012) Observer-based fault estimation and accomodation for dynamic systems, vol 436. Springer, Berlin

Zhang P, Ding SX (2008) An integrated trade-off design of observer based fault detection systems. Automatica 44(7):1886–1894

Zhong GX, Yang GH (2015) Robust control and fault detection for continuous-time switched systems subject to a dwell time constraint. Int J Robust Nonlinear Control 25(18):3799–3817

Part II
Strategies for Robust Integration of FE and FTC

Chapter 3
Sequential Integration of FE and FTC

3.1 Introduction

As discussed in Chap. 2, the FE-based FTC system involves the process of FE used to estimate the fault and FTC to compensate the fault effects. The combination of estimation and control leads to *bidirectional robustness interactions*, which breaks down the Separation Principle and gives rise to a requirement of integrating together the designs of FE and FTC functions. The design of each unit is challenging, while the interactions between them impose extra complexity. Under the nominal setting when the system is fault-free, a two-step strategy (following a framework similar to Fig. 3.1) is developed in Zhu and Pagilla (2007) for robust observer-based control for interconnected systems, where the robust controller is designed in the first step and then used to determine the robust observer in the second step. The difference between this two-step strategy and the separated design strategy in Luenberger (1971) is that the effects of control design, i.e. the *unidirectional robustness interaction*, is effectively taken into account in the observer design. This implies that the observer design is highly dependent on the controller.

Following this framework, a two-step FE-based FTC is developed in Shi and Patton (2015) for descriptor linear parameter varying system, where the FTC controller is designed in the first step and then used in the second step to determine the FE observer. However, system uncertainty is not considered in their work. In the spirit of this two-step strategy, a sequential strategy is proposed in this chapter for robust integration of FE and FTC, as shown in Fig. 3.1. The sequential manner reduces the design complexity and can achieve a suboptimal overall closed-loop system. It is worth noting that the effects of FE uncertainty on the FTC control system cannot be handled in this scheme. Moreover, the FE observer design is highly affected by the FTC controller designed in the first step, which will be discussed in detail in this chapter.

© The Author(s), under exclusive license to Springer Nature Switzerland AG 2021

J. Lan and R. J. Patton, *Robust Integration of Model-Based Fault Estimation and Fault-Tolerant Control*, Advances in Industrial Control,
https://doi.org/10.1007/978-3-030-58760-4_3

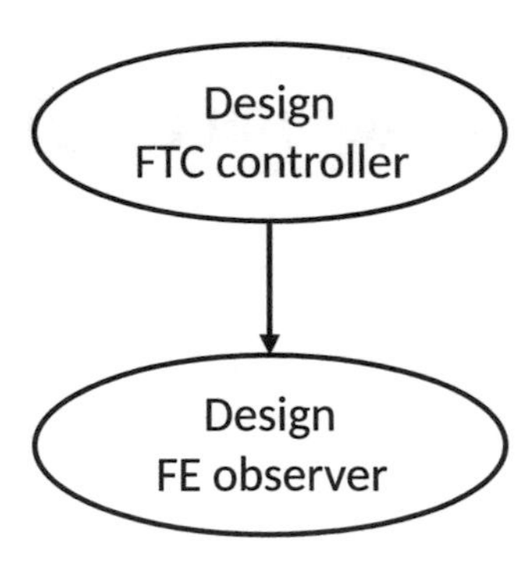

Fig. 3.1 Sequential integration of FE and FTC

3.2 Problem Description

Consider a class of uncertain linear systems described by

$$\begin{aligned} \dot{x} &= (A + \Delta A)x + Bu + Ff + Dd \\ y &= Cx \end{aligned} \tag{3.1}$$

where $x \in \mathbb{R}^n$, $u \in \mathbb{R}^m$, $y \in \mathbb{R}^p$, $f \in \mathbb{R}^q$, and $d \in \mathbb{R}^l$ are the state, control input, measured output, actuator fault, and external disturbance, respectively. The constant matrices A, B, F, D and C are known and of compatible dimensions. The matrix ΔA denotes the unknown system uncertainty. The system satisfies the assumptions below:

Assumption 3.1 The pair (A, B) is controllable and the triple (A, F, C) has no invariant zeros in the closed right-half complex plane. The actuator fault is matched, i.e. $\text{rank}[B\ F] = \text{rank}(B) = m$.

Assumption 3.2 The uncertainty matrix ΔA is norm-bounded (energy bounded) with the form $\Delta A = \mathcal{M}\mathcal{F}(t)\mathcal{N}$, where $\mathcal{M}$ and $\mathcal{N}$ are known constant matrices with appropriate dimensions, and $\mathcal{F}(t)$ is an unknown (time-varying) matrix satisfying $\mathcal{F}^\top(t)\mathcal{F}(t) \preceq I$.

Assumption 3.3 The fault f and disturbance d are norm-bounded. f has bounded first-order and second-order differentials.

Remark 3.1 The first part of Assumption 3.1 states some standard requirements for observer-based control systems, ensuring controllability of the system and observability of the state and fault. The rank condition $\text{rank}[B\ F] = \text{rank}(B)$ ensures that the fault f lies in the range space spanned by the input u so that the fault effects can be compensated through control actions, as described in Definition 1.7. Assumption 3.2 gives a general representation of unmatched uncertainty matrix used in robust control theory. Assumption 3.3 implies that the fault and disturbance considered here are norm-bounded with unknown upper bounds, which is rational in practical applications. The existence of differentials is required for the FE observer design.

This chapter aims to address the following problem:

Problem 3.1 Design a linear control law (FTC) to compensate the fault effects and stabilize the system state, by using the accurate estimate of state and fault obtained simultaneously from an observer (FE), in the presence of uncertainty and disturbance.

3.3 FE Observer and FTC Controller

This section describes designs of an observer to estimate both the system state x and fault f, and an FTC law to compensate the fault and stabilize the system (3.1).

3.3.1 FE Observer

By considering f and $\dot{f}$ as auxiliary state, the system (3.1) is augmented as

$$\begin{aligned} \dot{\bar{x}} &= \bar{A}\bar{x} + \Delta\bar{A}\bar{x} + \bar{B}u + \bar{D}\bar{d} \\ y &= \bar{C}\bar{x} \end{aligned} \tag{3.2}$$

with

$$\bar{x} = \begin{bmatrix} x \\ f \\ \dot{f} \end{bmatrix}, \ \bar{d} = \begin{bmatrix} d \\ \ddot{f} \end{bmatrix}, \ \bar{A} = \begin{bmatrix} A & F & 0 \\ 0 & 0 & I_q \\ 0 & 0 & 0 \end{bmatrix}, \ \Delta\bar{A} = \begin{bmatrix} \Delta A & 0 & 0 \\ 0 & 0 & 0 \\ 0 & 0 & 0 \end{bmatrix},$$

$$\bar{B} = \begin{bmatrix} B \\ 0 \\ 0 \end{bmatrix}, \ \bar{D} = \begin{bmatrix} D & 0 \\ 0 & 0 \\ 0 & I_q \end{bmatrix}, \ \bar{C} = [C \ 0 \ 0].$$

Remark 3.2 The reason of defining f and $\dot{f}$ as auxiliary state and $\ddot{f}$ as disturbance perturbing the augmented system is given below. If assuming $\dot{f} = 0$, then f can only be constant or slowly time-varying faults, e.g.. $f(t) = a$, $f(t) = e^{-b_1 t} + b_2$ and $f(t) = c\sin(\phi t)$, where a, b_1, b_2, c and ϕ are constants while b_1 and ϕ taking small values within (0, 1). In this chapter f is modelled as a second-order system which allows the proposed observer to estimate a wider range of faults. Moreover, if just considering f as auxiliary state with $\dot{f}$ as disturbance, then the augmented system may be affected by a larger disturbance $\dot{f}$ when $\dot{f} \geq \ddot{f}$. To cover more general faults, one can use a higher order system to model the fault, by regarding $f, \dot{f}, \ddot{f}, \ldots, f^{(s)}$, for an integer $s > 2$, as auxiliary state, see, for example, the results in Lan and Patton (2016). However, this will increase the observer dimension that is given by $(n + sq)$.

Since the augmented system (3.2) is to be used for designing the FE observer, its observability must be verified. This is shown in the proposition below.

Proposition 3.1 *Under Assumption 3.1, the augmented system (3.2) is observable.*

Proof It is stated in Assumption 3.1 that the triple (A, F, C) has no invariant zeros in the closed right-half complex plane. This is equivalent to

$$\text{rank}\begin{bmatrix} sI_n - A & F \\ C & 0 \end{bmatrix} = n + q, \ \forall s \in \mathbb{C}.$$

Under this condition, the augmented system (3.2) satisfies

$$\begin{aligned}
\text{rank}\begin{bmatrix} sI_{n+2q} - \bar{A} \\ \bar{C} \end{bmatrix} &= \text{rank}\begin{bmatrix} sI_n - A & -F & 0 \\ 0 & sI_q & -I_q \\ 0 & 0 & sI_q \\ C & 0 & 0 \end{bmatrix} \\
&= \text{rank}\left[\begin{array}{cc|c} sI_n - A & -F & 0 \\ C & 0 & 0 \\ \hline 0 & sI_q & -I_q \\ 0 & 0 & sI_q \end{array}\right] \\
&= n + 2q, \ \forall s \in \mathbb{C}.
\end{aligned}$$

Therefore, the pair $(\bar{A}, \bar{C})$ is observable, so is the augmented system (3.2). □

The state $\bar{x}$ is estimated by the augmented Luenberger state observer (ALSO):

$$\begin{aligned}
\dot{\hat{\bar{x}}} &= \bar{A}\hat{\bar{x}} + \bar{B}u + L(y - \hat{y}) \\
\hat{y} &= \bar{C}\hat{\bar{x}}
\end{aligned} \tag{3.3}$$

where $\hat{\bar{x}} \in \mathbb{R}^{n+2q}$ and $\hat{y} \in \mathbb{R}^p$ are the estimate of $\bar{x}$ and y, respectively. The observer gain $L \in \mathbb{R}^{(n+2q)\times p}$ is to be determined. The estimates of x and f can be calculated by $\hat{x} = [I_n \ 0_{n\times 2q}]\hat{\bar{x}}$ and $\hat{f} = [0_{q\times n} \ I_q \ 0_{q\times q}]\hat{\bar{x}}$, respectively.

Define the estimation error as $e = \bar{x} - \hat{\bar{x}}$, then subtracting (3.3) from (3.2) yields the error dynamics

$$\dot{e} = (\bar{A} - L\bar{C})e + \Delta\bar{A}\bar{x} + \bar{D}\bar{d}. \tag{3.4}$$

3.3.2 FTC Controller

To compensate the fault f and stabilize the state x using the available estimate $\hat{\bar{x}}$, the FTC controller is designed as

$$u = u_x + u_f \tag{3.5}$$

which consists of a state feedback baseline controller u_x and a fault compensator u_f. The baseline controller is designed as $u_x = K\hat{x}$ with a constant gain $K \in \mathbb{R}^{m\times n}$. Since the fault is matched, u_f is optimally designed as $u_f = -B^{\dagger}F\hat{f}$ such that $Bu_f + Ff = F(f - \hat{f})$. Hence, if the fault is accurately estimated, then $f - \hat{f} = 0$ and the fault can be completely compensated.

Substituting the controller (3.5) into (3.1) gives the closed-loop FTC system

$$\dot{x} = (A + BK)x + \Delta Ax + B_e e + Dd \tag{3.6}$$

where $B_e = [-BK\ F\ 0]$.

3.4 Integration of FE and FTC

Combining (3.4) and (3.6) gives the composite closed-loop system

$$\begin{aligned} \dot{x} &= (A + BK)x + \Delta Ax + B_e e + \tilde{D}\bar{d} \\ \dot{e} &= (\bar{A} - L\bar{C})e + \Delta\bar{A}\bar{x} + \bar{D}\bar{d} \\ z &= \mathrm{diag}(C_x x, C_e e) \end{aligned} \tag{3.7}$$

where $\tilde{D} = [D\ 0]$, $z \in \mathbb{R}^{2n+2q}$ is the performance output, and C_x and C_e are given constant weight matrices of appropriate dimensions.

In order to address Problem 3.1, the control gain K and the observer gain L should be designed to stabilize the system (3.7) in the presence of uncertainties and disturbances. One may spot that this design problem is very similar to the classical robust observer-based control problem studied in Lien (2004), except that here e includes also the fault estimation error. Nevertheless, the well-known robust control theory can be adopted for the FE-based FTC design problem studied here. Based on the composite closed-loop system (3.7), the following theorem is given.

Theorem 3.1 *Under Assumptions 3.1–3.3, the composite closed-loop system (3.7) is stable and satisfies the H_∞ performance $\|G_{z\bar{d}}\|_\infty < \gamma$, if the following optimization problem is feasible:*

$$\min_{K,L,P_0,Q,\gamma} \gamma \tag{3.8}$$

$$\text{s.t.} \quad \left[\begin{array}{c|cc} F_c & S_1 & S_2 \\ \hline \star & W_o & S_3 \\ \star & \star & \Lambda_o \end{array}\right] \prec 0 \tag{3.9}$$

$$P_0 = P_0^\top \succ 0,\ Q = Q^\top \succ 0,\ \gamma > 0 \tag{3.10}$$

with

$$F_c = \begin{bmatrix} \mathrm{He}(P_0(A+BK)) & P_0\tilde{D} & P_0\mathcal{M} & \mathcal{N}^\top & C_x^\top \\ \star & -\gamma I & 0 & 0 & 0 \\ \star & \star & -I & 0 & 0 \\ \star & \star & \star & -I & 0 \\ \star & \star & \star & \star & -\gamma I \end{bmatrix}, \; S_1 = \begin{bmatrix} P_0 B_e \\ (Q\bar{D})^\top \\ 0 \\ 0 \\ 0 \end{bmatrix}, \; S_2 = \begin{bmatrix} \mathcal{N}^\top & 0 & 0 \\ 0 & 0 & 0 \\ 0 & 0 & 0 \\ 0 & 0 & 0 \\ 0 & 0 & 0 \end{bmatrix},$$

$$W_o = \mathrm{He}(Q(\bar{A} - L\bar{C})), \; S_3 = [0 \; Q\bar{\mathcal{M}} \; C_e^\top],$$
$$\Lambda_o = \mathrm{diag}(-I, -I, -\gamma I), \; B_e = [-BK \; F \; 0].$$

Proof Consider the Lyapunov function $V_e = e^\top Q e$ for the error system in (3.7) with a s.p.d. matrix $Q \in \mathbb{R}^{(n+2q)\times(n+2q)}$. Under Assumption 3.2, it can be derived that

$$\mathrm{He}(e^\top Q \Delta \bar{A} \bar{x}) \le e^\top Q\bar{\mathcal{M}}(Q\bar{\mathcal{M}})^\top e + x^\top \mathcal{N}^\top \mathcal{N} x,$$

where $\bar{\mathcal{M}} = [\mathcal{M}^\top \; 0]^\top$. Hence, the time derivative of V_e along the error system is obtained as

$$\begin{aligned} \dot{V}_e \le\; & e^\top \left[\mathrm{He}(Q(\bar{A} - L\bar{C})) + Q\bar{\mathcal{M}}(Q\bar{\mathcal{M}})^\top\right] e \\ & + \mathrm{He}(e^\top Q\bar{D}\bar{d}) + x^\top \mathcal{N}^\top \mathcal{N} x. \end{aligned} \tag{3.11}$$

Consider the Lyapunov function $V_x = x^\top P_0 x$ for the closed-loop FTC system with a s.p.d. matrix $P_0 \in \mathbb{R}^{n\times n}$. Under Assumption 3.2, it is derived that

$$\mathrm{He}(x^\top P_0 \Delta A x) \le x^\top P_0\mathcal{M}(P_0\mathcal{M})^\top x + x^\top \mathcal{N}^\top \mathcal{N} x.$$

Hence, the derivative of V_x along the closed-loop FTC system is obtained as

$$\begin{aligned} \dot{V}_x \le\; & x^\top \left[\mathrm{He}(P_0(A+BK)) + P_0\mathcal{M}(P_0\mathcal{M})^\top + \mathcal{N}^\top\mathcal{N}\right] x \\ & + \mathrm{He}(x^\top P_0 B_e e + x^\top P_0 \tilde{D}\bar{d}). \end{aligned} \tag{3.12}$$

Define $V = V_x + V_e$, then combining (3.11) and (3.12) yields

$$\begin{aligned} \dot{V} \le\; & x^\top \left[\mathrm{He}(P_0(A+BK)) + P_0\mathcal{M}(P_0\mathcal{M})^\top + 2\mathcal{N}^\top\mathcal{N}\right] x \\ & + e^\top \left[\mathrm{He}(Q(\bar{A} - L\bar{C})) + Q\bar{\mathcal{M}}(Q\bar{\mathcal{M}})^\top\right] e \\ & + \mathrm{He}(x^\top P_0 B_e e + x^\top P_0\tilde{D}\bar{d} + e^\top Q\bar{D}\bar{d}). \end{aligned} \tag{3.13}$$

The H_∞ performance $\|G_{z\bar{d}}\|_\infty < \gamma$, where $\gamma > 0$, can be equivalently quantified by

$$J = \int_0^\infty \left(\frac{1}{\gamma} z(t)^\top z(t) - \gamma \bar{d}(t)^\top \bar{d}(t)\right) \mathrm{d}t < 0. \tag{3.14}$$

Under zero initial condition $z(0) = 0$, it holds that

$$
\begin{aligned}
J &= \int_0^{\infty} \Big(\frac{1}{\gamma} z(t)^\top z(t) - \gamma \bar{d}(t)^\top \bar{d}(t) + \dot{V}(t)\Big) \mathrm{d}t - \int_0^{\infty} \dot{V}(t) \mathrm{d}t \\
&= \int_0^{\infty} \Big(\frac{1}{\gamma} z(t)^\top z(t) - \gamma \bar{d}(t)^\top \bar{d}(t) + \dot{V}(t)\Big) \mathrm{d}t - V(\infty) + V(0) \\
&\le \int_0^{\infty} \Big(\frac{1}{\gamma} z(t)^\top z(t) - \gamma \bar{d}(t)^\top \bar{d}(t) + \dot{V}(t)\Big) \mathrm{d}t.
\end{aligned} \tag{3.15}
$$

A sufficient condition for (3.15) is given by

$$
\frac{1}{\gamma} z^\top z - \gamma \bar{d}^\top \bar{d} + \dot{V} < 0. \tag{3.16}
$$

Notice that $z^\top z = x^\top C_x^\top C_x x + e^\top C_e^\top C_e e$. Substituting (3.13) into (3.16) gives

$$
\begin{bmatrix} x \\ \bar{d} \\ e \end{bmatrix}^\top \begin{bmatrix} \Phi_{1,1} & \Phi_{1,2} \\ \star & \Phi_{2,2} \end{bmatrix} \begin{bmatrix} x \\ \bar{d} \\ e \end{bmatrix} < 0 \tag{3.17}
$$

and equivalently,

$$
\begin{bmatrix} \Phi_{1,1} & \Phi_{1,2} \\ \star & \Phi_{2,2} \end{bmatrix} \prec 0, \tag{3.18}
$$

where

$$
\Phi_{1,1} = \begin{bmatrix} \mathrm{He}(P_0(A+BK)) + P_0\mathcal{M}(P_0\mathcal{M})^\top + 2\mathcal{N}^\top\mathcal{N} + \frac{1}{\gamma} C_x^\top C_x & P_0\tilde{D} \\ \star & -\gamma I \end{bmatrix},
$$

$$
\Phi_{1,2} = \begin{bmatrix} P_0 B_e \\ (Q\bar{D})^\top \end{bmatrix}, \quad \Phi_{2,2} = \mathrm{He}(Q(\bar{A} - L\bar{C})) + Q\bar{\mathcal{M}}(Q\bar{\mathcal{M}})^\top + \frac{1}{\gamma} C_e^\top C_e.
$$

Applying Schur Complement (see Sect. 1.7.1) repeatedly to (3.18) and rearranging some entries, then it yields

$$
\begin{bmatrix}
\mathrm{He}(P_0(A+BK)) & P_0\tilde{D} & P_0\mathcal{M} & \mathcal{N}^\top & C_x^\top & P_0 B_e & \mathcal{N}^\top & 0 & 0 \\
\star & -\gamma I & 0 & 0 & 0 & (Q\bar{D})^\top & 0 & 0 & 0 \\
\star & \star & -I & 0 & 0 & 0 & 0 & 0 & 0 \\
\star & \star & \star & -I & 0 & 0 & 0 & 0 & 0 \\
\star & \star & \star & \star & -\gamma I & 0 & 0 & 0 & 0 \\
\star & \star & \star & \star & \star & \mathrm{He}(Q(\bar{A} - L\bar{C})) & 0 & Q\bar{\mathcal{M}} & C_e^\top \\
\star & \star & \star & \star & \star & \star & -I & 0 & 0 \\
\star & \star & \star & \star & \star & \star & \star & -I & 0 \\
\star & \star & \star & \star & \star & \star & \star & \star & -\gamma I
\end{bmatrix} \prec 0. \tag{3.19}
$$

Therefore, the design of K and L is formulated as the optimization problem (3.8).□

3.5 Feasibility Analysis

This section analyzes feasibility of the optimization problem (3.8). Before proceeding, the following lemma is extended from the Bounded Real Lemma (see Sect. 1.7.1).

Lemma 3.1 *Consider a linear system*

$$\begin{aligned} \dot{x} &= (A + \Delta A)x + Dd \\ z &= Cx \end{aligned} \tag{3.20}$$

where the uncertainty ΔA satisfies Assumption 3.2. This system is stable with the H_∞ performance $\|G_{zd}\|_\infty < \gamma$ if there exists a s.p.d. matrix $\mathcal{P}$ and a scalar $\gamma > 0$ such that

$$\begin{bmatrix} \mathrm{He}(\mathcal{P}A) & \mathcal{P}D & \mathcal{P}\mathcal{M} & \mathcal{N}^\top & C^\top \\ \star & -\gamma I & 0 & 0 & 0 \\ \star & \star & -I & 0 & 0 \\ \star & \star & \star & -I & 0 \\ \star & \star & \star & \star & -\gamma I \end{bmatrix} \prec 0. \tag{3.21}$$

3.5.1 Feasibility Under Ideal Case

We first analyze feasibility of the optimization problem (3.8) in the absence of uncertainty, i.e. $\Delta A = 0$.

Proposition 3.2 *Under the ideal case, the matrix inequality (3.21) is feasible if A is Hurwitz stable.*

Proof When $\Delta A = 0$, the matrix inequality (3.21) becomes

$$\begin{bmatrix} \mathrm{He}(\mathcal{P}A) & \mathcal{P}D & C^\top \\ \star & -\gamma I & 0 \\ \star & \star & -\gamma I \end{bmatrix} \prec 0. \tag{3.22}$$

Rewriting (3.22) into the compact form

$$\mathcal{F}_c = \begin{bmatrix} \mathcal{W}_c & R^\top \\ \star & \Gamma_\gamma \end{bmatrix} \prec 0 \tag{3.23}$$

with $\mathcal{W}_c = \mathrm{He}(\mathcal{P}A)$, $R^\top = [\mathcal{P}D \ C^\top]$ and $\Gamma_\gamma = \mathrm{diag}(-\gamma I, -\gamma I)$.

If A is Hurwitz stable, then there always exists a s.p.d. matrix P such that

$$\mathcal{W}_c = -Q \prec 0 \tag{3.24}$$

for any s.p.d. matrix Q. Applying Schur Complement to (3.23) gives

$$\mathcal{F}_c \prec 0 \iff \Gamma_\gamma \prec 0 \text{ and } \mathcal{W}_c - R^\top \Gamma_\gamma^{-1} R \prec 0. \tag{3.25}$$

Therefore, from the last condition of (3.25), $\mathcal{F}_c \prec 0$ is guaranteed by the existence of a large enough γ. □

With Proposition 3.2 at hand, the following statement can be given.

Lemma 3.2 *Under the ideal case, the optimization problem (3.8) is feasible if the pairs* (A, B) *and* (A, C) *are controllable and observable, respectively.*

Proof When $\Delta A = 0$, the columns and rows associated with $\mathcal{M}$, $\mathcal{N}$ and $\bar{\mathcal{M}}$ in (3.9) can be eliminated. The inequality (3.9) can then be rewritten into the compact form

$$\begin{bmatrix} F_c & S \\ \star & F_o \end{bmatrix} \prec 0 \tag{3.26}$$

with $S = [S_1 \; S_2]$ and $F_o = \begin{bmatrix} W_o & S_3 \\ \star & \Lambda_o \end{bmatrix}$.

If the pair (A, B) is controllable, then there exists a matrix K such that $(A + BK)$ is Hurwitz stable. Hence, it follows directly from Proposition 3.2 that there always exists a s.p.d. matrix P_0 and a large enough positive scalar γ satisfying

$$F_c \prec 0. \tag{3.27}$$

By using Schur Complement, it holds that

$$F_o = \begin{bmatrix} W_o & S_3 \\ \star & \Lambda_o \end{bmatrix} \prec 0 \iff \Lambda_o \prec 0 \text{ and } W_o - S_3 \Lambda_o^{-1} S_3^\top \prec 0. \tag{3.28}$$

Notice that $W_o \prec 0$ is satisfied if the pair (A, C) is observable. Hence, from the last condition of (3.28), $F_o \prec 0$ is guaranteed by the observability of (A, C) and the existence of a large enough γ.

Further applying Schur Complement to (3.26) gives

$$\begin{bmatrix} F_c & S \\ \star & F_o \end{bmatrix} \prec 0 \iff F_c \prec 0 \text{ and } F_o - S^\top F_c^{-1} S \prec 0. \tag{3.29}$$

By using (3.27), (3.28) and (3.29), the condition (3.26) is feasible for a large enough γ and so is the optimization problem (3.8). □

The result claimed in Lemma 3.2 coincides with the Separation Principle. It implies that the separately designed FE observer and FTC controller are able to achieve good closed-loop FE-based FTC system performance when being assembled together. However, as will be shown in the next subsection, this result is inapplicable to the uncertain case.

3.5.2 Feasibility Under Uncertain Case

This subsection further analyzes feasibility of the optimization problem (3.8) in the presence of uncertainty, i.e. $\Delta A \neq 0$.

Proposition 3.3 *Under the uncertain case, the matrix inequality (3.21) is feasible if there exists a s.p.d. matrix $\mathcal{P}$ such that*

$$\begin{bmatrix} \mathrm{He}(\mathcal{P}A) & \mathcal{P}\mathcal{M} & \mathcal{N}^\top \\ \star & -I & 0 \\ \star & \star & -I \end{bmatrix} \prec 0. \tag{3.30}$$

Proof The proof is sketched below. Rearranging some entries of (3.21) gives

$$\begin{bmatrix} \mathrm{He}(\mathcal{P}A) & \mathcal{P}\mathcal{M} & \mathcal{N}^\top & \mathcal{P}D & C^\top \\ \star & -I & 0 & 0 & 0 \\ \star & \star & -I & 0 & 0 \\ \star & \star & \star & -\gamma I & 0 \\ \star & \star & \star & \star & -\gamma I \end{bmatrix} \prec 0. \tag{3.31}$$

Rewriting (3.31) into the compact form

$$\mathcal{F}_c = \begin{bmatrix} \mathcal{W}_c & R^\top \\ \star & \Gamma_\gamma \end{bmatrix} \prec 0 \tag{3.32}$$

with

$$\mathcal{W}_c = \begin{bmatrix} \mathrm{He}(\mathcal{P}A) & \mathcal{P}\mathcal{M} & \mathcal{N}^\top \\ \star & -I & 0 \\ \star & \star & -I \end{bmatrix}, \ R^\top = [\mathcal{P}D \ C^\top], \ \Gamma_\gamma = \mathrm{diag}(-\gamma I, -\gamma I).$$

Applying Schur Complement to (3.32) gives

$$\mathcal{F}_c \prec 0 \iff \Gamma_\gamma \prec 0 \text{ and } \mathcal{W}_c - R^\top \Gamma_\gamma^{-1} R \prec 0. \tag{3.33}$$

Therefore, from the last condition of (3.33), $\mathcal{F}_c \prec 0$ is guaranteed provided (3.30) is satisfied and γ is large enough. □

Remark 3.3 Note that in the presence of uncertainty ΔA, Hurwitz stability of A is no longer the sufficient condition for $\mathcal{W}_c \prec 0$. In this case, by using Schur Complement, one has

$$\mathcal{W}_c \prec 0 \iff \mathrm{He}(\mathcal{P}A) + [\mathcal{P}\mathcal{M} \ \ \mathcal{N}^\top] \begin{bmatrix} \mathcal{M}^\top \mathcal{P} \\ \mathcal{N} \end{bmatrix} \prec 0.$$

The right-hand side condition is not guaranteed even if $\text{He}(\mathcal{P}A) \prec 0$, i.e. A is Hurwitz stable.

Now feasibility of the optimization problem (3.8) in the presence of uncertainty can be analyzed below.

Lemma 3.3 *Under the uncertain case, the optimization problem (3.8) is feasible if the following conditions hold*

$$\text{He}(\mathcal{P}A) + [\mathcal{P}\mathcal{M}\ \mathcal{N}^\top] \begin{bmatrix} \mathcal{M}^\top \mathcal{P} \\ \mathcal{N} \end{bmatrix} \prec 0, \tag{3.34}$$

$$F_o - \begin{bmatrix} S_1^\top \\ S_2^\top \end{bmatrix} F_c^{-1} [S_1\ S_2] \prec 0. \tag{3.35}$$

Proof Rewriting the inequality (3.9) into the compact form

$$\begin{bmatrix} F_c & S \\ \star & F_o \end{bmatrix} \prec 0 \tag{3.36}$$

with $S = [S_1\ S_2]$ and $F_o = \begin{bmatrix} W_o & S_3 \\ \star & \Lambda_o \end{bmatrix}$.

According to Proposition 3.3, if the inequality (3.34) holds, then there is a large enough γ such that

$$F_c \prec 0. \tag{3.37}$$

Further applying Schur Complement to (3.36) gives

$$\begin{bmatrix} F_c & S \\ \star & F_o \end{bmatrix} \prec 0 \iff F_c \prec 0 \text{ and } F_o - S^\top F_c^{-1} S \prec 0. \tag{3.38}$$

Considering (3.37) and (3.38), feasibility of the inequality (3.36) is guaranteed by the satisfaction of (3.34) and (3.35), and so is the optimization problem (3.8). □

As can be seen from (3.34) and (3.35), the presence of uncertainty ΔA leads to the existence of $\mathcal{M}$, $\mathcal{N}$ and $\bar{\mathcal{M}}$ and consequently S_2 and S_3, while the presence of estimation error leads to existence of S_1. Therefore, the results in Lemma 3.3 indicate that controllability and observability of the system (3.1) are not sufficient enough to ensure feasibility of the optimization problem (3.8). The feasible conditions also depend on the matrices $\mathcal{M}$ and $\mathcal{N}$ characterizing the uncertainty effects.

The above analysis gives the sufficient conditions (3.34) and (3.35) ensuring feasibility of the formulated optimization problem (3.8). However, the optimization problem is difficult to solve because the constraint (3.9) is a bilinear matrix inequality (BMI). This is due to the existing nonlinear terms $P_0 B K$ and QL of the decision variables P_0, K, Q and L. To overcome this difficulty, a sequential integration strategy is proposed in the next section for solving the optimization problem.

3.6 A Sequential Integration Strategy

Before presenting the sequential strategy, some preliminaries are provided below.

Let $P = P_0^{-1}$. Pre-and post-multiplying (3.9) with $\mathrm{diag}(P, I, I, I, I, I, I, I)$ and its transpose, respectively, then it yields

$$\left[\begin{array}{c|cc} F_c & S_1 & S_2 \\ \hline \star & W_o & S_3 \\ \star & \star & \Lambda_o \end{array}\right] \prec 0 \tag{3.39}$$

where

$$F_c = \begin{bmatrix} \mathrm{He}((A+BK)P) & \tilde{D} & \mathcal{M} & P\mathcal{N}^\top & PC_x^\top \\ \star & -\gamma I & 0 & 0 & 0 \\ \star & \star & -I & 0 & 0 \\ \star & \star & \star & -I & 0 \\ \star & \star & \star & \star & -\gamma I \end{bmatrix},\ S_1 = \begin{bmatrix} B_e \\ (Q\bar{D})^\top \\ 0 \\ 0 \\ 0 \end{bmatrix},\ S_2 = \begin{bmatrix} P\mathcal{N}^\top & 0 & 0 \\ 0 & 0 & 0 \\ 0 & 0 & 0 \\ 0 & 0 & 0 \\ 0 & 0 & 0 \end{bmatrix},$$

$W_o = \mathrm{He}(Q(\bar{A} - L\bar{C}))$, $S_3 = [0\ Q\bar{\mathcal{M}}\ C_e^\top]$,

$\Lambda_o = \mathrm{diag}(-I, -I, -\gamma I)$, $B_e = [-BK\ F\ 0]$.

Further define $M = KP$ and $Y = QL$, then the optimization problem (3.8) can be reformulated as

$$\min_{K,M,Y,P,Q,\gamma} \gamma \tag{3.40}$$

$$\text{s.t.} \quad \left[\begin{array}{c|cc} \hat{F}_c & \hat{S}_1 & \hat{S}_2 \\ \hline \star & \hat{W}_o & \hat{S}_3 \\ \star & \star & \hat{\Lambda}_o \end{array}\right] \prec 0 \tag{3.41}$$

$$P = P^\top \succ 0,\ Q = Q^\top \succ 0,\ \gamma > 0 \tag{3.42}$$

where

$$\hat{F}_c = \begin{bmatrix} \mathrm{He}(AP+BM) & \tilde{D} & \mathcal{M} & P\mathcal{N}^\top & PC_x^\top \\ \star & -\gamma I & 0 & 0 & 0 \\ \star & \star & -I & 0 & 0 \\ \star & \star & \star & -I & 0 \\ \star & \star & \star & \star & -\gamma I \end{bmatrix},\ \hat{S}_1 = \begin{bmatrix} B_e \\ (Q\bar{D})^\top \\ 0 \\ 0 \\ 0 \end{bmatrix},\ \hat{S}_2 = \begin{bmatrix} P\mathcal{N}^\top & 0 & 0 \\ 0 & 0 & 0 \\ 0 & 0 & 0 \\ 0 & 0 & 0 \\ 0 & 0 & 0 \end{bmatrix},$$

$\hat{W}_o = \mathrm{He}(Q\bar{A} - Y\bar{C})$, $\hat{S}_3 = [0\ Q\bar{\mathcal{M}}\ C_e^\top]$,

$\hat{\Lambda}_o = \mathrm{diag}(-I, -I, -\gamma I)$, $B_e = [-BK\ F\ 0]$.

The sequential strategy to solve the optimization problem (3.40) is summarized in Algorithm 3.1, where an extra decision variable Λ is introduced in Step 2 to ensure feasibility of the optimization problem (3.44). If simply setting $\Lambda = I$, then the matrices M, P and γ obtained from Step 1 may lead to infeasibility of Step 2.

Algorithm 3.1 Sequential strategy for robust integration of FE and FTC

Input: $A, B, F, D, C, \mathcal{M}, \mathcal{N}, \bar{A}, \bar{C}, \bar{D}, \tilde{D}, \bar{\mathcal{M}}, C_x, C_e$

Step 1: Solve the following optimization problem:

$$\min_{M,P,\gamma} \gamma \quad (3.43)$$
$$\text{s.t. } \hat{F}_c \prec 0,\ P = P^\top \succ 0,\ \gamma > 0.$$

Then the control gain is obtained as $K = MP^{-1}$.

Step 2: Solve the following optimization problem:

$$\min_{Y,Q,\Lambda} \text{trace}(\Lambda) \quad (3.44)$$
$$\text{s.t. } \left[\begin{array}{c|cc} \Lambda\hat{F}_c & \hat{S}_1 & \hat{S}_2 \\ \hline \star & \hat{W}_o & \hat{S}_3 \\ \star & \star & \hat{\Lambda}_o \end{array}\right] \prec 0,\ Q = Q^\top \succ 0,\ \Lambda \succ 0,$$

where $\Lambda = \text{diag}(\beta_1 I, \beta_2 I, \cdots, \beta_5 I)$, and M, P and γ are obtained from Step 1. Then the observer gain is obtained as $L = Q^{-1}Y$.

Output: K, L

It can be seen that solutions to the optimization problem in Steps 1 and 2 of Algorithm 3.1 induce the satisfaction of (3.34) and (3.35). According to Lemma 3.3, this thus ensures feasibility of the optimization problem (3.8). Using the sequential strategy in Algorithm 3.1 avoids the BMI difficulty and facilitates implementation of the proposed FE-based FTC design. However, it can be seen that the constraints of the two optimization problems in Algorithm 3.1 still depend on the uncertain matrices $\mathcal{M}$ and $\mathcal{N}$. Hence, feasibility of these two optimization problems cannot be guaranteed by controllability and observability of the system (3.1). Moreover, it is clear that solutions obtained from Step 1 directly affect those of Step 2, because the obtained variables K, P and γ appear at S_2 and Λ_o (which are equivalent to $\hat{S}_2$ and $\hat{\Lambda}_o$). This will be further analyzed through the tutorial example in Sect. 3.8.

3.7 Tutorial Example 1

Consider an aircraft system modified from the example in Sect. 5.7.1 of the book Edwards and Spurgeon (1998). The system is given in the form of (3.1) with

$$A = \begin{bmatrix} 0 & 0 & 1 & 0 & 0 \\ 0 & -0.154 & -0.0042 & 1.54 & 0 \\ 0 & 0.249 & -1 & -5.2 & 0 \\ 0.0386 & -0.996 & -0.0003 & -0.117 & 0 \\ 0 & 0.5 & 0 & 0 & -0.5 \end{bmatrix}, \; B = \begin{bmatrix} 0 & 0 \\ -3.72 & -0.16 \\ 1.685 & -5.6 \\ 0.1 & 0 \\ 0 & 0 \end{bmatrix},$$

$$F = \begin{bmatrix} 0 \\ -3.72 \\ 1.685 \\ 0.1 \\ 0 \end{bmatrix}, \; D = \begin{bmatrix} 0 & 0 \\ 1 & 1 \\ 0 & 1 \\ 1 & 0 \\ 0 & 0 \end{bmatrix}, \; C = \begin{bmatrix} 0 & 1 & 0 & 0 & 0 \\ 0 & 0 & 0 & 1 & 1 \\ 1 & 1 & 1 & 0 & 0 \end{bmatrix},$$

$$\Delta A = 0.05 \sin(0.1\pi t) \times A_p,$$

$$A_p = \begin{bmatrix} 0 & 0 & 0 & 0 & 0 \\ 0 & -0.154 & -0.0042 & 1.54 & 0 \\ 0 & 0.249 & -1 & -5.2 & 0 \\ 0.0386 & -0.996 & -0.0003 & -0.117 & 0 \\ 0 & 0.5 & 0 & 0 & -0.5 \end{bmatrix}.$$

The disturbance vector is given as $d(t) = [d_1(t)^\top \; d_2(t)^\top]^\top$, where $d_1(t)$ is a normally distributed random signal taking values within the interval $[-0.05, 0.05]$ and $d_2(t) = 0.01\cos(t)$. The actuator fault f is characterized by

$$f(t) = \begin{cases} 0, & 0\text{ s} \le t \le 30\text{ s} \\ 0.05t - 1.5, & 30\text{ s} < t \le 50\text{ s} \\ 1, & 50\text{ s} < t \le 80\text{ s} \\ 0, & 80\text{ s} < t \le 120\text{ s} \end{cases}.$$

It is verified that this example system satisfies Assumptions 3.1–3.3. According to Assumption 3.2, the matrices $\mathcal{M}$, $\mathcal{F}(t)$ and $\mathcal{N}$ are given as

$$\mathcal{M} = I_5, \; \mathcal{F}(t) = \sin(0.1\pi t) \times I_5, \; \mathcal{N} = 0.05 \times A_p.$$

Solving the optimization problem (3.43) gives

$$\gamma = 5.4888,$$

$$K = \begin{bmatrix} 3.4627 & 0.7450 & 4.3578 & -0.4691 & 0.0296 \\ 1.6998 & -2.4849 & 1.6535 & 1.9392 & 0.5749 \end{bmatrix},$$

$$P = \begin{bmatrix} 1.5658 & -0.0151 & -1.2540 & -0.0208 & -0.1164 \\ -0.0151 & 2.8847 & -0.0353 & 1.1121 & -0.0159 \\ -1.2540 & -0.0353 & 2.5967 & -0.0399 & 0.0262 \\ -0.0208 & 1.1121 & -0.0399 & 1.1790 & -0.5772 \\ -0.1164 & -0.0159 & 0.0262 & -0.5772 & 3.1034 \end{bmatrix}.$$

Further solving the optimization problem (3.44) gives

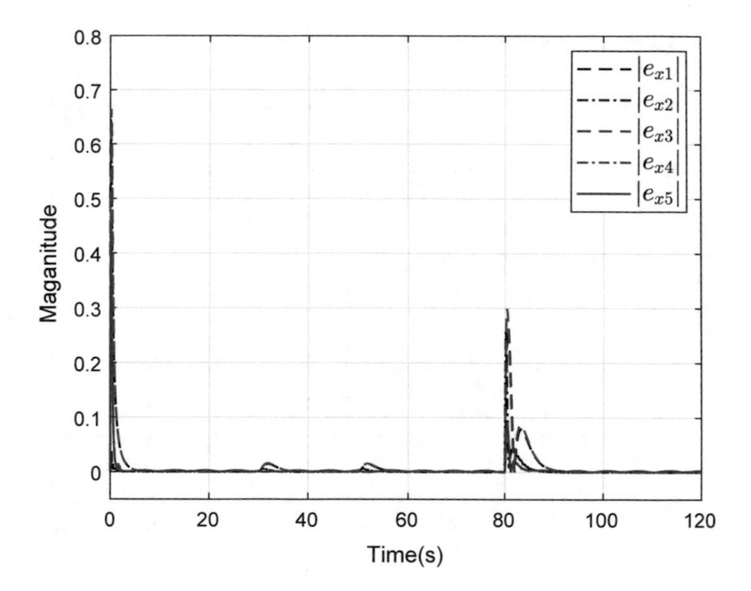

Fig. 3.2 State estimation performance

$$\Lambda = \text{diag}(16.9498,\ 6.4883,\ 19.9348,\ 3.8680,\ 24.4499),$$

$$L = \begin{bmatrix} -2.2137 & 1.3494 & -2.8677 \\ 10.2330 & 0.1861 & 0.1022 \\ -7.7698 & -4.0212 & 15.5234 \\ 3.9327 & 4.6845 & -8.1123 \\ -4.4726 & 1.9818 & 7.6228 \\ -5.5329 & 1.1892 & -2.5049 \\ -2.9187 & 0.4365 & -0.3781 \end{bmatrix}.$$

The closed-loop system is simulated under initial conditions $x(0) = [0.5\ 0\ 0.5\ 0]^\top$ and $\hat{\bar{x}}(0) = 0_{7\times 1}$.

From the results depicted in Figs. 3.2 and 3.3, it can be seen that the proposed observer achieves accurate state and fault estimation.

The closed-loop system state responses under either the baseline controller $u = Kx$ or the FTC controller $u = K\hat{x} + K_f\hat{f}$ are shown in Fig. 3.4 and 3.5, respectively. From these results, it is observed that the state are unstable using the baseline controller when the fault presents, but always maintained stable using the proposed FE-based FTC design.

The above results show that the proposed sequential strategy is effective in achieving stable closed-loop FE-based FTC system with good estimation and state response.

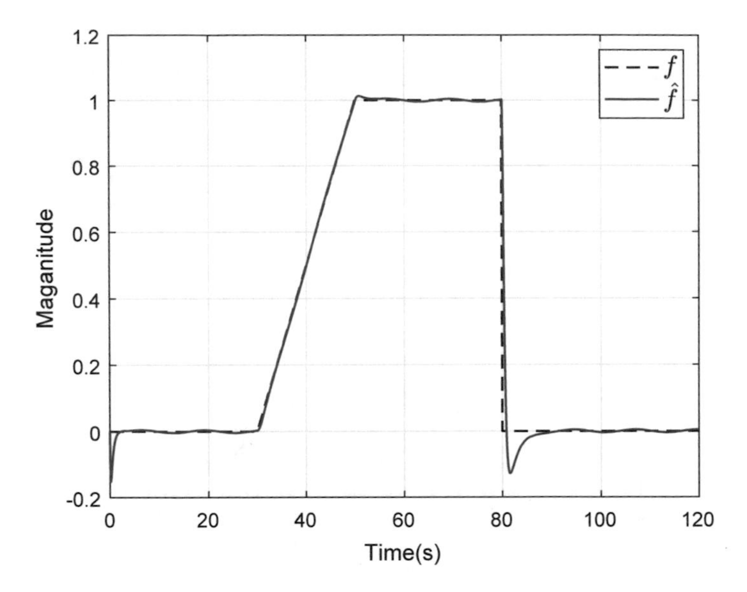

Fig. 3.3 Fault estimation performance

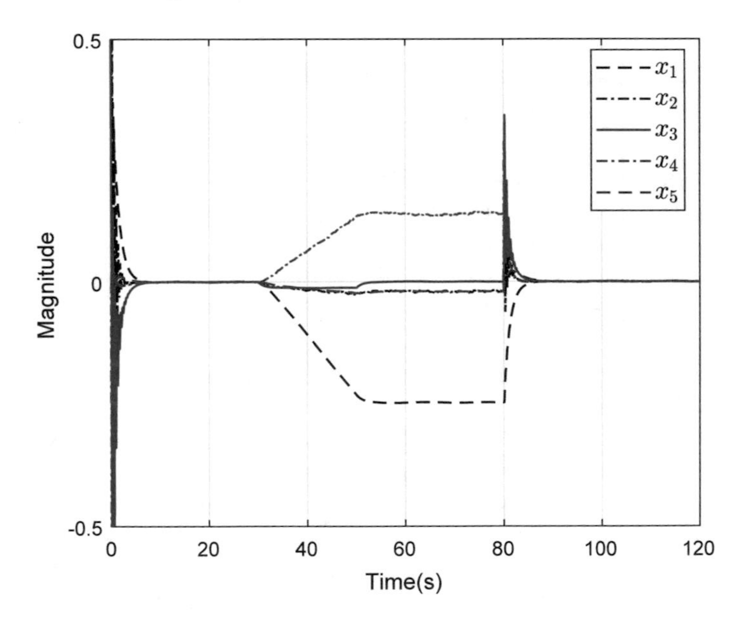

Fig. 3.4 State response under baseline controller $u = Kx$

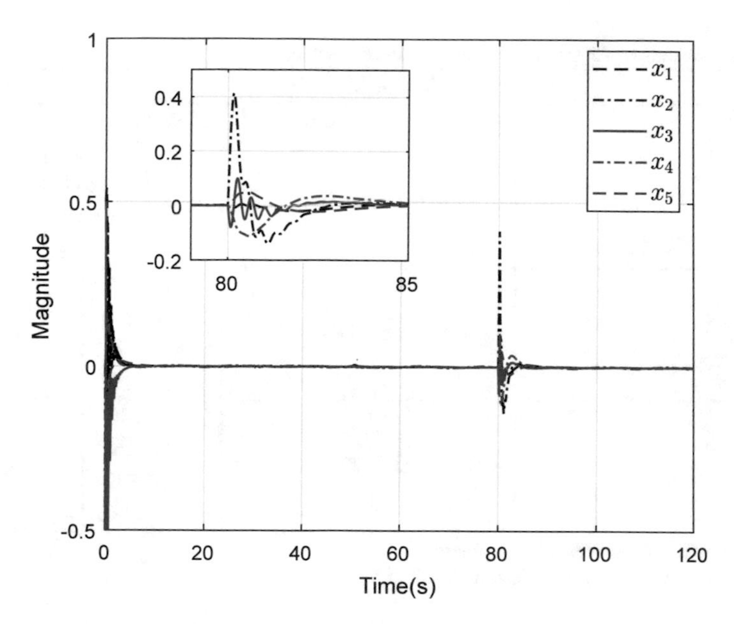

Fig. 3.5 State response under FTC controller $u = K\hat{x} + K_f\hat{f}$

3.8 Tutorial Example 2

This section presents another tutorial example to further investigate the following aspects of the proposed sequential strategy:

- How does it perform in achieving robust integration of FE and FTC?
- How does the solution of Step 1 in Algorithm 3.1 affects that of Step 2?

Consider again the motivating example (2.22) in Sect. 2.4.2. The optimization problems (3.43) and (3.44) are solved by manually specifying different values for γ. In order to find feasible solution to problem (3.44) under small values of γ, the minimization problem (min trace(Λ)) is replaced by the relaxed existence problem (Find trace(Λ)).

The experimental results are reported in Table 3.1. It is seen that the optimum γ gives good control performance, but infeasible observer design. The increase of γ results in a smaller L but may degrade the control performance. These results suggest that a compromise should be made to balance the FTC and FE performances in real implementation of the sequential strategy. A trial and error test may be required to obtain desirable controller and observer gains.

The closed-loop FE-based FTC system is simulated with the gains obtained from the case $\gamma = 40$, and under the same fault, disturbance and initial conditions as in Sect. 2.4.2. The uncertainty is set as $\Delta A = 0.1A$.

The results depicted in Figs. 3.6 and 3.7 show that the proposed FE-based FTC design achieves accurate estimation and stable state response in the presence of

Table 3.1 Results of Algorithm 3.1 under different values of γ

γ	Eigenvalues of $(A + BK)$	L
0.6233*	$-5.6316 \times 10^6, -8.87 \times 10^4$	Infeasible
31**	$-17.7147 \pm 6.9805i$	$10^4 \times \begin{bmatrix} 0.0237 \\ 0.1935 \\ -1.5812 \\ -0.0083 \end{bmatrix}$
40	$-18.2471 \pm 6.8584i$	$\begin{bmatrix} 88.4205 \\ 191.2280 \\ -961.7882 \\ -18.4192 \end{bmatrix}$
60	$-18.4697 \pm 8.1272i$	$\begin{bmatrix} 73.2044 \\ 115.2566 \\ -465.3388 \\ -12.7182 \end{bmatrix}$

*The optimum γ to optimization problem (3.43)
**The smallest value of γ that permits feasible L

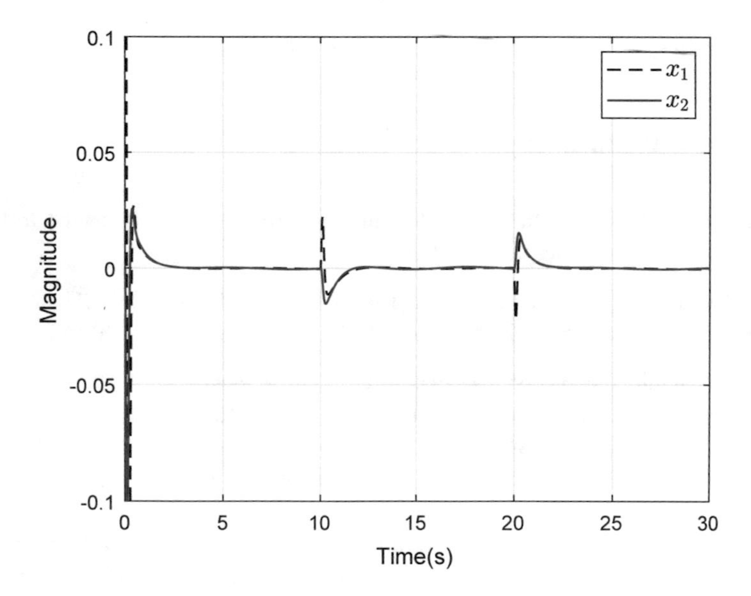

Fig. 3.6 State and fault estimation performance

uncertainty. Recalling here that in Sect. 2.4.2, combination of the separately designed FE and FTC causes system instability even for a much smaller uncertainty $\Delta A = 0.018\,A$. This demonstrates that the proposed sequential strategy can achieve robust integration of FE and FTC, and thus is superior over the separated strategy where the *bidirectional robustness interactions* are ignored.

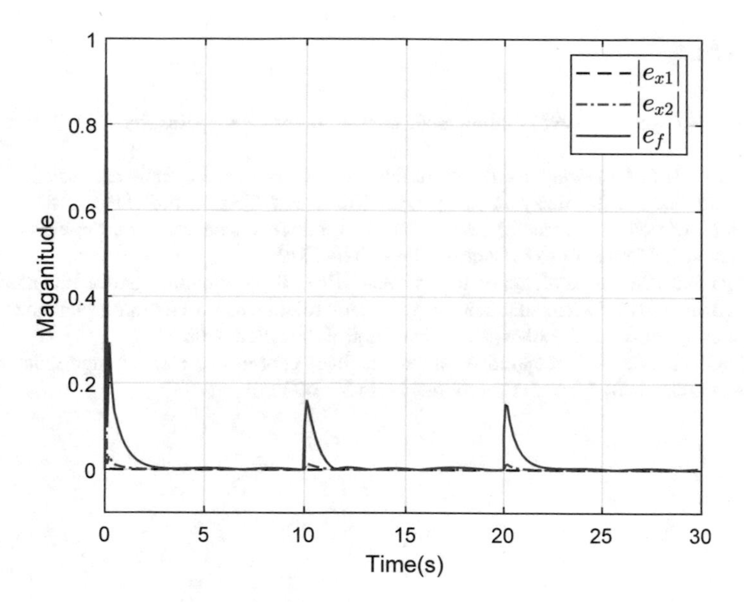

Fig. 3.7 FE-based FTC system state response

3.9 Notes

The sequential strategy avoids the difficulty in solving the optimization problem (3.8) with the BMI constraint (3.9), which reduces the design complexity and facilitates the real implementation. However, it is restrictive in the following aspects:

- The matrices M, P and γ are fixed in Step 2 of Algorithm 3.1, which reduces the design freedom.
- The solution of Step 2 is highly dependent on that of Step 1 and thus the obtained observer may have poor performance. In some cases, the predetermined FTC controller may lead to infeasibility of the FE design, as shown in Sect. 3.8.
- In practice, the obtained observer gains may be high gains and are very sensitive to system noises. The closed-loop transient performance may be undesired, e.g. with big overshoot, oscillation and long settling time.

To avoid obtaining the high gains, extra constraints can be imposed on the magnitudes of K and L, e.g. following the methods described in Zhu and Pagilla (2007). In terms of improving the transient performance, additional constraints, e.g. pole placement constraint (see Sect. 1.7.1), can be incorporated when solving the optimization problems.

References

Edwards C, Spurgeon S (1998) Sliding mode control: theory and applications. CRC Press, Bocca Raton

Lan J, Patton RJ (2016) Integrated design of fault-tolerant control for nonlinear systems based on fault estimation and T-S fuzzy modeling. IEEE Trans. Fuzzy Syst. 25(5):1141–1154

Lien CH (2004) Robust observer-based control of systems with state perturbations via LMI approach. IEEE Trans. Autom. Control 49(8):1365–1370

Luenberger D (1971) An introduction to observers. IEEE Trans. Autom. Control 16(6):596–602

Shi F, Patton RJ (2015) Fault estimation and active fault tolerant control for linear parameter varying descriptor systems. Int. J. Robust Nonlinear Control 25(5):689–706

Zhu Y, Pagilla PR (2007) Decentralized output feedback control of a class of large-scale interconnected systems. IMA J. Math. Control Inf. 24(1):57–69

Chapter 4
Iterative Integration of FE and FTC

4.1 Introduction

A sequential strategy (see Fig. 4.1(1)) for robust integration of FE and FTC is proposed in Chap. 3, by taking into account only the *unidirectional robustness interaction*, i.e. effects of FTC uncertainty on the FE observer. This chapter further presents an iterative integration strategy (see Fig. 4.1(2)) to account for the *bidirectional robustness interactions* through an iteration manner. Under this strategy, the FTC controller and FE observer are synthesized in an iterative loop. The strategy builds on the sequential strategy in Chap. 3, while actively using the FE observer obtained in Step 2 to improve the FTC controller design in Step 1.

The iteration starts with an initial feasible FTC controller, which is then used to design the associated FE observer. After that, the FE observer is fixed and the FTC controller is refined by adding an extra state feedback controller component. The iteration continues until targeting the prescribed accuracy of the robust control performance index. The resulting FTC controller is the sum of the initial controller and the extra state feedback controller components generated at each iteration. It is proved that the iterative procedure finitely converges to a (local) optima under a prescribed stopping criterion. The iteration manner is able to improve results of the sequential strategy, but it can only obtain suboptimal solutions to the robust integration of FE and FTC.

An augmented state unknown input observer (ASUIO) is developed in this chapter to estimate the state and fault simultaneously. There are many existing FE observers, e.g. adaptive observers (Jiang et al. 2006; Kabore and Wang 2001), SMO (Edwards and Tan 2006, ASO (Gao and Ding 2007), UIO (Gao et al. 2016), ALSO (see Chap. 3), moving horizon estimator (Feng and Patton 2014), and combined SMO and ASO (Shi et al. 2015). However, the adaptive observers estimate faults with finite errors. Moreover, in order to estimate time-varying faults the observer has a Proportional-Integral structure with carefully chosen learning rate. The canonical form SMO (Edwards and Tan 2006) requires several state transformations as well as a priori knowledge of the

© The Author(s), under exclusive license to Springer Nature Switzerland AG 2021

J. Lan and R. J. Patton, *Robust Integration of Model-Based Fault Estimation and Fault-Tolerant Control*, Advances in Industrial Control,
https://doi.org/10.1007/978-3-030-58760-4_4

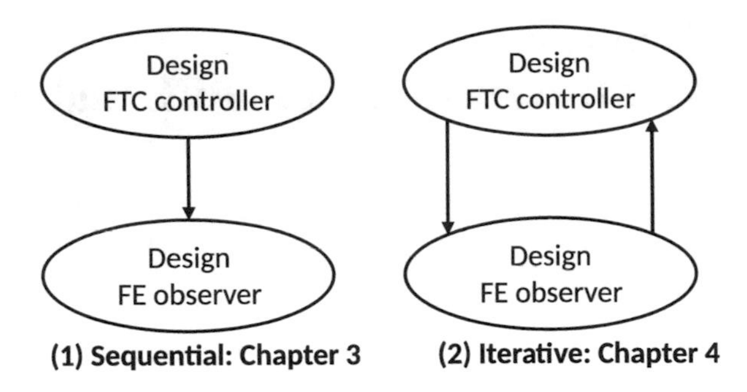

Fig. 4.1 Sequential and iterative integration of FE and FTC

upper bounds of the faults. The ASO reconstructs faults in a polynomial form with a priori knowledge of their orders. The moving horizon estimation is a complex online optimization problem. The existing UIOs are obtained after satisfying a well-known matching condition (see Sect. 1.7.3). By combining ASO and UIO, an ASUIO not requiring the matching condition is proposed in this chapter. The proposed ASUIO do not require state transformation and fault information (upper bounds and fault characteristics), or online optimization. Moreover, the ASUIO has more design freedom than the ALSO in Chap. 3.

4.2 Problem Description

Consider a linear system with the form

$$\begin{aligned} \dot{x} &= (A + \Delta A)x + Bu + Ff + Dd \\ y &= Cx \end{aligned} \tag{4.1}$$

where $x \in \mathbb{R}^n$, $u \in \mathbb{R}^m$, $y \in \mathbb{R}^p$, $f \in \mathbb{R}^q$, and $d \in \mathbb{R}^l$ are the state, control input, measured output, actuator fault, and external disturbance, respectively. The constant matrices A, B, F, D and C are known and of compatible dimensions. The matrix ΔA denotes the unknown system uncertainty. The proposed design relies on the following assumptions:

Assumption 4.1 The pair (A, B) is controllable. The triple (A, F, C) has no invariant zeros in the closed right-half complex plane. The actuator fault is matched, i.e. $\text{rank}[B\ F] = \text{rank}(B) = m$.

Assumption 4.2 The uncertainty matrix ΔA is norm-bounded (energy bounded) with the form $\Delta A = \mathcal{M}\mathcal{F}(t)\mathcal{N}$, where $\mathcal{M}$ and $\mathcal{N}$ are known matrices with appropriate dimensions, and $\mathcal{F}(t)$ is an unknown matrix satisfying $\mathcal{F}^\top(t)\mathcal{F}(t) \preceq I$.

Assumption 4.3 The fault f and disturbance d are norm-bounded. Moreover, f has norm-bounded first-order and second-order time differentials.

4.3 FE Observer and FTC Controller

4.3.1 FE Observer

Define f and $\dot{f}$ as auxiliary state, then the system (4.1) is augmented as

$$\begin{aligned} \dot{\bar{x}} &= \bar{A}\bar{x} + \Delta\bar{A}\bar{x} + \bar{B}u + \bar{D}\bar{d} \\ y &= \bar{C}\bar{x} \end{aligned} \tag{4.2}$$

with

$$\bar{x} = \begin{bmatrix} x \\ f \\ \dot{f} \end{bmatrix}, \ \bar{d} = \begin{bmatrix} d \\ \ddot{f} \end{bmatrix}, \ \bar{A} = \begin{bmatrix} A & F & 0 \\ 0 & 0 & I_q \\ 0 & 0 & 0 \end{bmatrix}, \ \Delta\bar{A} = \begin{bmatrix} \Delta A & 0 & 0 \\ 0 & 0 & 0 \\ 0 & 0 & 0 \end{bmatrix},$$
$$\bar{B} = \begin{bmatrix} B \\ 0 \\ 0 \end{bmatrix}, \ \bar{D} = \begin{bmatrix} D & 0 \\ 0 & 0 \\ 0 & I_q \end{bmatrix}, \ \bar{C} = [C \ 0 \ 0].$$

In order to estimate $\bar{x}$, the following ASUIO is proposed:

$$\begin{aligned} \dot{\xi} &= N\xi + Gu + Ly \\ \hat{\bar{x}} &= \xi + Hy, \end{aligned} \tag{4.3}$$

where $\xi \in \mathbb{R}^{n+2q}$ is the observer state and $\hat{\bar{x}} \in \mathbb{R}^{n+2q}$ is the estimate of $\bar{x}$. The matrices N, G, L and H are of appropriate dimensions and to be designed. The estimates of the original system state x and actuator fault f can be calculated by $\hat{x} = [I_n \ 0_{n\times 2q}]\hat{\bar{x}}$ and $\hat{f} = [0_{q\times n} \ I_q \ 0_{q\times q}]\hat{\bar{x}}$, respectively.

Remark 4.1 The use of ASUIO for simultaneous state and fault estimation is an extension and further development of the conventional UIO (see Sect. 1.7.3), which has been widely used for state estimation and residual generation with unknown input (disturbance) decoupling.

Define the estimation error as $e = \bar{x} - \hat{\bar{x}}$. By using the definition of $\hat{\bar{x}}$ in (4.3), it gives

$$\bar{x} = e + \xi + Hy, \ e = \Xi\bar{x} - \xi \tag{4.4}$$

with $\Xi = I_{n+2q} - H\bar{C}$.

Let $L = L_1 + L_2$. By using (4.2), (4.3) and (4.4), the estimation error system can be derived as

$$
\begin{aligned}
\dot{e} &= \Xi\dot{\bar{x}} - \dot{\xi} \\
&= \Xi\left(\bar{A}\bar{x} + \Delta\bar{A}\bar{x} + \bar{B}u + \bar{D}\bar{d}\right) - (N\xi + Gu + Ly) \\
&= \Xi\bar{A}\bar{x} - N\xi - Ly + (\Xi\bar{B} - G)u + \Xi\Delta\bar{A}\bar{x} + \Xi\bar{D}\bar{d} \\
&= (\Xi\bar{A} - L_1\bar{C})\bar{x} - N\xi - L_2 y + (\Xi\bar{B} - G)u + \Xi\Delta\bar{A}\bar{x} + \Xi\bar{D}\bar{d} \\
&= (\Xi\bar{A} - L_1\bar{C})(e + \xi + Hy) - N\xi - L_2 y + (\Xi\bar{B} - G)u + \Xi\Delta\bar{A}\bar{x} + \Xi\bar{D}\bar{d} \\
&= (\Xi\bar{A} - L_1\bar{C})e + (\Xi\bar{A} - L_1\bar{C} - N)\xi + [(\Xi\bar{A} - L_1\bar{C})H - L_2]y \\
&\quad + (\Xi\bar{B} - G)u + \Xi\Delta\bar{A}\bar{x} + \Xi\bar{D}\bar{d}.
\end{aligned} \tag{4.5}
$$

Define the following matrix equations:

$$\Xi\bar{A} - L_1\bar{C} = N \tag{4.6}$$

$$\Xi\bar{B} = G \tag{4.7}$$

$$(\Xi\bar{A} - L_1\bar{C})H = L_2. \tag{4.8}$$

Substituting (4.6)–(4.8) into (4.5) gives the estimation error system

$$\dot{e} = (\Xi\bar{A} - L_1\bar{C})e + \Xi\Delta\bar{A}\bar{x} + \Xi\bar{D}\bar{d}. \tag{4.9}$$

It is worth noting that once L_1 and H (in Ξ) are designed, all the observer gains can then be determined using (4.6)–(4.8), for which the details are given in Remark 4.2.

Remark 4.2 Once H is obtained, one can calculate $\Xi = I_{n+2q} - H\bar{C}$. Then substituting Ξ and L_1 into (4.6) gives N. Substituting Ξ into (4.7) gives G. Substituting Ξ, L_1 and H into (4.8) gives L_2. Finally, one can calculate $L = L_1 + L_2$. It can be seen that although there are four matrices (N, G, L, H) to determine, by using the matrix equations (4.6)–(4.8), only two (L_1, H) need to be designed. This reduces the design complexity dramatically without imposing any conservativeness.

Before proceeding to the design of L_1 and H to stabilize the error system (4.9), it is necessary to check its stabilizability. This can be achieved by verifying the observability of the pair $(\Xi\bar{A}, \bar{C})$, as shown in the proposition below.

Proposition 1 *Under Assumption 4.1, there exists a s.p.d. matrix Q such that*

$$\mathrm{He}\left(Q(\Xi\bar{A} - L_1\bar{C})\right) \prec 0. \tag{4.10}$$

Proof By using the definitions of $\bar{A}$ and $\bar{C}$ in (4.2) and Proposition 3.1, the pair $(\bar{A}, \bar{C})$ is observable under Assumption 4.1. Hence, there always exists a matrix H such that

$$\mathrm{rank}\begin{bmatrix} \bar{C} \\ \Xi\bar{A} \end{bmatrix} = \mathrm{rank}\left\{\begin{bmatrix} I & 0 \\ 0 & I - H\bar{C} \end{bmatrix}\begin{bmatrix} \bar{C} \\ \bar{A} \end{bmatrix}\right\} = n + 2q. \text{ (Full rank)} \tag{4.11}$$

By inspection, an obvious solution to (4.11) is $H = 0$. This implies that there always exists a matrix H such that the pair $(\Xi\bar{A}, \bar{C})$ is observable and consequently the statement (4.10) follows. □

The following example is used to further illustrate the results in Proposition 1.

Example 4.1 Consider a system in the form of (4.1) with the following matrices

$$A = \begin{bmatrix} -24 & -12 \\ 4.4379 & -2.2189 \end{bmatrix}, \quad F = \begin{bmatrix} 20 \\ 0 \end{bmatrix}, \quad C = \begin{bmatrix} 1 & 0 \\ 0 & 1 \end{bmatrix}.$$

For this example system, the related dimensions are $n = 2$ and $q = 1$. Since $\text{rank}(C) = 2$, the pair (A, C) is observable and $\text{rank}\begin{bmatrix} C \\ A \end{bmatrix} = 2$. Hence, it can be derived that

$$\text{rank}\begin{bmatrix} \bar{C} \\ \bar{A} \end{bmatrix} = \text{rank}\begin{bmatrix} 1 & 0 & 0 & 0 \\ 0 & 1 & 0 & 0 \\ -24 & -12 & 20 & 0 \\ 4.4379 & -2.2189 & 0 & 0 \\ 0 & 0 & 0 & 1 \\ 0 & 0 & 0 & 0 \end{bmatrix} = 4 = n + 2q.$$

This implies that the pair $(\Xi\bar{A}, \bar{C})$ is observable and the matrix H satisfying (4.11) always exists.

4.3.2 *FTC Controller*

To compensate the fault f and stabilize the state x, the FTC controller is designed as

$$u = u_x + u_f \tag{4.12}$$

which consists of a state feedback baseline controller u_x and a fault compensator u_f. The baseline controller is designed as $u_x = K\hat{x}$ with a constant gain $K \in \mathbb{R}^{m \times n}$. Since the fault f is matched, u_f can be optimally designed as $u_f = -B^{\dagger}F\hat{f}$ such that $Bu_f + Ff = F(f - \hat{f})$. This implies that given accurate fault estimation, i.e. $f - \hat{f} = 0$, then the fault can be completely compensated.

Substituting (4.12) into (4.1) gives the closed-loop FTC system

$$\dot{x} = (A + BK)x + \Delta Ax + B_e e + Dd \tag{4.13}$$

where $B_e = [-BK \;\; F \;\; 0]$.

4.4 Integration of FE and FTC

Combining the estimation error system (4.9) with the FTC system (4.13) gives the composite closed-loop system

$$\begin{aligned} \dot{x} &= (A + BK)x + \Delta Ax + B_e e + \tilde{D}\bar{d} \\ \dot{e} &= (\Xi\bar{A} - L_1\bar{C})e + \Xi\Delta\tilde{A}x + \Xi\bar{D}\bar{d} \\ z &= \text{diag}(C_x x, C_e e) \end{aligned} \tag{4.14}$$

where $\tilde{D} = [D\ 0]$, $\Delta\tilde{A} = [\Delta A^\top\ 0]^\top$, and $z \in \mathbb{R}^{2n+2q}$ is the performance output with given weights C_x and C_e.

Lemma 4.1 *Under Assumptions 4.1–4.3, the composite closed-loop system (4.14) is stable and satisfies the H_∞ performance $\|G_{z\bar{d}}\|_\infty < \gamma$, if the following optimization problem is feasible*

$$\min_{K,H,L_1,P_0,Q,\gamma} \quad \gamma \tag{4.15}$$

$$\text{s.t.} \quad \left[\begin{array}{c|cc} F_c & S_1 & S_2 \\ \hline \star & W_o & S_3 \\ \star & \star & \Lambda_o \end{array}\right] \prec 0 \tag{4.16}$$

$$P_0 = P_0^\top \succ 0,\ Q = Q^\top \succ 0,\ \gamma > 0,$$

where the block matrices F_c, S_1, S_2, W_o, S_3 and Λ_o are given as

$$F_c = \begin{bmatrix} \text{He}(P_0(A + BK)) & P_0\tilde{D} & P_0\mathcal{M} & \mathcal{N}^\top & C_x^\top \\ \star & -\gamma I & 0 & 0 & 0 \\ \star & \star & -I & 0 & 0 \\ \star & \star & \star & -I & 0 \\ \star & \star & \star & \star & -\gamma I \end{bmatrix},$$

$$S_1 = \begin{bmatrix} P_0 B_e \\ (Q\Xi\bar{D})^\top \\ 0 \\ 0 \\ 0 \end{bmatrix},\ S_2 = \begin{bmatrix} \mathcal{N}^\top & 0 & 0 \\ 0 & 0 & 0 \\ 0 & 0 & 0 \\ 0 & 0 & 0 \\ 0 & 0 & 0 \end{bmatrix},$$

$$W_o = \text{He}(Q(\Xi\bar{A} - L_1\bar{C})),\ S_3 = [0\ Q\Xi\bar{\mathcal{M}}\ C_e^\top],$$
$$\Lambda_o = \text{diag}(-I, -I, -\gamma I),\ B_e = [-BK\ F\ 0].$$

Proof The proof is similar to that of Theorem 3.1 and it is not repeated here. □

Feasibility analysis for the optimization problem (4.15) can be performed following the content in Sect. 3.5. Notice that condition (4.16) is a BMI due to the existence of nonlinear terms P_0BK, QH and QL_1 of the decision variables. This imposes solvability difficulty on the optimization problem. To overcome it, an iterative strategy is developed in the next section.

4.5 An Iterative Integration Strategy

This section describes an iterative strategy to solve the optimization problem (4.15). For this purpose, the baseline control gain K in (4.12) is redesigned as

$$K = \left(K_0 + \sum_{i=1}^{j} \Delta K^{(i)}\right) \tag{4.17}$$

where K_0 is the baseline control gain such that the fault-free system (4.1) is robustly stable against the disturbance and uncertainty. The extra components $\Delta K^{(i)}$ are the enhanced control gains used to compensate the impacts from the estimation error. They will be determined at iteration step i, $i = 1, 2, \ldots, j$, with an integer $j \geq 1$.

Designs of the baseline control gain and the enhanced control gain will be described in Sects. 4.5.1 and 4.5.2, respectively. Finally. the complete iterative algorithm will be described in Sect. 4.5.3.

4.5.1 *Baseline Control Gain Design*

Substituting $u_x = K_0x$ into (4.1) yields the fault-free closed-loop control system

$$\begin{aligned} \dot{x} &= (A + BK_0)x + \Delta Ax + Dd \\ z_0 &= C_0x \end{aligned} \tag{4.18}$$

where $z_0 \in \mathbb{R}^n$ is the performance output with a given weight matrix $C_0 \in \mathbb{R}^{n\times n}$.

The design of baseline control gain K_0 is formulated as below.

Lemma 4.2 *Under Assumptions 4.1–4.3, the closed-loop system (4.18) is stable with H_∞ performance $\|G_{z_0d}\|_\infty < \gamma_0$ if the following optimization problem is feasible*

$$\text{Find } \gamma_0 \tag{4.19}$$

$$\text{s.t.} \begin{bmatrix} \text{He}(AP_0 + BM_0) & D & \mathcal{M} & P_0\mathcal{N}^\top & P_0C_0^\top \\ \star & -\gamma_0 I & 0 & 0 & 0 \\ \star & \star & -I & 0 & 0 \\ \star & \star & \star & -I & 0 \\ \star & \star & \star & \star & -\gamma_0 I \end{bmatrix} \prec 0 \tag{4.20}$$

$$P_0 = P_0^\top \succ 0,\ \gamma_0 > 0.$$

Then the baseline control gain is obtained as $K_0 = M_0 P_0^{-1}$.

Proof The proof is based on Lemma 3.1 in Chap. 3, by replacing $(A + \Delta A)$, $\mathcal{P}$ and γ in (3.21) with $(A + \Delta A + BK_0)$, P and γ_0, respectively. Let $P_0 = P^{-1}$ and $M_0 = K_0 P_0$, then pre- and post-multiplying (3.21), respectively, with $\mathrm{diag}(P_0, I, I, I, I)$ and its transpose gives (4.20). □

The objective in (4.19) can be either "Find" or "min". However, the optimal baseline control gain K_0 associated with the optimum γ_0 may make the iterative algorithm converge to local optima too early, which results in less desirable solutions.

4.5.2 Enhanced Control Gain Design

Substituting (4.12) with K defined in (4.17) into (4.14) gives the closed-loop system

$$\begin{aligned} \dot{x} &= \Big[A + B\Big(K_0 + \sum_{i=1}^{j} \Delta K^{(i)}\Big)\Big]x + \Delta A x + B_e e + \tilde{D}\bar{d} \\ \dot{e} &= (\Xi\bar{A} - L_1\bar{C})e + \Xi\Delta\tilde{A}x + \Xi\bar{D}\bar{d} \\ z &= \mathrm{diag}(C_x x, C_e e) \end{aligned} \tag{4.21}$$

where

$$\tilde{D} = [D\ 0],\ \Delta\tilde{A} = [\Delta A^\top\ 0]^\top,$$
$$B_e = \Big[-B\Big(K_0 + \sum_{i=1}^{j} \Delta K^{(i)}\Big)\ F\ 0\Big].$$

The signal $z \in \mathbb{R}^{2n+2q}$ is the performance output with given weights C_x and C_e.

According to Lemma 4.1, the following theorem is given.

Theorem 1 *Under Assumptions 4.1–4.3, the gains* $\Delta K^{(i)}$, $i = 1, 2, \ldots, j$, H *and* L_1 *can be determined via solving the following optimization problem:*

$$\min_{M^{(i)}, X, Y, P, Q, \gamma} \gamma \tag{4.22}$$

$$\text{s.t.} \quad \left[\begin{array}{c|cc} \tilde{F}_c & \tilde{S}_1 & \tilde{S}_2 \\ \hline \star & \tilde{W}_o & \tilde{S}_3 \\ \star & \star & \tilde{\Lambda}_o \end{array}\right] \prec 0 \tag{4.23}$$

$$P = P^\top \succ 0,\ Q = Q^\top \succ 0,\ \gamma > 0 \tag{4.24}$$

where the block matrices $\tilde{F}_c$, $\tilde{S}_1$, $\tilde{S}_2$, $\tilde{W}_o$, $\tilde{S}_3$ *and* $\tilde{\Lambda}_o$ *are given as*

$$\tilde{F}_c = \begin{bmatrix} \text{He}\left[AP + B\left(K_0 P + \sum_{i=1}^{j} M^{(i)}\right)\right] & \tilde{D} & \mathcal{M} & P\mathcal{N}^\top & PC_x^\top \\ \star & -\gamma I & 0 & 0 & 0 \\ \star & \star & -I & 0 & 0 \\ \star & \star & \star & -I & 0 \\ \star & \star & \star & \star & -\gamma I \end{bmatrix},$$

$$\tilde{S}_1 = \begin{bmatrix} \tilde{B}_e \\ (Q\bar{D} - X\bar{C}\bar{D})^\top \\ 0 \\ 0 \\ 0 \end{bmatrix}, \quad \tilde{S}_2 = \begin{bmatrix} P\mathcal{N}^\top & 0 & 0 & -B\left(K_0 P + \sum_{i=1}^{j} M^{(i)}\right) & 0 \\ 0 & 0 & 0 & 0 & 0 \\ 0 & 0 & 0 & 0 & 0 \\ 0 & 0 & 0 & 0 & 0 \\ 0 & 0 & 0 & 0 & I \end{bmatrix},$$

$$\tilde{W}_o = \text{He}\left[(Q - X\bar{C})\bar{A} - Y\bar{C}\right], \; \tilde{S}_3 = [0 \; (Q - X\bar{C})\bar{\mathcal{M}} \; C_e^\top \; 0 \; 0],$$
$$\tilde{\Lambda}_o = \text{diag}(-I, -I, -\gamma I, -P, -P), \; \tilde{B}_e = [0 \; F \; 0].$$

Then the enhanced control gains and observer gains are obtained as

$$\Delta K^{(i)} = M^{(i)} P^{-1}, \; i = 1, 2, \ldots, j,$$
$$H = Q^{-1} X, \; L_1 = Q^{-1} Y.$$

Proof Replacing K in (4.16) with $(K_0 + \sum_{i=1}^{j} \Delta K^{(i)})$. Let $P = P_0^{-1}$. Pre- and post-multiplying (4.16), respectively, with $\text{diag}(P, I, I, I, I, I, I, I, I)$ and its transpose yields

$$\begin{bmatrix} \hat{F}_c & S_1 & S_2 \\ \star & W_o & S_3 \\ \star & \star & \Lambda_o \end{bmatrix} \prec 0, \tag{4.25}$$

where

$$\hat{F}_c = \begin{bmatrix} \text{He}\left[AP + B\left(K_0 + \sum_{i=1}^{j} \Delta K^{(i)}\right) P\right] & \tilde{D} & \mathcal{M} & P\mathcal{N}^\top & PC_x^\top \\ \star & -\gamma I & 0 & 0 & 0 \\ \star & \star & -I & 0 & 0 \\ \star & \star & \star & -I & 0 \\ \star & \star & \star & \star & -\gamma I \end{bmatrix},$$

$$S_1 = \begin{bmatrix} B_e \\ (Q\Xi\bar{D})^\top \\ 0 \\ 0 \\ 0 \end{bmatrix}, \quad S_2 = \begin{bmatrix} P\mathcal{N}^\top & 0 & 0 \\ 0 & 0 & 0 \\ 0 & 0 & 0 \\ 0 & 0 & 0 \\ 0 & 0 & 0 \end{bmatrix},$$

$$W_o = \text{He}\left[Q(\Xi\bar{A} - L_1\bar{C})\right], \; S_3 = [0 \;\; Q\Xi\bar{\mathcal{M}} \;\; C_e^\top],$$

$$\Lambda_o = \text{diag}(-I, -I, -\gamma I), \; B_e = \left[-B\left(K_0 + \sum_{i=1}^{j} \Delta K^{(i)}\right) F \; 0\right].$$

Further define $M^{(i)} = \Delta K^{(i)} P$, $i = 1, 2, \ldots, j$, $X = QH$ and $Y = QL_1$, then the optimization problem (4.15) becomes

$$\min_{\Delta K^{(i)}, M^{(i)}, X, Y, P, Q, \gamma} \gamma \tag{4.26}$$

$$\text{s.t.} \quad \left[\begin{array}{c|cc} \hat{F}_c & \hat{S}_1 & \hat{S}_2 \\ \hline \star & \hat{W}_o & \hat{S}_3 \\ \star & \star & \hat{\Lambda}_o \end{array}\right] \prec 0 \tag{4.27}$$

$$P = P^\top \succ 0,\ Q = Q^\top \succ 0,\ \gamma > 0,$$

where

$$\hat{F}_c = \begin{bmatrix} \text{He}\left[AP + B\left(K_0 P + \sum_{i=1}^{j} M^{(i)}\right)\right] & \tilde{D} & \mathcal{M} & P\mathcal{N}^\top & PC_x^\top \\ \star & -\gamma I & 0 & 0 & 0 \\ \star & \star & -I & 0 & 0 \\ \star & \star & \star & -I & 0 \\ \star & \star & \star & \star & -\gamma I \end{bmatrix},$$

$$\hat{S}_1 = \begin{bmatrix} B_e \\ (Q\bar{D} - X\bar{C}\bar{D})^\top \\ 0 \\ 0 \\ 0 \end{bmatrix},\ \hat{S}_2 = \begin{bmatrix} P\mathcal{N}^\top & 0\ 0 \\ 0 & 0\ 0 \\ 0 & 0\ 0 \\ 0 & 0\ 0 \\ 0 & 0\ 0 \end{bmatrix},$$

$$\hat{W}_o = \text{He}\left[(Q - X\bar{C})\bar{A} - Y\bar{C}\right],$$

$$\hat{S}_3 = [0\ (Q - X\bar{C})\bar{\mathcal{M}}\ C_e^\top],$$

$$\hat{\Lambda}_o = \text{diag}(-I, -I, -\gamma I),$$

$$B_e = \left[-B\left(K_0 + \sum_{i=1}^{j} \Delta K^{(i)}\right) F\ 0\right].$$

The coexistence of decision variables $M^{(i)}$ and $\Delta K^{(i)}$ in (4.27) is undesirable. Hence, a further treatment is made below to remove $\Delta K^{(i)}$. Define

$$V_1 = \left[-\left(B\left(K_0 + \sum_{i=1}^{j} \Delta K^{(i)}\right)\right)^\top 0\,0\,0\,0\,0\,0\,0\,0\right]^\top,\ V_2^\top = [0\,0\,0\,0\,I\,0\,0\,0\,0].$$

By using Young inequality (see Sect. 1.7.1), the following relation holds

$$\text{He}(V_1 V_2^\top) \preceq (V_1 P)P^{-1}(V_1 P)^\top + V_2^\top P^{-1} V_2. \tag{4.28}$$

Applying Schur Complement (see Sect. 1.7.1) to the term $-B(K_0 + \sum_{i=1}^{j} \Delta K^{(i)})$ in (4.27) and using (4.28) with $M^{(i)} = \Delta K^{(i)} P$, then the optimization problem (4.26) is converted into (4.22). □

4.5.3 Iterative Algorithm

This section presents the iterative algorithm based on the designs of the baseline control gain and the enhanced control gain described in Sects. 4.5.1 and 4.5.2, respectively.

At iteration j the control gain K can be presented as

$$K = K_{\text{known}}^{(j-1)} + \Delta K^{(j)} \tag{4.29}$$

where the part $K_{\text{known}}^{(j-1)} = (K_0 + \sum_{i=1}^{j-1} \Delta K^{(i)})$ is determined at iteration $(j-1)$ and treated known at iteration j, while $\Delta K^{(j)}$ is a decision variable. In this case, the optimization problem (4.22) becomes

$$\min_{M^{(j)}, X^{(j)}, Y^{(j)}, P^{(j)}, Q^{(j)}, \gamma^{(j)}} \quad \gamma^{(j)} \tag{4.30}$$

$$\text{s.t.} \quad \left[\begin{array}{c|cc} \tilde{F}_c^{(j)} & \tilde{S}_1^{(j)} & \tilde{S}_2^{(j)} \\ \hline \star & \tilde{W}_o^{(j)} & \tilde{S}_3^{(j)} \\ \star & \star & \tilde{\Lambda}_o^{(j)} \end{array}\right] \prec 0 \tag{4.31}$$

$$P^{(j)} = (P^{(j)})^\top \succ 0, \;\; Q^{(j)} = (Q^{(j)})^\top \succ 0, \;\; \gamma^{(j)} > 0 \tag{4.32}$$

where

$$\tilde{F}_c^{(j)} = \begin{bmatrix} \text{He}\left[AP^{(j)} + B\left(K_{\text{known}}^{(j-1)} P^{(j)} + M^{(j)}\right)\right] & \tilde{D} & \mathcal{M} & P^{(j)} \mathcal{N}^\top & P^{(j)} C_x^\top \\ \star & -\gamma^{(j)} I & 0 & 0 & 0 \\ \star & \star & -I & 0 & 0 \\ \star & \star & \star & -I & 0 \\ \star & \star & \star & \star & -\gamma^{(j)} I \end{bmatrix},$$

$$\tilde{S}_1^{(j)} = \begin{bmatrix} \tilde{B}_e \\ (Q^{(j)} \bar{D} - X^{(j)} \bar{C} \bar{D})^\top \\ 0 \\ 0 \\ 0 \end{bmatrix}, \; \tilde{S}_2^{(j)} = \begin{bmatrix} P^{(j)} \mathcal{N}^\top & 0 \; 0 & -B\left(K_{\text{known}}^{(j-1)} P^{(j)} + M^{(j)}\right) & 0 \\ 0 & 0 \; 0 & 0 & 0 \\ 0 & 0 \; 0 & 0 & 0 \\ 0 & 0 \; 0 & 0 & 0 \\ 0 & 0 \; 0 & 0 & I \end{bmatrix},$$

$\tilde{W}_o^{(j)} = \text{He}\left[(Q^{(j)} - X^{(j)} \bar{C}) \bar{A} - Y^{(j)} \bar{C}\right]$, $\tilde{S}_3^{(j)} = [0 \; (Q^{(j)} - X^{(j)} \bar{C}) \bar{\mathcal{M}} \; C_e^\top \; 0 \; 0]$,

$\tilde{\Lambda}_o^{(j)} = \text{diag}(-I, -I, -\gamma^{(j)} I, -P^{(j)}, -P^{(j)})$, $\tilde{B}_e = [0 \; F \; 0]$.

Then the design gains are obtained as

$$\Delta K^{(j)} = M^{(j)}(P^{(j)})^{-1},\ H^{(j)} = (Q^{(j)})^{-1}X,\ L_1^{(j)} = (Q^{(j)})^{-1}Y^{(j)}.$$

Based on Lemma 4.2 and the optimization problem (4.30), the iterative algorithm is presented below, where ϵ is a prescribed positive scalar.

Algorithm 4.1 Iterative strategy for robust integration of FE and FTC

Input: $A, B, F, D, C, \mathcal{M}, \mathcal{N}, \bar{A}, \bar{C}, \bar{D}, \tilde{D}, \bar{\mathcal{M}}, C_x, C_e, K_0, P_0, \gamma_0, \epsilon$

Initialization: Set $j = 1$, $K_{\text{known}}^{(0)} = K_0$, $P^{(0)} = P_0$, $\gamma^{(0)} = \gamma_0$

while ($j \geq 1$) **do**

Step 1: Solve the optimization problem $\mathcal{P}_1$:

$$\min_{X^{(j)}, Y^{(j)}, Q^{(j)}, \Lambda^{(j)}} \quad \text{trace}(\Lambda^{(j)})$$

$$\text{s.t.} \quad \left[\begin{array}{c|cc} \Lambda^{(j)} \tilde{F}_c^{(j)} & \tilde{S}_1^{(j)} & \tilde{S}_2^{(j)} \\ \hline \star & \tilde{W}_o^{(j)} & \tilde{S}_3^{(j)} \\ \star & \star & \tilde{\Lambda}_o^{(j)} \end{array}\right] \prec 0,$$

$$Q^{(j)} = (Q^{(j)})^\top \succ 0,\ \Lambda^{(j)} = \text{diag}(\beta_1 I, \beta_2 I, \cdots, \beta_5 I) \succ 0,$$
$$M^{(j)} = 0,\ \gamma^{(j)} = \gamma^{(j-1)},\ P^{(j)} = P^{(j-1)}.$$

Calculate the observer gains $H^{(j)} = (Q^{(j)})^{-1}X^{(j)}$ and $L_1^{(j)} = (Q^{(j)})^{-1}Y^{(j)}$.

Step 2: Solve the optimization problem $\mathcal{P}_2$:

$$\min_{M^{(j)}, P^{(j)}, \gamma^{(j)}} \gamma^{(j)}$$

$$\text{s.t.} \quad \left[\begin{array}{c|cc} \Lambda^{(j)} \tilde{F}_c^{(j)} & \tilde{S}_1^{(j)} & \tilde{S}_2^{(j)} \\ \hline \star & \tilde{W}_o^{(j)} & \tilde{S}_3^{(j)} \\ \star & \star & \tilde{\Lambda}_o^{(j)} \end{array}\right] \prec 0,\ P^{(j)} = (P^{(j)})^\top \succ 0,\ \gamma > 0.$$

Calculate the enhanced control gain $\Delta K^{(j)} = M^{(j)}(P^{(j)})^{-1}$.

if $|\gamma^{(j)} - \gamma^{(j-1)}| < \epsilon$ **then**

Set $j^* = j$ and stop

else

Set $j = j + 1$, $K_{\text{known}}^{(j)} = K_{\text{known}}^{(j-1)} + \Delta K^{(j)}$

end if

end while

Output: $j^*, \gamma^*, K = K_0 + \sum_{i=1}^{j^*} \Delta K^{(i)}, H = H^{(j^*)}, L_1 = L_1^{(j^*)}$

For both the theoretical and practical concerns, it is necessary and important to analyze convergence of the iterative algorithm. This is provided in the proposition below.

Proposition 2 *The gain sequence $\{\gamma^{(j)}\}_0^\infty$ converges to a local minimum γ^*.*

Proof By construction, Algorithm 4.1 creates a series of positive scalars $\gamma^{(j)}$. Furthermore, it holds that $\gamma^{(j+1)} \leq \gamma^{(j)}$. This is because the solution to problem $\mathcal{P}_2$ at iteration $j + 1$ is always one feasible solution to this problem at iteration j by setting $M^{(j+1)} = 0$, $P^{(j+1)} = P^{(j)}$ and $\gamma^{(j+1)} = \gamma^{(j)}$. Therefore, the sequence $\{\gamma^{(j)}\}_{j=0}^\infty$ is non-increasing and bounded below by zero. Let γ^* be the greatest lower bound of the sequence, then by definition, one has

$$\gamma^{(j)} \geq \gamma^*, \ j = 0, 1, 2, \ldots, \infty.$$

However, for every $\varepsilon > 0$, there exists an integer N such that $\gamma^N < \gamma^* + \varepsilon$, otherwise γ^* is not the greatest lower bound of $\{\gamma^{(j)}\}_{j=0}^{\infty}$. Since the sequence is non-increasing, for all $j \geq N$, it holds that

$$\gamma^* - \varepsilon \leq \gamma^{(j)} \leq \gamma^* + \varepsilon.$$

This means that $\{\gamma^{(j)}\}_{j=0}^{\infty}$ converges to γ^* and it is thus a Cauchy sequence (Rudin 1964). Therefore, for any $\varepsilon > 0$, there exists an integer N_c such that for all $j > N_c$, $|\gamma^{(j+1)} - \gamma^{(j)}| < \varepsilon$. This implies that Algorithm 4.1 will terminate in finite iterations and find an arbitrarily close approximation to the true local minima γ^*. □

Remark 4.3 It is shown by Proposition 2 that Algorithm 4.1 will converge provided any K_0 such that system (4.1) is robustly stable. However, the selection of K_0 will affect the convergence speed. This will be discussed in Sect. 4.6 using experimental results.

4.6 Tutorial Example

Consider an aircraft system modified from the example in Sect. 5.7.1 of the book Edwards and Spurgeon (1998). The system is given in the form of (4.1) with the following matrices

$$A = \begin{bmatrix} 0 & 0 & 1 & 0 & 0 \\ 0 & -0.154 & -0.0042 & 1.54 & 0 \\ 0 & 0.249 & -1 & -5.2 & 0 \\ 0.0386 & -0.996 & -0.0003 & -0.117 & 0 \\ 0 & 0.5 & 0 & 0 & -0.5 \end{bmatrix}, \ B = \begin{bmatrix} 0 & 0 \\ -3.72 & -0.16 \\ 1.685 & -5.6 \\ 0.1 & 0 \\ 0 & 0 \end{bmatrix},$$

$$F = \begin{bmatrix} 0 \\ -3.72 \\ 1.685 \\ 0.1 \\ 0 \end{bmatrix}, \ D = \begin{bmatrix} 0 & 0 \\ 1 & 1 \\ 0 & 1 \\ 1 & 0 \\ 0 & 0 \end{bmatrix}, \ C = \begin{bmatrix} 0 & 1 & 0 & 0 & 0 \\ 0 & 0 & 0 & 1 & 1 \\ 1 & 1 & 1 & 0 & 0 \end{bmatrix}, \ \Delta A = 0.05 \sin(0.1\pi t) \times A_p,$$

$$A_p = \begin{bmatrix} 0 & 0 & 0 & 0 & 0 \\ 0 & -0.154 & -0.0042 & 1.54 & 0 \\ 0 & 0.249 & -1 & -5.2 & 0 \\ 0.0386 & -0.996 & -0.0003 & -0.117 & 0 \\ 0 & 0.5 & 0 & 0 & -0.5 \end{bmatrix}.$$

According to Assumption 4.2, the matrices $\mathcal{M}$, $\mathcal{F}(t)$ and $\mathcal{N}$ can be given as

$$\mathcal{M} = I_5,\ \mathcal{F}(t) = \sin(0.1\pi t) \times I_5,\ \mathcal{N} = 0.05 \times A_p.$$

The disturbance is $d(t) = [d_1(t)\ d_2(t)]^{\top}$, with $d_1(t)$ taking random values within $[-0.05, 0.05]$ and $d_2(t) = 0.01\cos(t)$. The actuator fault f is characterized by

$$f(t) = \begin{cases} 0, & 0\text{ s} \le t \le 50\text{ s} \\ 1, & 50\text{ s} < t \le 100\text{ s} \\ 2, & 100\text{ s} < t \le 150\text{ s} \\ 1, & 150\text{ s} < t \le 200\text{ s} \\ 0, & 200\text{ s} < t \le 250\text{ s} \end{cases}.$$

4.6.1 Design of FE Observer and FTC Controller

The optimization problem in Lemma 4.2 is solved by minimizing γ_0 and the obtained initial gains are as follows:

$$\begin{aligned}
\gamma_0 &= 5.4888, \\
K_0 &= \begin{bmatrix} 3.4627 & 0.7450 & 4.3578 & -0.4691 & 0.0296 \\ 1.6998 & -2.4849 & 1.6535 & 1.9392 & 0.5749 \end{bmatrix}, \\
P_0 &= \begin{bmatrix} 1.5658 & -0.0151 & -1.2540 & -0.0208 & -0.1164 \\ -0.0151 & 2.8847 & -0.0353 & 1.1121 & -0.0159 \\ -1.2540 & -0.0353 & 2.5967 & -0.0399 & 0.0262 \\ -0.0208 & 1.1121 & -0.0399 & 1.1790 & -0.5772 \\ -0.1164 & -0.0159 & 0.0262 & -0.5772 & 3.1034 \end{bmatrix}.
\end{aligned}$$

Algorithm 4.1 is run with $\epsilon = 1.0\text{e-}5$. It terminates in 26 steps and converges to the local optimum with $\gamma^* = 2.0962$. The evolution of γ and Λ are depicted in Figs. 4.2 and 4.3, respectively. It is seen that the variables γ and Λ_i, $i = 1, 2, 3, 4, 5$, all converge. The obtained controller and observer gains are as follows:

$$\begin{aligned}
K &= \begin{bmatrix} -0.0377 & 0.9384 & -0.0326 & -0.6833 & 0.0916 \\ 0.5744 & 0.2613 & 0.5285 & -1.0634 & 0.0404 \end{bmatrix}, \\
N &= \begin{bmatrix} -0.7522 & -0.2906 & 0.2488 & 0.1602 & -0.2070 & -0.2408 & 0 \\ 0.2838 & -0.5022 & 0.2838 & 0.0606 & 0.0612 & -0.0010 & 0 \\ 0.2497 & -0.2768 & -0.7512 & -0.1822 & 0.1856 & 0.2424 & 0 \\ 0.0190 & -0.0379 & 0.0231 & -3.6491 & 3.2397 & 1.6727 & 0 \\ 0.0021 & -0.0842 & -0.0020 & 3.1454 & -3.7410 & -1.6715 & 0 \\ -0.0093 & 0.0396 & 0.0180 & -0.6913 & 0.7204 & -2.3713 & 1 \\ -0.0176 & 0.0322 & -0.0002 & -0.1585 & 0.1669 & -1.6936 & 0 \end{bmatrix},
\end{aligned}$$

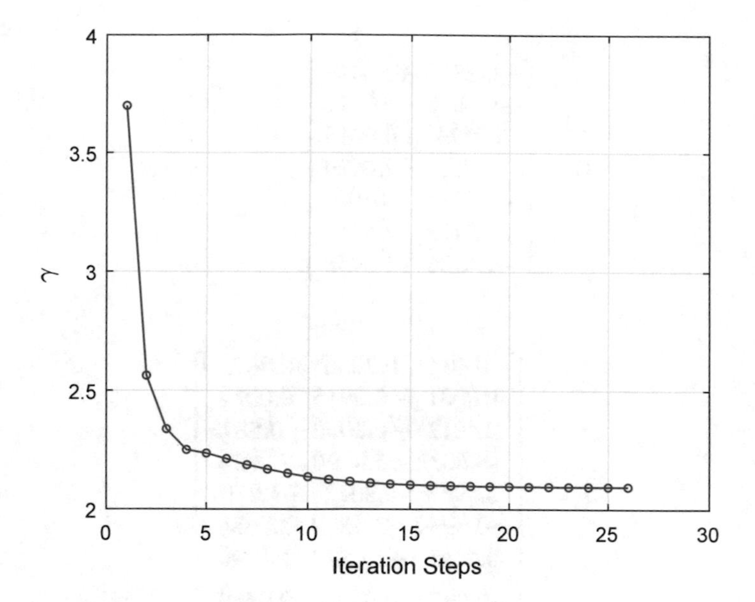

Fig. 4.2 Evolution of gain γ

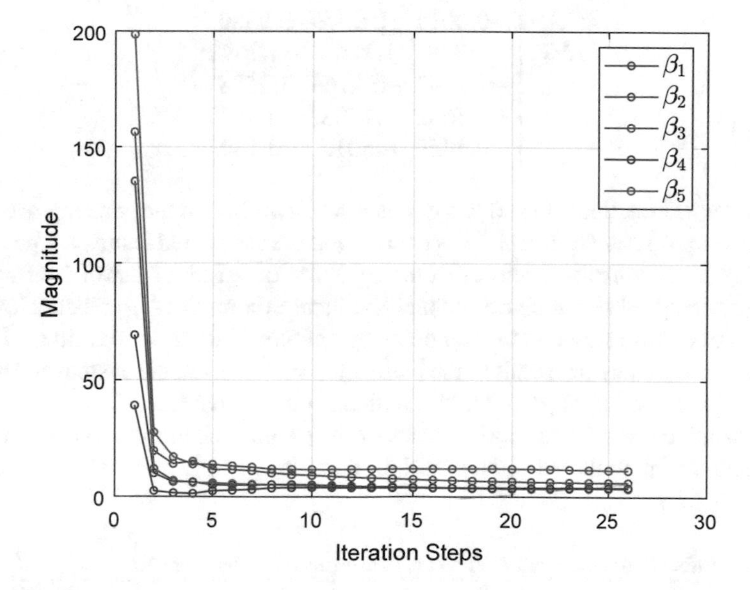

Fig. 4.3 Evolution of gain Λ

$$G = \begin{bmatrix} -0.2408 & 0.3416 \\ -0.0010 & -0.0011 \\ 0.2424 & -0.3415 \\ 1.6727 & -7.0420 \\ -1.6715 & 7.0396 \\ -2.3713 & -2.6911 \\ -1.6936 & -1.0958 \end{bmatrix},$$

$$L = \begin{bmatrix} -0.4027 & 0.2056 & 0.5873 \\ 0.0001 & -0.0015 & 0.0012 \\ 0.4012 & -0.2050 & -0.5873 \\ -8.6032 & -5.8090 & 7.8734 \\ 8.5998 & 5.8072 & -7.8707 \\ -0.6969 & -2.5549 & 2.6684 \\ 0.4665 & -1.4523 & 1.1790 \end{bmatrix},$$

$$H = \begin{bmatrix} -0.0979 & 0.0275 & 0.0620 \\ 0.9998 & 0.0003 & -0.0002 \\ -0.9016 & -0.0279 & 0.9380 \\ 1.1389 & 1.1162 & -1.2542 \\ -1.1383 & -0.1163 & 1.2538 \\ -0.3662 & 0.7908 & -0.4570 \\ -0.3429 & 0.5016 & -0.1807 \end{bmatrix}.$$

In order to investigate the effects of baseline controller on the convergence speed of Algorithm 4.1, five different cases are run and the associated results are presented in Table 4.1. The results confirm that the selection of baseline controller affects the convergence speed of the algorithm and the local minimum γ^* to reach. However, it is in a stochastic way due to the nonconvex nature of iterative algorithm. Hence, the minimum γ_0 may not result in the lowest γ^*. For the example system considered here, the lowest value of γ^* is 2.0958, which is obtained in Case 5.

It is worth comparing the performance of the proposed iterative strategy with that of the sequential strategy in Chap. 3. Note that the example system studied here is

Table 4.1 Effect of baseline controller on the convergence of Algorithm 4.1

Case	Baseline gain γ_0	Iteration number j	Optimum gain γ^*
1	5.4888	26	2.0962
2	6.3606	72	2.1182
3	6.7558	26	3.2696
4	7.6874	17	2.5617
5	8.0368	27	2.0958

the same as the one in Sect. 3.7 for the sequential strategy, except that a different fault signal is simulated. The performance index obtained in Sect. 3.7 is $\gamma = 5.4888$, which is the same as γ_0 obtained from the baseline control design in Sect. 4.6.1. As seen from Table 4.1, it is obvious that the optimum gain γ^* obtained by the iterative strategy is much lower than $\gamma = 5.4888$. Therefore, the proposed iterative strategy can achieve better robust performance than the sequential strategy in Chap. 3.

4.6.2 Performance of FE and FTC

The closed-loop FE-based FTC system is simulated under the initial conditions $x(0) = [0.5\ 0\ 0.5\ 0\ 0]^\top$ and $\hat{\bar{x}}(0) = 0_{7\times 1}$.

The state and fault estimation errors are shown in Figs. 4.4 and 4.5, respectively. It is seen in Fig. 4.4 that the absolute state estimation errors e_{xi}, $i = 1, 2, 3, 4, 5$, are very small though with overshoots at transients. As shown in Fig. 4.5, the estimated actuator fault $\hat{f}$ is almost the same as the actual fault f, though with some overshoots at transients. The above results demonstrate that the proposed observer achieves accurately simultaneous estimation of system state and actuator fault.

The closed-loop system performances under the baseline controller $u = K_0 x$ and the FTC controller $u = K\hat{x} + K_f \hat{f}$ are shown in Figs. 4.6 and 4.7, respectively. It is seen from Fig. 4.6 that under the baseline controller the system state x_i, $i = 1, 2, 3, 4, 5$, are stable without actuator fault f, in the time intervals [0, 50] s and [200, 250] s. However, they are unstable when the actuator fault f occurs in the time

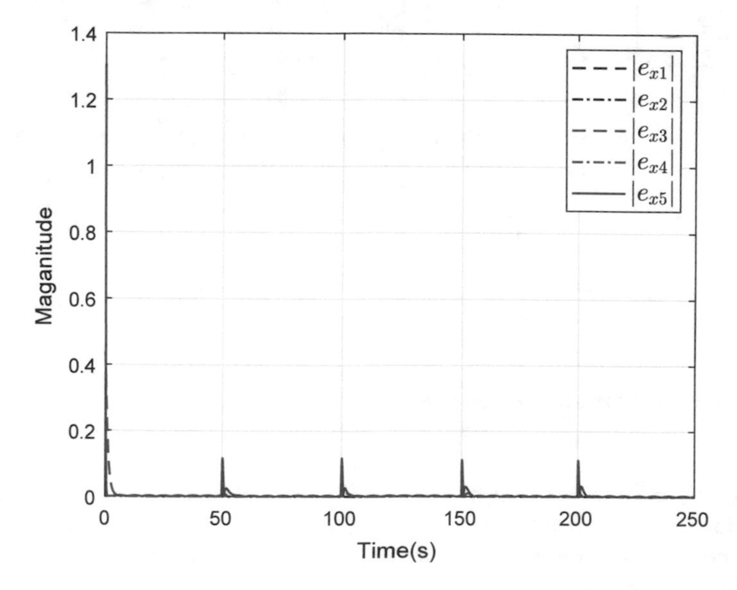

Fig. 4.4 State estimation performance

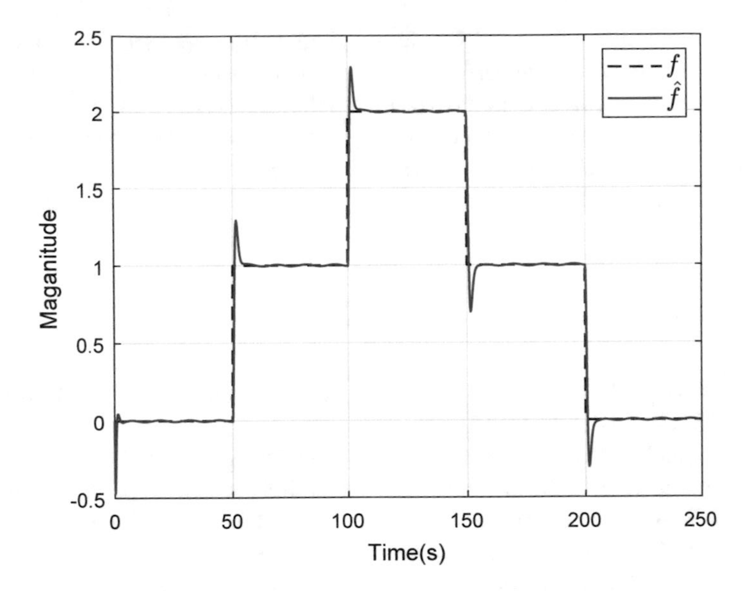

Fig. 4.5 Fault Estimation performance

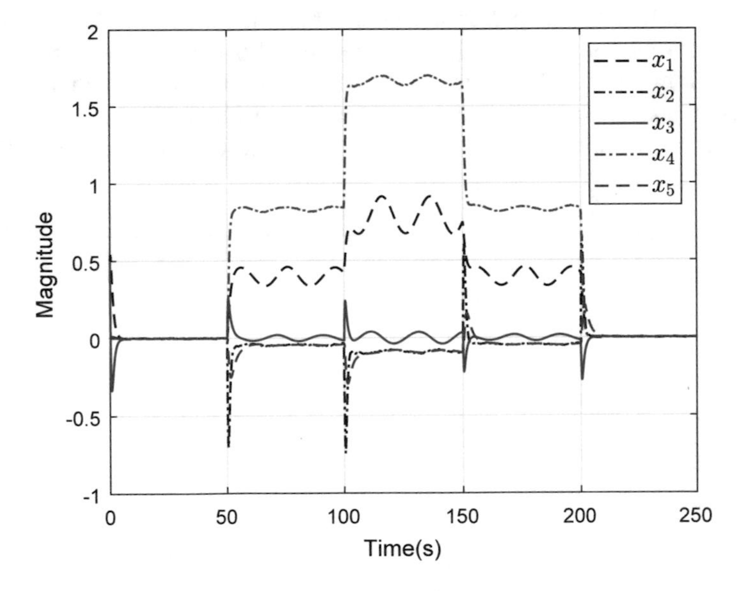

Fig. 4.6 State response under baseline controller $u = K_0 x$

interval (50, 200) s. It is seen from Fig. 4.7 that the system state are always stable in the entire time interval [0, 250] s under FTC. The above results confirm that the FTC controller can ensure system stability in the presence of fault, but the baseline controller alone cannot.

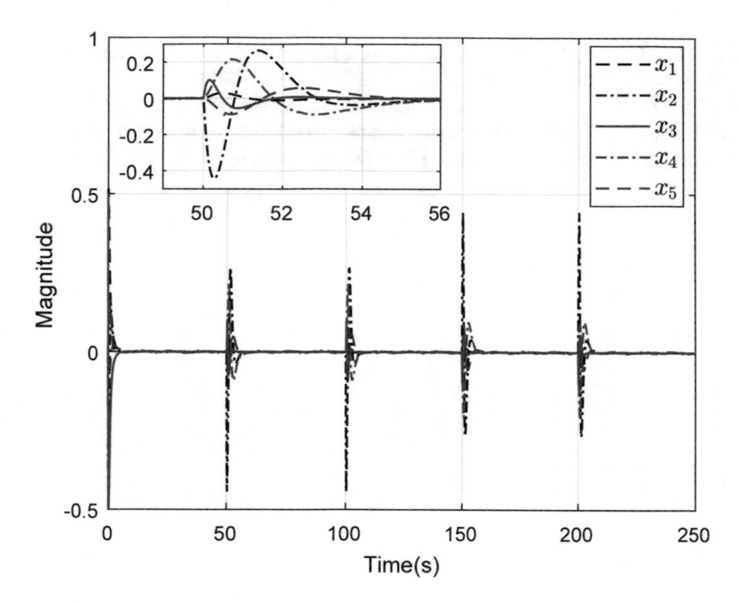

Fig. 4.7 State response under FTC controller $u = K\hat{x} + K_f\hat{f}$

4.7 Notes

Compared with the sequential strategy in Chap. 3, the iterative strategy proposed in this chapter takes into account of the *bidirectional robustness interactions* between FTC system and FE observer. Hence, it can achieve a more robust closed-loop system, as demonstrated by the simulation results. However, similar to the sequential strategy, the iterative strategy can normally generate local optimal solutions to the robust integration design. This is due to the nonlinear nature of the iterative procedure.

References

Edwards C, Spurgeon S (1998) Sliding mode control: theory and applications. CRC Press
Edwards C, Tan CP (2006) Sensor fault tolerant control using sliding mode observers. Control Eng Pract 14(8):897–908
Feng X, Patton R (2014) Active fault tolerant control of a wind turbine via fuzzy MPC and moving horizon estimation. IFAC Proc Vol 47(3):3633–3638
Gao Z, Ding SX (2007) Actuator fault robust estimation and fault-tolerant control for a class of nonlinear descriptor systems. Automatica 43(5):912–920
Gao Z, Liu X, Chen MZ (2016) Unknown input observer-based robust fault estimation for systems corrupted by partially decoupled disturbances. IEEE Trans Ind Electron 63(4):2537–2547
Jiang B, Staroswiecki M, Cocquempot V (2006) Fault accommodation for nonlinear dynamic systems. IEEE Trans Autom Control 51(9):1578–1583

Kabore R, Wang H (2001) Design of fault diagnosis filters and fault-tolerant control for a class of nonlinear systems. IEEE Trans Autom Control 46(11):1805–1810
Rudin W (1964) Principles of mathematical analysis, vol 3. McGraw-hill, New York
Shi P, Liu M, Zhang L (2015) Fault-tolerant sliding mode observer synthesis of Markovian jump systems using quantized measurements. IEEE Trans Ind Electron 62(9):5910–5918

Chapter 5
Simultaneous Integration of FE and FTC

5.1 Introduction

It has been shown in Chaps. 3 and 4 that the integration of FE and FTC considered in this book can be formulated as a robust observer-based control problem. Its solution lies on solving H_∞ optimization problems with BMI constraints that are difficult to solve using off-the-shelf solvers, e.g. the LMI toolbox. To obviate the BMI problem, sequential (see Fig. 5.1 (1)) and iterative strategies (see Fig. 5.1 (2)) are proposed in Chaps. 3 and 4, respectively. However, they can only obtain a suboptimal solution to the overall FE-based FTC system design.

This chapter aims to develop a simultaneous integration strategy (see Fig. 5.1 (3)) to obtain optimal FTC controller and FE observer gains in one shot based on a fully LMI formulation. The main results in this chapter are summarized below:

- *Reduced-order and full-order augmented state unknown input observers (ASUIOs) are proposed to achieve FE.* In this chapter, reduced-order and full-order ASUIOs without the well-known matching condition (see Sect. 1.7.3) are proposed to achieve, respectively: (1) time-varying fault estimation for the state feedback case and (2) simultaneous estimation of time-varying faults and system state for the output feedback case. A new property of the reduced-order ASUIO for FE is that the estimation of system state is not necessary. This results in an observer with lower dimension than the one in Chap. 4. Both actuator and sensor faults are studied, while only the former is considered in Chaps. 3 and 4. Moreover, the faults considered in this chapter can be either additive or multiplicative faults, while only the former is discussed in Chaps. 3 and 4.
- *Both adaptive state and output feedback sliding mode FTC controllers are developed.* Considering its salient feature of robustness to uncertainty and disturbance, sliding mode control (SMC) is combined with the linear FTC controllers used in Chaps. 3 and 4 to enhance FTC system robustness. Sliding mode FTC controllers are developed for both the state and output feedback cases. Moreover, adaptive

© The Author(s), under exclusive license to Springer Nature Switzerland AG 2021

J. Lan and R. J. Patton, *Robust Integration of Model-Based Fault Estimation and Fault-Tolerant Control*, Advances in Industrial Control,
https://doi.org/10.1007/978-3-030-58760-4_5

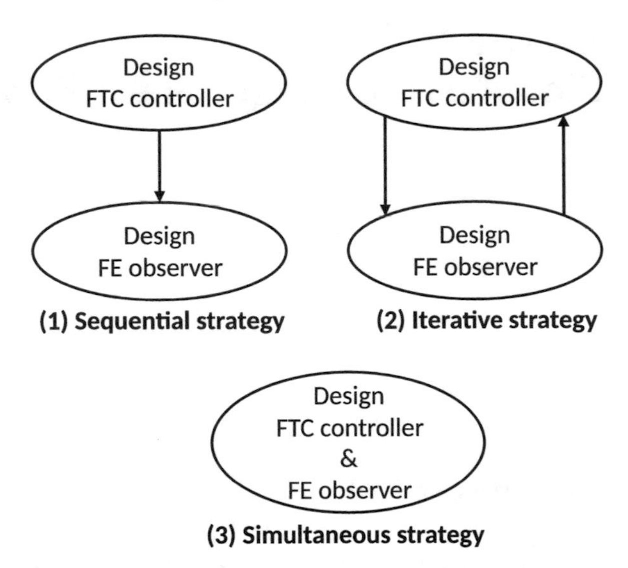

Fig. 5.1 Sequential, iterative and simultaneous integration of FE and FTC

gains are incorporated with SMC to avoid using knowledge of the unknown upper bounds of fault, uncertainty and disturbance.

- *A simultaneous strategy is proposed for robust integration of FE and FTC*. For systems with additive or multiplicative faults, it is shown that the robust integration of FE and FTC can be converted into a robust observer-based control problem solved using a single-step LMI formulation.

5.2 Problem Description

Consider a class of linear systems represented by

$$\begin{aligned} \dot{x} &= (A + \Delta A)x + Bu + F_a f_a + Dd \\ y &= Cx + F_s f_s \end{aligned} \tag{5.1}$$

where $x \in \mathbb{R}^n, u \in \mathbb{R}^m, f_a \in \mathbb{R}^q, d \in \mathbb{R}^l, y \in \mathbb{R}^p$, and $f_s \in \mathbb{R}^{q_1}$ are the state, control input, actuator fault, external disturbance, measured output, and sensor fault, respectively. The constant matrices A, B, F_a, D, C and F_s are known and of compatible dimensions. The matrix ΔA represents the unknown system uncertainty. This system satisfies the assumptions below.

Assumption 5.1 The pair (A, B) is controllable, and the actuator fault f_a is matched, i.e. $\text{rank}[B\ F_a] = \text{rank}(B) = m$.

Assumption 5.2 $\text{rank}(F_a) = q$, $\text{rank}(F_s) = q_1$, and the quadruple $(A, [F_a\ 0], C, [0\ F_s])$ has no invariant zeros in the closed right-half complex plane.

Assumption 5.3 The uncertainty matrix ΔA is norm-bounded (energy bounded) and has the form: $\Delta A = M_0 F_0(t) N_0$, where M_0 and N_0 are known matrices with appropriate dimensions, and $F_0(t)$ is an unknown matrix satisfying $F_0^\top(t) F_0(t) \preceq I$.

Assumption 5.4 The faults and disturbance satisfy $\|f_a\| \leq \bar{f}_a$, $\|f_s\| \leq \bar{f}_s$, and $\|d\| \leq d_0$ with unknown positive scalars $\bar{f}_a$, $\bar{f}_s$, and d_0, respectively. Moreover, f_a and f_s have norm-bounded first-order time derivatives.

This chapter aims to address the following problem:

problem 5.1 For the system (5.1), design together the following FE and FTC functions to guarantee closed-loop system stability: (1) *FE observer*: to estimate the faults for the state feedback case, and simultaneously the faults and system state for the output feedback case; (2) *Adaptive sliding mode FTC controller*: to compensate the fault effects and ensure closed-loop system stability by using state or output feedback.

5.3 Integration of FE and FTC: State Feedback

Provided that all the system state variables are available, then only the fault estimation is needed for the FTC design. In this case, a reduced-order ASUIO is proposed below to estimate only the faults.

5.3.1 Reduced-Order ASUIO-Based FE Design

By defining the faults f_a and f_s as auxiliary state, then the system (5.1) can be augmented as

$$\begin{aligned} \dot{\bar{x}} &= \bar{A}\bar{x} + \bar{B}u + \Delta\bar{A}\bar{x} + \bar{D}\bar{d} \\ y &= \bar{C}\bar{x} \end{aligned} \tag{5.2}$$

where

$$\bar{x} = \begin{bmatrix} x \\ f_a \\ f_s \end{bmatrix}, \ \bar{d} = \begin{bmatrix} d \\ \dot{f}_a \\ \dot{f}_s \end{bmatrix}, \ \bar{A} = \begin{bmatrix} A & F_a & 0 \\ 0 & 0 & 0 \\ 0 & 0 & 0 \end{bmatrix}, \ \Delta\bar{A} = \begin{bmatrix} \Delta A & 0 & 0 \\ 0 & 0 & 0 \\ 0 & 0 & 0 \end{bmatrix},$$

$$\bar{B} = \begin{bmatrix} B \\ 0 \\ 0 \end{bmatrix}, \ \bar{D} = \begin{bmatrix} D & 0 & 0 \\ 0 & I_q & 0 \\ 0 & 0 & I_{q_1} \end{bmatrix}, \ \bar{C} = [C \ 0 \ F_s].$$

According to Assumption 5.2, it can be verified that the augmented system (5.2) is observable. Define $z = L\bar{x}$ with $L = [0 \ I_{q+q_1}] \in \mathbb{R}^{(q+q_1)\times(n+q+q_1)}$, then a reduced-order ASUIO to estimate the faults f_a and f_s is designed as

$$\begin{aligned} \dot{\xi}_s &= M\xi_s + Gu + Ry \\ \hat{z} &= \xi_s + Hy \end{aligned} \tag{5.3}$$

where $\xi_s \in \mathbb{R}^{q+q_1}$ and $\hat{z} \in \mathbb{R}^{q+q_1}$ are the observer state and the estimate of z, respectively. The matrices M, G, R and H are of appropriate dimensions and to be designed.

Define $\varepsilon = \xi_s - T\bar{x}$, then the estimation error system is derived as

$$\begin{aligned} \dot{\varepsilon} &= M\varepsilon + (MT + R\bar{C} - T\bar{A})\bar{x} + (G - T\bar{B})u - T\Delta\bar{A}\bar{x} - T\bar{D}\bar{d} \\ e_s &= \varepsilon + (T + H\bar{C} - L)\bar{x}. \end{aligned} \tag{5.4}$$

The existence conditions of an asymptotically stable observer (5.3) are given below.

Theorem 5.1 *There exists an asymptotically stable observer (5.3) for the system (5.2) when $\Delta\bar{A}\bar{x} = 0$ and $\bar{d} = 0$, if the following conditions hold*

$$M \text{ is Hurwitz} \tag{5.5}$$

$$MT + R\bar{C} - T\bar{A} = 0 \tag{5.6}$$

$$T + H\bar{C} - L = 0 \tag{5.7}$$

$$G - T\bar{B} = 0. \tag{5.8}$$

Proof With (5.6)–(5.8) and $\Delta\bar{A}\bar{x} = 0$ and $\bar{d} = 0$, the error system (5.4) becomes

$$\begin{aligned} \dot{\varepsilon} &= M\varepsilon, \\ e_s &= \varepsilon. \end{aligned}$$

Since M is Hurwitz, it holds that $\lim_{t\to\infty} e_s(t) = 0$. □

In the following, a parametrization method is used to solve the matrix equations (5.5)–(5.8). Before that, the solvability of these matrix equations is proved using Lemmas 5.1 and 5.2. Before presenting these lemmas, the following full row rank matrix to be used by them are defined:

$$S = [L^{\dagger}\ (I_{n+q+q_1} - L^{\dagger}L)] = [S_1\ S_2] \tag{5.9}$$

where the matrices S_1 and S_2 satisfy $S_2S_1 = 0$ and $\text{rank}(S_1) + \text{rank}(S_2) = \text{rank}(S) = n + q + q_1$.

Lemma 5.1 *There exist a solution Λ to the matrix equation*

$$\Lambda\Omega = \Psi \tag{5.10}$$

where $\Omega = \begin{bmatrix} \bar{C}S_2 \\ \bar{C}\bar{A}S_2 \end{bmatrix}$ *and* $\Psi = L\bar{A}S_2$, *if the following rank condition holds*

$$\text{rank} \begin{bmatrix} L\bar{A} \\ \bar{C} \\ \bar{C}\bar{A} \\ L \end{bmatrix} = \text{rank} \begin{bmatrix} \bar{C} \\ \bar{C}\bar{A} \\ L \end{bmatrix}. \tag{5.11}$$

Proof Post-multiplying both sides of (5.11) with the matrix $[S_2 \; S_1]$ gives

$$\text{rank} \begin{bmatrix} L\bar{A}S_2 & L\bar{A}S_1 \\ \bar{C}S_2 & \bar{C}S_1 \\ \bar{C}\bar{A}S_2 & \bar{C}\bar{A}S_1 \\ LS_2 & LS_1 \end{bmatrix} = \text{rank} \begin{bmatrix} \bar{C}S_2 & \bar{C}S_1 \\ \bar{C}\bar{A}S_2 & \bar{C}\bar{A}S_1 \\ LS_2 & LS_1 \end{bmatrix}. \tag{5.12}$$

By using the definitions of S_1 and S_2 in (5.9), it gives $LS_2 = 0$ and $LS_1 = I_{n+q+q_1}$. Hence, (5.12) is equivalent to

$$\text{rank} \left[\begin{array}{c|c} L\bar{A}S_2 & L\bar{A}S_1 \\ \bar{C}S_2 & \bar{C}S_1 \\ \bar{C}\bar{A}S_2 & \bar{C}\bar{A}S_1 \\ \hline 0 & I_{n+q+q_1} \end{array}\right] = \text{rank} \left[\begin{array}{c|c} \bar{C}S_2 & \bar{C}S_1 \\ \bar{C}\bar{A}S_2 & \bar{C}\bar{A}S_1 \\ \hline 0 & I_{n+q+q_1} \end{array}\right]. \tag{5.13}$$

The above equation implies that

$$\text{rank} \begin{bmatrix} L\bar{A}S_2 \\ \bar{C}S_2 \\ \bar{C}\bar{A}S_2 \end{bmatrix} = \text{rank} \begin{bmatrix} \bar{C}S_2 \\ \bar{C}\bar{A}S_2 \end{bmatrix}. \tag{5.14}$$

Therefore, the following condition holds

$$\text{rank} \begin{bmatrix} \Psi \\ \Omega \end{bmatrix} = \text{rank}(\Omega).$$

This guarantees feasibility of the matrix equation (5.10). $\square$

Lemma 5.2 *The pair* (M_2, M_1), *with* $M_1 = L\bar{A}S_1 - \Psi\Omega^{\dagger}\Gamma$, $M_2 = (I_{2p} - \Omega\Omega^{\dagger})\Gamma$, *and* $\Gamma = \begin{bmatrix} \bar{C}S_1 \\ \bar{C}\bar{A}S_1 \end{bmatrix}$, *is detectable if*

$$\text{rank} \begin{bmatrix} sL - L\bar{A} \\ \bar{C} \\ \bar{C}\bar{A} \end{bmatrix} = \text{rank} \begin{bmatrix} \bar{C} \\ \bar{C}\bar{A} \\ L \end{bmatrix}, \; \forall s \in \mathbb{C}, \; \text{Re}(s) \geq 0. \tag{5.15}$$

Proof Post-multiplying the left-hand side of (5.15) by the full row rank matrix $[S_1\ S_2]$ gives

$$\begin{aligned}
&\operatorname{rank}\left(\begin{bmatrix} sL - L\bar{A} \\ \bar{C} \\ \bar{C}\bar{A} \end{bmatrix} [S_1\ S_2]\right) \\
&= \operatorname{rank}\begin{bmatrix} sI_{q+q_1} - L\bar{A}S_1 & -\Psi \\ \Gamma & \Omega \end{bmatrix} \\
&= \operatorname{rank}\left(\begin{bmatrix} I_{q+q_1} & \Psi\Omega^\dagger \\ 0 & (I_{2p} - \Omega\Omega^\dagger) \\ 0 & \Omega\Omega^\dagger \end{bmatrix} \begin{bmatrix} sI_{q+q_1} - L\bar{A}S_1 & -\Psi \\ \Gamma & \Omega \end{bmatrix}\right) \\
&= \operatorname{rank}\begin{bmatrix} sI_{q+q_1} - M_1 & 0 \\ M_2 & 0 \\ \Omega\Omega^\dagger\Gamma & \Omega \end{bmatrix} \\
&= \operatorname{rank}\begin{bmatrix} sI_{q+q_1} - M_1 \\ M_2 \end{bmatrix} + \operatorname{rank}(\Omega). \qquad (5.16)
\end{aligned}$$

Similarly, the right-hand side of (5.15) is equivalent to

$$\begin{aligned}
\operatorname{rank}\left(\begin{bmatrix} \bar{C} \\ \bar{C}\bar{A} \\ L \end{bmatrix} [S_1\ S_2]\right) &= \operatorname{rank}\begin{bmatrix} \bar{C}S_1 & \bar{C}S_2 \\ \bar{C}\bar{A}S_1 & \bar{C}\bar{A}S_2 \\ I_{q+q_1} & 0 \end{bmatrix} \\
&= q + q_1 + \operatorname{rank}(\Omega). \qquad (5.17)
\end{aligned}$$

Comparing (5.16) with (5.17) gives

$$\operatorname{rank}\begin{bmatrix} sI_{q+q_1} - M_1 \\ M_2 \end{bmatrix} = q + q_1.$$

This implies detectability of the pair (M_2, M_1). □

Since $L\bar{A} = 0$, the sufficient condition (5.11) is always satisfied and thus the matrix equation (5.10) is always solvable. With this result at hand, it is ready to give the parameterized solutions of the observer gains. It follows from (5.7) that $T = L - H\bar{C}$. Substituting this into (5.6) yields

$$M(L - H\bar{C}) + R\bar{C} - (L - H\bar{C})\bar{A} = 0.$$

Define $T_1 = R - MH$, it follows that

$$L\bar{A} - ML = [T_1\ H]\begin{bmatrix} \bar{C} \\ \bar{C}\bar{A} \end{bmatrix}. \qquad (5.18)$$

Post-multiplying both sides of (5.18) by S yields

$$M = L\bar{A}S_1 - [T_1 \ H]\begin{bmatrix} \bar{C}S_1 \\ \bar{C}\bar{A}S_1 \end{bmatrix} \tag{5.19}$$

$$L\bar{A}S_2 = [T_1 \ H]\begin{bmatrix} \bar{C}S_2 \\ \bar{C}\bar{A}S_2 \end{bmatrix}. \tag{5.20}$$

The equation (5.20) can be rearranged as

$$[T_1 \ H]\Omega = \Psi. \tag{5.21}$$

According to Lemma 5.1, the matrix equation (5.21) is solvable with the general solution

$$[T_1 \ H] = \Psi\Omega^{\dagger} + Z(I_{2p} - \Omega\Omega^{\dagger}) \tag{5.22}$$

where $Z \in \mathbb{R}^{(q+q_1)\times 2p}$ is a design matrix.

It follows from (5.19) and (5.22) that

$$M = M_1 - ZM_2, \ H = H_1 + ZH_2 \tag{5.23}$$

where $H_1 = \Psi\Omega^{\dagger}\Gamma_1$, $H_2 = (I_{2p} - \Omega\Omega^{\dagger})\Gamma_1$ and $\Gamma_1 = [0 \ I_p]^{\top}$.

The matrices M_1 and M_2 in (5.23) are known from Lemma 5.2. Since $L\bar{A} = 0$, the sufficient condition (5.15) is always satisfied and thus the pair (M_2, M_1) is detectable. Hence, one can design the matrix Z such that M is Hurwitz. Then, the gain H can be obtained. Further using $T_1 = R - MH$ gives R and using Theorem 5.1 gives the gain G.

However, since there exists uncertainty and disturbance in the system (5.1), i.e. the terms $\Delta\bar{A}\bar{x} \neq 0$ and $\bar{d} \neq 0$, the error system (5.4) should be made robustly stable.

Define $\bar{H}_1 = H_1\bar{C} - L$. Substituting $M = M_1 - ZM_2$ and $T = L - H\bar{C}$ into (5.9) gives the estimation error system

$$\dot{e}_s = (M_1 - ZM_2)e_s + (\bar{H}_1 + ZH_2\bar{C})(\Delta\bar{A}\bar{x} + \bar{D}\bar{d}). \tag{5.24}$$

Hence, by designing the matrix Z such that (5.24) is robustly stable, a robustly stable observer (5.3) for the system (5.1) can be built.

The proposed reduced-order ASUIO (with an order of $(q + q_1)$) is interesting in three respects: (1) The traditional UIOs (Chen and Patton 1999; Odgaard and Stoustrup 2012; Xiong and Saif 2003) decouple the disturbance upon satisfaction of the rank condition $\text{rank}(\bar{C}\bar{D}) = \text{rank}(\bar{D})$, which is restrictive and often cannot be satisfied. H_∞ optimization is employed in this chapter to attenuate the disturbance and the matrix Z is designed using LMI tools; (2) In contrast to the FE used in the majority of existing FE (and Chaps. 3 and 4) with full-order $(n + q + q_1)$, the reduced-order ASUIO obtains the fault estimation without extra effort to estimate the available system state; (3) Notice that (5.11) and (5.15) are two sufficient conditions for the existence of a solution to Theorem 5.1 and the proposed reduced-order ASUIO. Since $L\bar{A} = 0$, these two conditions are always satisfied for all s with $\text{Re}(s) > 0$.

5.3.2 State Feedback Sliding Mode FTC Design

This section describes the design of an FTC controller to compensate the fault effects and stabilize the system (5.1), by using the concept of SMC. The general aim of SMC is to achieve robust insensitivity to matched uncertainty acting within the control channels, via a combination of linear and switched feedback (Edwards and Spurgeon 1998). The SMC must be designed to reach a sliding surface and the switching operation is designed to keep the system motion in the sliding manifold.

Since it is assumed that all the state variables are available, the switching function for (5.1) using the system state is defined as

$$s_1 = N_1 x \tag{5.25}$$

where $s_1 \in \mathbb{R}^m$, $N_1 = B^\dagger - Y_1(I_n - BB^\dagger)$ and $B^\dagger = (B^\top B)^{-1}B^\top$. $Y_1 \in \mathbb{R}^{m\times n}$ is a design matrix introduced here to allow more design freedom, compared with the use of a fixed gain $N_1 = B^\dagger$.

The first step of SMC design is to determine the controller u to ensure satisfaction of the reachability condition $s_1^\top \dot{s}_1 \leq 0$ (Edwards and Spurgeon 1998). To this end, differentiating s_1 with respect to time gives

$$\dot{s}_1 = N_1(A + \Delta A)x + u + N_1 F_a f_a + N_1 D d. \tag{5.26}$$

Design the control input u as

$$u = u_{l_1} + u_{n_1} \tag{5.27}$$

with a linear feedback component u_{l_1} and a nonlinear component u_{n_1}. The linear feedback component is designed as $u_{l_1} = -K_s x - E_1 \hat{f}_a$ with constant gains $K_s \in \mathbb{R}^{m\times n}$ and $E_1 = B^\dagger F_a$. The nonlinear component is designed as $u_{n_1} = -\varrho_{s_1}\operatorname{sign}(s_1)$ with a design parameter ϱ_{s_1}.

To prove satisfaction of the reachability condition $s_1^\top \dot{s}_1 \leq 0$, the following Lyapunov function is used:

$$V_{s_{10}} = \frac{1}{2} s_1^\top s_1.$$

By using (5.26) and (5.27), the time derivative of $V_{s_{10}}$ is

$$\begin{aligned} \dot{V}_{s_{10}} &= s_1^\top \left[N_1(A + \Delta A)x + u + N_1 F_a f_a + N_1 D d\right] \\ &= s_1^\top \left[(N_1 A - K_s + N_1 \Delta A)x + E_1 e_{f_a} + N_1 D d - \varrho_{s_1}\operatorname{sign}(s_1)\right] \\ &\leq (\eta_{s_1} - \varrho_{s_1})\|s_1\| \end{aligned} \tag{5.28}$$

where η_{s_1} is an unknown scalar satisfying $\eta_{s_1} \geq (\|N_1 A - K_s\| + \|N_1 M_0\|\|N_0\|)$ $\|x\| + \|E_1\|\|e_{f_a}\| + \|N_1 D\| d_0$.

Design $\varrho_{s_1} = \hat{\eta}_{s_1} + \varepsilon_{s_1}$, where ε_{s_1} is a positive scalar and $\hat{\eta}_{s_1}$ is introduced to estimate the unknown scalar η_{s_1}. The scalar $\hat{\eta}_{s_1}$ is updated by

$$\dot{\hat{\eta}}_{s_1} = \sigma_1 \|s_1\|, \ \hat{\eta}_{s_1}(0) = 0 \tag{5.29}$$

where $\sigma_1 > 0$ is a design constant.

Define the estimation error of η_{s_1} as $\tilde{\eta}_{s_1} = \eta_{s_1} - \hat{\eta}_{s_1}$. Consider a Lyapunov function

$$V_{s_1} = V_{s_{10}} + \frac{1}{2\sigma_1}\tilde{\eta}_{s_1}^2.$$

It follows from (5.28) and (5.29) that

$$\begin{aligned} \dot{V}_{s_1} &= \dot{V}_{s_{10}} - \frac{1}{\sigma_1}\tilde{\eta}_{s_1}\dot{\hat{\eta}}_{s_1} \\ &\leq (\eta_{s_1} - \varrho_{s_1})\|s_1\| - \tilde{\eta}_{s_1}\|s_1\| \\ &\leq -\varepsilon_{s_1}\|s_1\| \\ &\leq 0. \end{aligned} \tag{5.30}$$

Since V_{s_1} is positive definite, it follows from (5.30) and the Barbalat's Lemma (see Sect. 1.7.1) that $V_{s_1}(t) \leq V_{s_1}(0)$. Therefore, $s_1(t)$ and $\tilde{\eta}_{s_1}(t)$ are bounded. This means that the designed controller (5.27) can maintain the sliding motion around the sliding surface $s_1 = 0$. Moreover, in the case of zero initial condition (i.e. $V_{s_1}(0) = 0$), it holds that $\lim_{t\to\infty} s_1(t) = 0$. In such case, the sliding surface $s_1 = 0$ is reachable and the ideal sliding motion is maintained.

The next step is to analyze the system stability corresponding to the sliding mode. By setting $\dot{s}_1 = 0$, it follows from (5.26) that the equivalent control input of u can be defined as

$$u_{eq_1} = -\left[N_1(A + \Delta A)x + N_1 Dd\right] + u_{l_1}. \tag{5.31}$$

Substituting (5.31) into (5.1) gives the equivalent closed-loop control system

$$\dot{x} = (\Theta_1 A - BK_s)x + \Theta_1 \Delta A x + F_1 e_s + \Theta_1 Dd \tag{5.32}$$

where $\Theta_1 = I_n - BN_1$ and $F_1 = [F_a \ 0]$.

Therefore, the system (5.1) is maintained on the sliding mode with the equivalent control (5.31) by designing K_s to stabilize (5.32). The closed-loop system (5.32) contains the uncertainty ΔAx and disturbance d, which must be minimized to achieve a suitable degree of robustness. This is achieved using H_∞ optimization, whose details are given in the next section.

Remark 5.1 It is worth noting that the equivalent control input u_{eq_1} is slightly different from the conventional form given below (Edwards and Spurgeon 1998):

$$u_{eq_1} = -\left[N_1(A + \Delta A)x + N_1 F_a f_a + N_1 Dd\right] + u_{l_1}. \tag{5.33}$$

The reason for choosing the particular form (5.31) is that under this u_{eq_1}, the sliding dynamics become

$$\dot{s}_1 = -K_s x - E_1 e_{fa} \tag{5.34}$$

which is asymptotically stable provided that x and e_{fa} are asymptotically stable.

If instead applying the conventional form (5.33), the sliding dynamics become

$$\dot{s}_1 = -K_s x - E_1 \hat{f}_a \tag{5.35}$$

and the resulting equivalent closed-loop control system is given as

$$\dot{x} = (\Theta_1 A - BK_s)x + \Theta_1 \Delta A x - F_a \hat{f}_a + \Theta_1 D d. \tag{5.36}$$

It can be seen that both (5.35) and (5.36) are perturbed by $\hat{f}_a$, which leads to unstable sliding dynamics even when x and e_{fa} are asymptotically stable. This also imposes more disturbance ($\hat{f}_a$) on the equivalent closed-loop control system.

5.3.3 Simultaneous Integration of FE and FTC

With the reduced-order FE observer presented in Sect. 5.3.1 and the state feedback FTC controller presented in Sect. 5.3.2, the proposed state feedback FE-based FTC system is outlined in Fig. 5.2. This section describes a simultaneous integration strategy to design the observer and controller gains.

The composite closed-loop system consisting of (5.24) and (5.32) is given as

$$\begin{aligned}
\dot{x} &= (\Theta_1 A - BK_s)x + \Theta_1 \Delta A x + F_1 e_s + D_1 \bar{d} \\
\dot{e}_s &= (M_1 - ZM_2)e_s + (\bar{H}_1 + ZH_2\bar{C})(\Delta\bar{A}\bar{x} + \bar{D}\bar{d}) \\
y_c &= y - F_s \hat{f}_s \\
z_s &= \mathrm{diag}(C_{sx}x, C_{se}e_s)
\end{aligned} \tag{5.37}$$

where $D_1 = [\Theta_1 D \; 0]$ and z_s is the performance output with given weights C_{sx} and C_{se} of compatible dimensions.

Theorem 5.2 *Under Assumptions 5.1–5.4, the composite closed-loop system (5.37) is stable with H_∞ performance $\|G_{z_s\bar{d}}\|_\infty < \gamma_s$, if the following optimization problem is feasible:*

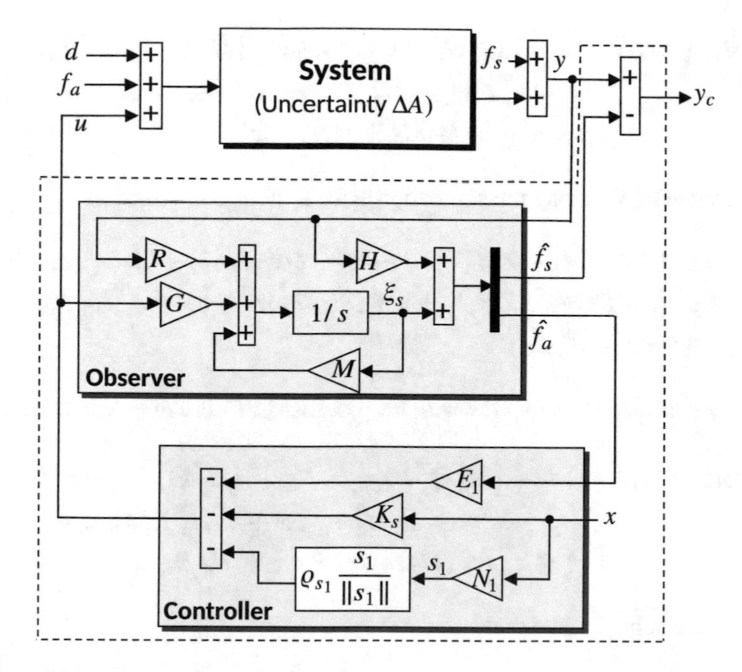

Fig. 5.2 Integrated FE and FTC design: state feedback case

$$\min \bar{\gamma}_s \tag{5.38}$$

$$\text{s.t.} \quad PB = B\hat{P} \tag{5.39}$$

$$\begin{bmatrix} \chi_{11} & \chi_{12} & \chi_{13} & \chi_{14} & 0 & C_{sx}^\top & 0 \\ \star & \chi_{22} & \chi_{23} & 0 & \chi_{25} & 0 & C_{se}^\top \\ \star & \star & -\bar{\gamma}_s I & 0 & 0 & 0 & 0 \\ \star & \star & \star & -I & 0 & 0 & 0 \\ \star & \star & \star & \star & -I & 0 & 0 \\ \star & \star & \star & \star & \star & -I & 0 \\ \star & \star & \star & \star & \star & \star & -I \end{bmatrix} \prec 0 \tag{5.40}$$

$$P = P^\top \succ 0, \; Q = Q^\top \succ 0, \; \bar{\gamma}_s > 0, \tag{5.41}$$

where

$$\chi_{11} = \text{He}(P\Theta_1 A - BR_1) + 2N_0^\top N_0, \; \chi_{12} = PF_1, \; \chi_{13} = PD_1, \; \chi_{14} = P\Theta_1 M_0,$$
$$\chi_{22} = \text{He}(QM_1 - R_2 M_2), \; \chi_{23} = (Q\bar{H}_1 + R_2 H_2 \bar{C})\bar{D}, \; \chi_{25} = (Q\bar{H}_1 + R_2 H_2 \bar{C})\bar{M}_0.$$

Then the gains are obtained as: $\gamma_s = \sqrt{\bar{\gamma}_s}$, $K_s = \hat{P}^{-1} R_1$, $Z = Q^{-1} R_2$.

Proof Consider the Lyapunov function $V_{e_s} = e_s^\top Q e_s$ with a s.p.d. matrix Q. Define $W_1 = \bar{H}_1 + ZH_2\bar{C}$ and $\bar{M}_0 = [M_0^\top \; 0]^\top$. According to Assumption 5.3, it is derived that

$$\begin{aligned}\mathrm{He}(e_s^\top QW\Delta\bar{A}\bar{x}) &= -\left[\bar{M}_0^\top W_1^\top Qe_s - F_0N_0x\right]^\top\left[\bar{M}_0^\top W_1^\top Qe_s - F_0N_0x\right] \\ &\quad + e_s^\top QW_1\bar{M}_0\bar{M}_0^\top W_1^\top Qe_s + x^\top N_0^\top F_0^\top F_0N_0x \\ &\le e_s^\top QW_1\bar{M}_0\bar{M}_0^\top W_1^\top Qe_s + x^\top N_0^\top N_0x.\end{aligned}$$

The derivative of V_{e_s} along the estimation error system is obtained as

$$\begin{aligned}\dot{V}_{e_s} &= e_s^\top\left[\mathrm{He}(Q(M_1 - ZM_2))\right]e_s + \mathrm{He}(e_s^\top QW_1\Delta\bar{A}\bar{x}) + \mathrm{He}(e_s^\top QW_1\bar{D}\bar{d}) \\ &\le e_s^\top\left[\mathrm{He}(Q(M_1 - ZM_2)) + QW_1\bar{M}_0\bar{M}_0^\top W_1^\top Q\right]e_s + x^\top N_0^\top N_0x \\ &\quad + \mathrm{He}(e_s^\top QW_1\bar{D}\bar{d}).\end{aligned} \tag{5.42}$$

Further consider $V_{xs} = x^\top Px$ with a s.p.d. matrix P. It can be derived that

$$\begin{aligned}\mathrm{He}(x^\top P\Theta_1\Delta Ax) &= -\left[M_0^\top\Theta_1^\top Px - F_0N_0x\right]^\top\left[M_0^\top\Theta_1^\top Px - F_0N_0x\right] \\ &\quad + x^\top P\Theta_1M_0M_0^\top\Theta_1^\top Px + x^\top N_0^\top F_0^\top F_0N_0x \\ &\le x^\top P\Theta_1M_0M_0^\top\Theta_1^\top Px + x^\top N_0^\top N_0x.\end{aligned}$$

Similarly, it can be derived that

$$\begin{aligned}\dot{V}_{xs} &= x^\top\left[\mathrm{He}\left(P(\Theta_1A - BK_s)\right) + P\Theta_1M_0M_0^\top\Theta_1^\top P + N_0^\top N_0\right]x \\ &\quad + \mathrm{He}\left(x^\top PF_1e + x^\top PD_1\bar{d}\right).\end{aligned} \tag{5.43}$$

The H_∞ performance $\|G_{z_s\bar{d}}\|_\infty < \gamma_s$ can be represented by

$$J = \int_0^\infty\left(z_s(t)^\top z_s(t) - \gamma_s^2\bar{d}(t)^\top\bar{d}(t)\right)\mathrm{d}t < 0. \tag{5.44}$$

Define $V_s = V_{xs} + V_{es}$, then under zero initial conditions, it holds that

$$\begin{aligned}J &= \int_0^\infty\left(z_s(t)^\top z_s(t) - \gamma_s^2\bar{d}(t)^\top\bar{d}(t)\right)\mathrm{d}t - \int_0^\infty\dot{V}_s(t)\mathrm{d}t \\ &= \int_0^\infty\left(z_s(t)^\top z_s(t) - \gamma_s^2\bar{d}(t)^\top\bar{d}(t)\right)\mathrm{d}t - V_s(\infty) + V_s(0) \\ &\le \int_0^\infty\left(z_s(t)^\top z_s(t) - \gamma_s^2\bar{d}(t)^\top\bar{d}(t) + \dot{V}_s(t)\right)\mathrm{d}t.\end{aligned} \tag{5.45}$$

It can be seen from (5.45) that a sufficient condition for (5.44) is

$$z_s^\top z_s - \gamma_s^2\bar{d}^\top\bar{d} + \dot{V}_s < 0. \tag{5.46}$$

Substituting (5.42) and (5.43) into (5.46) gives

$$\begin{bmatrix} e \\ e_s \\ \bar{d} \end{bmatrix}^\top \begin{bmatrix} J_{11} & \chi_{12} & \chi_{13} \\ \star & J_{22} & \chi_{23} \\ \star & \star & -\gamma_s^2 I \end{bmatrix} \begin{bmatrix} e \\ e_s \\ \bar{d} \end{bmatrix} < 0$$

and equivalently,

$$\begin{bmatrix} J_{11} & \chi_{12} & \chi_{13} \\ \star & J_{22} & \chi_{23} \\ \star & \star & -\gamma_s^2 I \end{bmatrix} \prec 0, \tag{5.47}$$

where $J_{11} = \chi_{11} + P\Theta_1 M_0 M_0^\top \Theta_1^\top P + C_{sx}^\top C_{sx}$, $\chi_{11} = \text{He}(P(\Theta_1 A - BK_x)) + 2N_0^\top N_0$, $\chi_{12} = PF_1$, $\chi_{13} = PD_1$, $J_{22} = \chi_{22} + C_{se}^\top C_{se} + QW_1\bar{M}_0\bar{M}_0^\top W_1^\top Q$, $\chi_{22} = \text{He}(Q(M_1 - ZM_2))$, and $\chi_{23} = QW_1\bar{D}$.

By using Schur Complement (see Sect. 1.7.1), the inequality (5.47) becomes

$$\begin{bmatrix} \chi_{11} & \chi_{12} & \chi_{13} & P\Theta_1 M_0 & 0 & C_{sx}^\top & 0 \\ \star & \chi_{22} & \chi_{23} & 0 & QW_1\bar{M}_0 & 0 & C_{se}^\top \\ \star & \star & -\gamma_s^2 I & 0 & 0 & 0 & 0 \\ \star & \star & \star & -I & 0 & 0 & 0 \\ \star & \star & \star & \star & -I & 0 & 0 \\ \star & \star & \star & \star & \star & -I & 0 \\ \star & \star & \star & \star & \star & \star & -I \end{bmatrix} \prec 0. \tag{5.48}$$

The inequality (5.48) is BMI and cannot be solved directly using linear optimization tools. Therefore, it is further converted into the LMI (5.40) by defining $\bar{\gamma}_s = \gamma_s^2$, $PB = B\hat{P}$, $R_1 = \hat{P}K_s$ and $R_2 = QZ$. This results in the optimization problem (5.38). □

The equality constraint (5.39) is difficult to solve using off-the-shelf linear optimization solvers. It can be converted into the LMI (5.50) by using the method described in Corless and Tu (1998). In this case, the optimization problem (5.38) can be reformulated as the following multi-objective optimization problem:

$$\min \alpha_s \bar{\gamma}_s + \beta_s \tag{5.49}$$

$$\text{s.t.} \begin{bmatrix} \beta_s I & PB - B\hat{P} \\ \star & \beta_s I \end{bmatrix} \succ 0, \ \beta_s \geq 0 \tag{5.50}$$

$$(5.39),\ (5.40),\ (5.41) \tag{5.51}$$

where $\alpha_s \geq 0$ is a given weight. This multi-objective optimization problem can then be directly solved using off-the-shelf linear optimization solvers, such as the LMI control toolbox (Gahinet et al. 1995) and MOSEK (Mosek 2018).

5.4 Integration of FE and FTC: Output Feedback

Section 5.3 presents a state feedback integrated FE/FTC strategy with the assumption that state variables are fully available. However, this is often not the case in practical applications. Hence, this section considers an output feedback integrated FE/FTC strategy, for which purpose one more assumption is made.

Assumption 5.5 For the system (5.1), $\mathrm{rank}(CB) = \mathrm{rank}(B)$.

Since the state variables are unavailable, both the state and faults need to be estimated. To achieve this, a full-order observer is designed in Sect. 5.4.1.

5.4.1 Full-Order ASUIO-Based FE Design

The augmented state $\bar{x}$ in (5.2) is estimated by the following full-order ASUIO:

$$\begin{aligned} \dot{\xi}_o &= M_o \xi_o + G_o u + L_o y \\ \hat{\bar{x}} &= \xi_o + H_o y \end{aligned} \tag{5.52}$$

where $\xi_o \in \mathbb{R}^{n+q+q_1}$ and $\hat{\bar{x}} \in \mathbb{R}^{n+q+q_1}$ denote the observer state and the estimate of $\bar{x}$, respectively. The matrices M_o, G_o, L_o and H_o are of appropriate dimensions and to be determined.

Define the estimation error as $e_o = \bar{x} - \hat{\bar{x}}$. By using (5.2) and (5.52), the estimation error system is derived as

$$\begin{aligned} \dot{e}_o =& (\Xi \bar{A} - L_1 \bar{C}) e_o + (\Xi \bar{A} - L_1 \bar{C} - M_o) \xi_o + (\Xi \bar{B} - G_o) u \\ &+ [(\Xi \bar{A} - L_1 \bar{C}) H_o - L_2] y + \Xi \Delta \bar{A} \bar{x} + \Xi \bar{D} \bar{d} \end{aligned} \tag{5.53}$$

where $\Xi = I_{n+q+q_1} - H_o \bar{C}$ and $L_o = L_1 + L_2$.

In the absence of uncertainty and disturbance, it follows from Theorem 5.1 that sufficient conditions of an asymptotically stable observer (5.53) are given as

$$M_o \text{ is Hurwitz} \tag{5.54}$$

$$\Xi \bar{A} - L_1 \bar{C} - M_o = 0 \tag{5.55}$$

$$\Xi \bar{B} - G_o = 0 \tag{5.56}$$

$$(\Xi \bar{A} - L_1 \bar{C}) H_o - L_2 = 0. \tag{5.57}$$

By using (5.55)–(5.57), the estimation error system (5.53) becomes

$$\dot{e}_o = (\Xi \bar{A} - L_1 \bar{C}) e_o + \Xi \Delta \bar{A} \bar{x} + \Xi \bar{D} \bar{d}. \tag{5.58}$$

Therefore, by designing H_0 and L_1 such that (5.58) is robustly stable, the observer (5.52) can then be built with robustness to system uncertainty and disturbance.

5.4.2 Output Feedback Sliding Mode FTC Design

A switching function for the system (5.1) using the system output information is designed as

$$s_2 = N_2 y_c \tag{5.59}$$

where $N_2 = (CB)^\dagger - Y_2(I_p - CB(CB)^\dagger)$ with a design matrix $Y_2 \in \mathbb{R}^{m\times p}$ and $(CB)^\dagger = \left((CB)^\top CB\right)^{-1}(CB)^\top$. $y_c = y - F_s\hat{f}_s = Cx + F_s e_{f_s}$ is the system output with the sensor fault f_s compensated, and e_{f_s} is the estimation error of f_s.

Differentiating s_2 with respect to time gives

$$\dot{s}_2 = N_2 C\left[(A+\Delta A)x + F_a f_a + Dd\right] + N_2 F_s \dot{e}_{f_s} + u. \tag{5.60}$$

Design the FTC controller as

$$u = u_{l_2} + u_{n_2} \tag{5.61}$$

with a linear component u_{l_2} and a nonlinear component u_{n_2}. The linear component is designed as $u_{l_2} = -K_o\hat{x} - E_2\hat{f}_a$, where $K_o \in \mathbb{R}^{m\times n}$ and $E_2 = B^\dagger F_a$ are design matrices, and $\hat{x}$ and $\hat{f}_a$ are the estimates of state and actuator fault, respectively. The nonlinear component u_{n_2} is designed as $u_{n_2} = -\varrho_{s_2}\mathrm{sign}(s_2)$ with a time-varying scalar ϱ_{s_2} to be determined.

To analyze satisfaction of the reachability condition $s_2^\top\dot{s}_2 \le 0$, the following Lyapunov function is used:

$$V_{s20} = \frac{1}{2}s_2^\top s_2.$$

By using (5.60) and (5.61), the time derivative of V_{s20} is derived as

$$\begin{aligned}
\dot{V}_{s20} &= s_2^\top\left[N_2C\left((A+\Delta A)x + F_a f_a + Dd\right) + N_2F_s\dot{e}_{f_s} + u\right]\\
&= s_2^\top\left[\delta(x,e,d) - \varrho_{s_2}\mathrm{sign}(s_2)\right]\\
&\le \left(\|\delta(x,e,d)\| - \varrho_{s_2}\right)\|s_2\|\\
&\le \left(\eta_{s_2} - \varrho_{s_2}\right)\|s_2\|
\end{aligned} \tag{5.62}$$

where

$$\delta(x,e,d) = (N_2CA + N_2C\Delta A - K_o)x + K_o e_x + E_2 e_{f_a} + N_2CDd + N_2F_s\dot{e}_{f_s},$$

and η_{s_2} is an unknown scalar satisfying $\eta_{s_2} \ge \|\delta(x,e,d)\|$.

Define $\varrho_{s_2} = \hat{\eta}_{s_2} + \varepsilon_{s_2}$, where ε_{s_2} is a positive design scalar and $\hat{\eta}_{s_2}$ is used to estimate η_{s_2} with an update law

$$\dot{\hat{\eta}}_{s_2} = \sigma_2 \|s_2\|,\ \hat{\eta}_{s_2}(0) = 0 \tag{5.63}$$

where $\sigma_2 > 0$ is a design constant.

Define the estimation error of η_{s_2} as $\tilde{\eta}_{s_2} = \eta_{s_2} - \hat{\eta}_{s_2}$. Consider the Lyapunov function

$$V_{s_2} = V_{s_{20}} + \frac{1}{2\sigma_2}\tilde{\eta}_{s_2}^2.$$

By using (5.62) and (5.63), it can be derived that

$$\begin{aligned}\dot{V}_{s_2} &= \dot{V}_{s_{20}} - \frac{1}{\sigma_2}\tilde{\eta}_{s_2}\dot{\hat{\eta}}_{s_2} \\ &\leq (\eta_{s_2} - \varrho_{s_2})\|s_2\| - \tilde{\eta}_{s_2}\|s_2\| \\ &\leq -\varepsilon_{s_2}\|s_2\| \\ &\leq 0.\end{aligned} \tag{5.64}$$

Since V_{s_2} is positive definite, it follows from (5.64) and the Barbalat's Lemma that $V_{s_2}(t) \leq V_{s_2}(0)$. Therefore, $s_2(t)$ and $\tilde{\eta}_{s_2}(t)$ are bounded. This means that the designed controller (5.61) can maintain the sliding motion around $s_2 = 0$. Moreover, in the case of zero initial condition (i.e. $V_{s_2}(0) = 0$), it holds that $\lim_{t\to\infty} s_2(t) = 0$. In such case, the sliding surface $s_2 = 0$ is reachable and the ideal sliding motion is maintained.

The next step is to analyze the system stability corresponding to the sliding mode. It follows from (5.60) that the equivalent control input can be defined as

$$u_{eq_2} = -N_2 C\left[(A + \Delta A)x + Dd\right] - N_2 F_s \dot{e}_{f_s} + u_{l_2}. \tag{5.65}$$

Substituting (5.65) into (5.1) gives the equivalent closed-loop control system

$$\dot{x} = (\Theta_2 A - BK_o)x + F_2 e_o + \Theta_2 \Delta A x + \bar{D}_2 \bar{d}_2 \tag{5.66}$$

where

$$\begin{aligned}&\Theta_2 = I_n - BN_2C,\ F_2 = [BK_o\ F_a\ 0],\\ &\bar{D}_2 = [\Theta_2 D\ \ -BN_2F_s],\ \bar{d}_2 = [d^\top\ \dot{e}_{f_s}^\top]^\top.\end{aligned}$$

Therefore, the system (5.1) is maintained on the sliding surface with the equivalent control (5.65) by designing K_o such that (5.66) is stable. The closed-loop control system (5.66) contains the uncertainty ΔAx and disturbance $\bar{d}$, which must be minimized to achieve a suitable degree of robustness. This is achieved using H_∞ optimization with the details given in the next section.

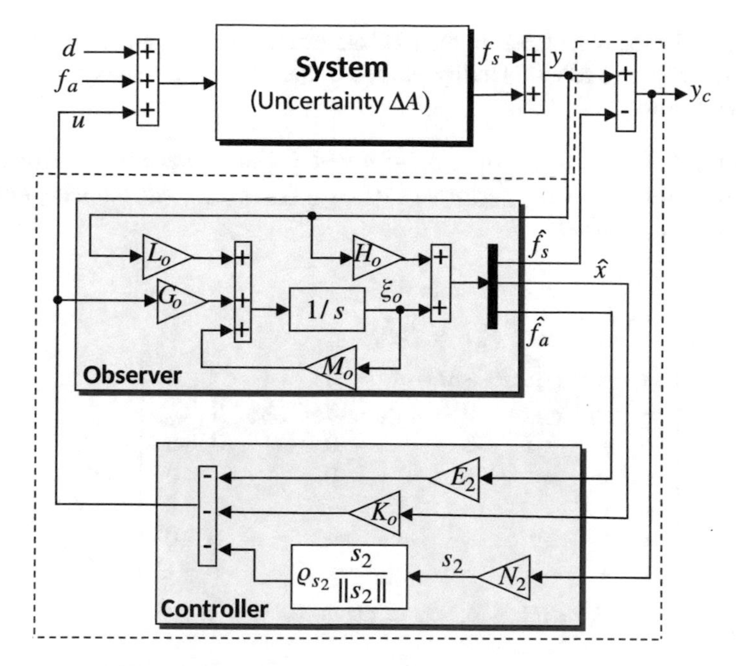

Fig. 5.3 Integrated FE and FTC design: output feedback case

5.4.3 Simultaneous Integration of FE and FTC

With the full-order FE observer presented in Sect. 5.4.1 and the output feedback FTC controller presented in Sect. 5.4.2, the proposed output feedback FE-based FTC system is outlined in Fig. 5.3. This section describes a simultaneous integration strategy to design the observer and controller gains.

The composite closed-loop system consisting of (5.58) and (5.66) is given as

$$\begin{aligned}
\dot{x} &= (\Theta_2 A - BK_o)x + F_2 e_o + \Theta_2 \Delta A x + D_2 \tilde{d}_2 \\
\dot{e}_o &= (\Xi \bar{A} - L_1 \bar{C})e_o + \Xi \Delta \bar{A} \bar{x} + \Xi \bar{D}_1 \tilde{d}_2 \\
y_c &= y - F_s \hat{f}_s \\
z_o &= \mathrm{diag}(C_{ox} x, C_{oe} e_o)
\end{aligned} \tag{5.67}$$

where

$$\tilde{d}_2 = [\bar{d}^\top \; \dot{e}_{fs}^\top]^\top, \; \bar{D}_1 = [\bar{D}\; 0], \; D_2 = [\Theta_2 D \; 0 \; 0 \; - BN_2 F_s],$$

and z_o is the performance output with given weights C_{ox} and C_{oe}.

The simultaneous synthesis of the FTC controller gain K_o and FE observer gains H and L_1 to ensure robust stability of the composite closed-loop system (5.67) is described below.

Theorem 5.3 *Under Assumptions 5.1–5.4 and 5.5, the composite closed-loop system (5.67) is stable with H_∞ performance $\|G_{z_o\tilde{d}_2}\|_\infty < \gamma_o$, if the following optimization problem is feasible:*

$$\min \bar{\gamma}_o \tag{5.68}$$

$$\text{s.t.} \quad P_o B = B\hat{P}_o \tag{5.69}$$

$$\begin{bmatrix} \chi_{11} & \chi_{12} & \chi_{13} & P\Theta_2 M_0 & 0 & C_{ox}^\top & 0 \\ \star & \chi_{22} & \chi_{23} & 0 & (Q_o - X_2\bar{C})\bar{M}_0 & 0 & C_{oe}^\top \\ \star & \star & -\bar{\gamma}_o I & 0 & 0 & 0 & 0 \\ \star & \star & \star & -I & 0 & 0 & 0 \\ \star & \star & \star & \star & -I & 0 & 0 \\ \star & \star & \star & \star & \star & -I & 0 \\ \star & \star & \star & \star & \star & \star & -I \end{bmatrix} \prec 0 \tag{5.70}$$

$$P_o = P_o^\top \succ 0, \; Q_o = Q_o^\top \succ 0, \; \bar{\gamma}_o \geq 0 \tag{5.71}$$

where

$$\chi_{11} = \text{He}(P_o\Theta_2 A - BX_1) + 2N_0^\top N_0, \; \chi_{12} = [BX_1 \; P_o F_a \; 0], \; \chi_{13} = P_o D_2,$$
$$\chi_{22} = \text{He}(Q_o\bar{A} - X_2\bar{C}\bar{A} - X_3\bar{C}), \; \chi_{23} = (Q_o - X_2\bar{C})\bar{D}_1.$$

Then the gains are obtained as: $\gamma_o = \sqrt{\bar{\gamma}_o}$, $K_o = \hat{P}_o^{-1}X_1$, $H = Q_o^{-1}X_2$, $L_1 = Q_o^{-1}X_3$.

Proof Consider the Lyapunov function $V_{e_o} = e_o^\top Q e_o$ with a s.p.d. matrix $Q_o \in \mathbb{R}^{(n+q+q_1)\times(n+q+q_1)}$. Define $\bar{M}_0 = [M_0^\top \; 0]^\top$. According to Assumption 5.3, it is derived that

$$\text{He}(e_o^\top Q_o \Xi \Delta\bar{A}\bar{x}) \leq e_o^\top Q_o \Xi \bar{M}_0 \bar{M}_0^\top \Xi^\top Q_o e_o + x^\top N_0^\top N_0 x.$$

The derivative of V_{e_o} along the estimation error system is obtained as

$$\begin{aligned} \dot{V}_{e_o} &= e_o^\top \left[\text{He}(Q_o(\Xi\bar{A} - L_1\bar{C}))\right] e_o + \text{He}(e_o^\top Q_o \Xi\Delta\bar{A}\bar{x}) + \text{He}(e_o^\top Q_o \Xi \bar{D}_1 \tilde{d}_2) \\ &\leq e_o^\top \left[\text{He}(Q_o(\Xi\bar{A} - L_1\bar{C})) + Q_o\Xi\bar{M}_0\bar{M}_0^\top \Xi^\top Q_o\right] e_o + x^\top N_0^\top N_0 x \\ &\quad + \text{He}(e_o^\top Q_o \Xi \bar{D}_1 \tilde{d}_2). \end{aligned} \tag{5.72}$$

Consider $V_{xo} = x^\top P_o x$ with a s.p.d. matrix $P_o \in \mathbb{R}^{n\times n}$. It can be derived that

$$\mathrm{He}(x^\top P_o \Theta_2 \Delta A x) \leq x^\top (P_o \Theta_2 M_0 M_0^\top \Theta_2^\top P_o + N_0^\top N_0) x.$$

Then the derivative of V_{xo} is obtained as

$$\begin{aligned} \dot{V}_{xo} \leq x^\top & \left[\mathrm{He}\left(P_o(\Theta_2 A - BK_o)\right) + P_o \Theta_2 M_0 M_0^\top \Theta_2^\top P_o + N_0^\top N_0\right] x \\ & + \mathrm{He}\left(x^\top P_o F_2 e_o + x^\top P_o D_2 \tilde{d}_2\right). \end{aligned} \tag{5.73}$$

The H_∞ performance $\|G_{z_o \tilde{d}_2}\|_\infty < \gamma_o$ can be represented by

$$J = \int_0^\infty \left(z_o(t)^\top z_o(t) - \gamma_o^2 \tilde{d}_2(t)^\top \tilde{d}_2(t)\right) \mathrm{d}t < 0. \tag{5.74}$$

Define $V_o = V_{xo} + V_{eo}$, then under zero initial conditions, it holds that

$$J \leq \int_0^\infty \left(z_o(t)^\top z_o(t) - \gamma_o^2 \tilde{d}_2(t)^\top \tilde{d}_2(t) + \dot{V}_o(t)\right) \mathrm{d}t. \tag{5.75}$$

It can be seen from (5.75) that a sufficient condition of (5.74) is

$$z_o^\top z_o - \gamma_o^2 \tilde{d}_2^\top \tilde{d}_2 + \dot{V}_o < 0. \tag{5.76}$$

Substituting (5.72) and (5.73) into (5.76) gives

$$\begin{bmatrix} x \\ e_o \\ \tilde{d}_2 \end{bmatrix}^\top \begin{bmatrix} J_{11} & \chi_{12} & \chi_{13} \\ \star & J_{22} & \chi_{23} \\ \star & \star & -\gamma_o^2 I \end{bmatrix} \begin{bmatrix} x \\ e_o \\ \tilde{d}_2 \end{bmatrix} < 0 \tag{5.77}$$

and equivalently,

$$\begin{bmatrix} J_{11} & \chi_{12} & \chi_{13} \\ \star & J_{22} & \chi_{23} \\ \star & \star & -\gamma_o^2 I \end{bmatrix} \prec 0 \tag{5.78}$$

where

$$\begin{aligned} & J_{11} = \chi_{11} + P_o \Theta_2 M_0 M_0^\top \Theta_2^\top P_o + C_{ox}^\top C_{ox}, \\ & \chi_{11} = \mathrm{He}\left(P_o(\Theta_2 A - BK_o)\right) + 2N_0^\top N_0, \; \chi_{12} = P_o F_2, \; \chi_{13} = P_o D_2, \\ & J_{22} = \chi_{22} + Q_o \Xi \bar{M}_0 \bar{M}_0^\top \Xi^\top Q_o + C_{oe}^\top C_{oe}, \; \chi_{22} = \mathrm{He}(Q_o(\Xi \bar{A} - L_1 \bar{C})), \\ & \chi_{23} = Q_o \Xi \bar{D}_1. \end{aligned}$$

By applying Schur Complement to the inequality (5.78), it gives

$$\begin{bmatrix} \chi_{11} & \chi_{12} & \chi_{13} & P_o\Theta_2 M_0 & 0 & C_{ox}^\top & 0 \\ \star & \chi_{22} & \chi_{23} & 0 & Q_o\Xi\bar{M}_0 & 0 & C_{oe}^\top \\ \star & \star & -\gamma_o^2 I & 0 & 0 & 0 & 0 \\ \star & \star & \star & -I & 0 & 0 & 0 \\ \star & \star & \star & \star & -I & 0 & 0 \\ \star & \star & \star & \star & \star & -I & 0 \\ \star & \star & \star & \star & \star & \star & -I \end{bmatrix} \prec 0. \tag{5.79}$$

The BMI (5.79) is further converted into the LMI (5.70) by defining $\bar{\gamma}_o = \gamma_o^2$, $P_o B = B\hat{P}_o$, $X_1 = \hat{P}_o K_o$, $X_2 = Q_o H$ and $X_3 = Q_o L_1$. This results in the optimization problem (5.68). □

Similar to the state feedback case, Theorem 5.3 can be further converted into the following multi-objective optimization problem:

$$\min \alpha_o \bar{\gamma}_o + \beta_o \tag{5.80}$$

$$\text{s.t.} \begin{bmatrix} \beta_o I & P_o B - B\hat{P}_o \\ \star & \beta_o I \end{bmatrix} \succ 0, \; \beta_o \geq 0 \tag{5.81}$$

$$(5.69), \; (5.70), \; (5.71) \tag{5.82}$$

where $\alpha_o \geq 0$ is a given weight.

5.5 Integration of FE and FTC: Multiplicative Faults

Sections 5.3 and 5.4 (and also Chaps. 3 and 4) focus on the robust integration of FE and FTC for systems with actuator and/or sensor faults. These faults are added to the system state and output, i.e. additive faults, causing changes in the mean values of the system state and outputs. Besides additive faults, multiplicative faults which are defined as component faults (even some kinds of actuator and sensor faults are in the form of multiplicative faults, e.g. partial loss of actuator effectiveness) also need to be handled, because they affect the stability and degrade the performance of the post-fault system. Several works have been published on some topics related to multiplicative faults, e.g. multiplicative fault modelling and diagnosis (Ding 2008; Li et al. 2020), and multiplicative fault estimation (Gao and Duan 2012; Rotondo et al. 2016; Tan and Edwards 2004; Wang and Daley 1996). However, none of them address the integration of FE and FTC in the presence of system uncertainties and disturbances. Motivating by the above background, this section presents an integration strategy of FE and FTC for linear systems with multiplicative faults, based on the results in Sects. 5.3–5.4.

Consider an uncertain linear system with the form of

$$\begin{aligned} \dot{x} &= (A + \Delta A)x + Bu + F_m f_m + Dd \\ y &= Cx \end{aligned} \tag{5.83}$$

where x, u, d, y, A, B, D and C are defined the same as in (5.1). The constant matrix $F_m \in \mathbb{R}^{n \times q_m}$ indicates the distribution of the following fictitious fault:

$$f_m = B_m \sum_{i=1}^{q_m} \theta_i \phi_i(A, B, x, u) \tag{5.84}$$

where $B_m = F_m^{\dagger} - (F_m^{\dagger} F_m - I_{q_m})W$ with some matrix $W \in \mathbb{R}^{q_m \times n}$. The scalar functions $\theta_i \in \mathbb{R}, i = 1, 2, \ldots, q_m$, are time-varying and present the multiplicative faults. The functions $\phi_i(A, B, x, u) \in \mathbb{R}^{n \times 1}, i = 1, 2, \ldots, q_m$, are known and related to A, B, x and u.

The formulation (5.84) can represent a wide class of multiplicative faults occurring at the system matrices, see, for example,

$$\sum_{i=1}^{q_A} \theta_{Ai} A_i x = F_{m_A}\left(B_{m_A} \sum_{i=1}^{q_A} \theta_{Ai} A_i x\right) = F_{m_A} f_{m_A},$$
$$\sum_{i=1}^{q_B} \theta_{Bi} B_i u = F_{m_B}\left(B_{m_B} \sum_{i=1}^{q_B} \theta_{Bi} B_i u\right) = F_{m_B} f_{m_B},$$
$$\sum_{i=1}^{q_A} \theta_{Ai} A_i x + \sum_{i=1}^{q_B} \theta_{Bi} B_i u = F_m\left(B_m \sum_{i=1}^{q_A} \theta_{Ai} A_i x + B_m \sum_{i=1}^{q_B} \theta_{Bi} B_i u\right) = F_m f_m,$$

where $A_i, i = 1, 2, \ldots, q_A$, and $B_i, i = 1, 2, \ldots, q_B$, are known matrices related to A and B.

In the literature (Ding 2008; Gao and Duan 2012; Rotondo et al. 2016; Tan and Edwards 2004; Wang and Daley 1996), the effort was put into the estimation of $\theta_i, i = 1, 2, \ldots, q_m$. However, few works have considered FTC design for systems with multiplicative faults. If the aim is to achieve acceptable closed-loop system performance, then the purpose of FTC design is to compensate for the effect of the multiplicative faults, whatever their sources or sizes. This can be achieved even if the fictitious multiplicative fault f_m cannot reflect the real fault location and size. In this respect, the integrated FE/FTC design of the system (5.83) with multiplicative fault can be achieved through the designs proposed in Sects. 5.3–5.4 with minor modification, by estimating and compensating the fictitious multiplicative fault f_m with the chosen F_m satisfying $\text{rank}[B\ F_m] = \text{rank}(B) = m \le n$.

The system (5.1) is required to satisfy $\text{rank}[B\ F_a] = \text{rank}(B)$ as in Definition 1.7. However, when $\text{rank}[B\ F_a] \neq \text{rank}(B)$ but $\text{rank}(B) = m \le n$, the actuator fault f_a can be handled in the following way: Denote $F_a f_a = (BB^{\dagger} + B^{\perp}B^{\perp^{\dagger}})F_a f_a$ where $B^{\perp} \in \mathbb{R}^{n \times (n-m)}$ spans the null space of B and $BB^{\dagger} + B^{\perp}B^{\perp^{\dagger}} = I_n$. Using the proposed design strategy, the matched part $BB^{\dagger}F_a f_a$ of the actuator fault can be estimated and compensated, while the unmatched part $B^{\perp}B^{\perp^{\dagger}}F_a f_a$ can be treated as disturbance.

5.6 Tutorial Example 1: System with Additive Fault

Considering the stabilization control for a DC motor characterized by

$$\begin{aligned} \dot{x} &= (A + \Delta A)x + Bu + Dd \\ y &= Cx \end{aligned} \tag{5.85}$$

with the state vector $x = [i_a\ w]^\top$, control input $u = v_a$, disturbance $d = -\frac{T_l}{J_i}$, output y, and the system matrices

$$A = \begin{bmatrix} -\frac{R_a}{L_a} & -\frac{K_v}{L_a} \\ \frac{K_m}{J_i} & -\frac{B_0}{J_i} \end{bmatrix},\ B = \begin{bmatrix} \frac{1}{L_a} \\ 0 \end{bmatrix},\ D = \begin{bmatrix} 0 \\ 1 \end{bmatrix},\ C = \begin{bmatrix} 1 & 0 \\ 0 & 1 \end{bmatrix},\ \Delta A = \begin{bmatrix} 0 & \sigma_v \\ \sigma_m & 0 \end{bmatrix},$$

where i_a, w and v_a are the armature current, angular velocity and armature voltage, respectively. R_a is the armature resistance and L_a is the inductance. K_v and K_m are the voltage and motor constants which are supposed to have parameter variations $|\sigma_v| \leq 0.06$ and $|\sigma_m| \leq 0.06$, respectively. J_i is the moment of inertia and B_0 is the friction coefficient. T_l is the unknown load torque. The control design aims to regulate the output y to zero.

The simulations use the following nominal parameters of the DC motor (Bélanger 1995): $R_a = 1.2$, $L_a = 0.05$, $K_v = 0.6$, $K_m = 0.6$, $J_i = 0.1352$ and $B_0 = 0.3$.

The parameter variations and disturbance are assumed to be $\sigma_v = \sigma_m = -0.01$ and $d = 0.01 \sin(t)$, respectively. Denote $|\sigma_v| \leq \alpha_v$ and $|\sigma_m| \leq \alpha_m$ with two positive scalars α_v and α_m. According to Assumption 5.3, the following matrices are chosen to characterize the uncertainty:

$$M_0 = \begin{bmatrix} 1 & 0 \\ 0 & 1 \end{bmatrix},\ F_0 = \begin{bmatrix} \frac{\sigma_v}{\alpha_v} & 0 \\ 0 & \frac{\sigma_m}{\alpha_m} \end{bmatrix},\ N_0 = \begin{bmatrix} 0 & \alpha_v \\ \alpha_m & 0 \end{bmatrix},$$

where $\alpha_v = 0.01$ and $\alpha_m = 0.01$.

There may be additive faults during the operation of the DC motor system (Isermann 2011). Consider here an offset fault of the armature current and angular velocity sensors, i.e. sensor fault f_s, and a voltage fault of v_a, i.e. actuator fault f_a. The model (5.85) can be rewritten in the form of (5.1) and given as

$$\begin{aligned} \dot{x} &= (A + \Delta A)x + Bu + F_a f_a + Dd \\ y &= Cx + F_s f_s \end{aligned} \tag{5.86}$$

where

$$F_a = \begin{bmatrix} \frac{1}{10L_a} \\ 0 \end{bmatrix}, \ f_a = \begin{cases} 0, & 0\,\text{s} \le t \le 2\,\text{s} \\ 0.05t - 0.1, & 2\,\text{s} < t \le 10\,\text{s} \\ 0.4, & 10\,\text{s} < t \le 15\,\text{s} \\ 0.2, & 15\,\text{s} < t \le 25\,\text{s} \end{cases},$$

$$F_s = \begin{bmatrix} -1 \\ 2 \end{bmatrix}, \ f_s = \begin{cases} 0, & 0\,\text{s} \le t \le 1\,\text{s} \\ 0.1\sin(0.5t - 0.5), & t > 1\,\text{s} \end{cases}.$$

5.6.1 Case 1: State Feedback

It is verified that Assumptions 5.1–5.4 are satisfied for the system (5.86). Given $C_{sx} = I_2$, $C_{se} = 0.1 \times I_2$, $Y_1 = [0.5\ 0.5]$ and $\alpha_s = 10$, solving the optimization problem (5.49) gives the optimal solution:

$$\begin{aligned} \beta_s &= 3.7747 \times 10^{-8}, \ \gamma_s = 0.505, \\ K_x &= [4.1845\ 2.5489], \\ M &= \begin{bmatrix} -2.4116 & -0.0007 \\ -0.0065 & -4.4439 \end{bmatrix}, \\ G &= \begin{bmatrix} -24.1164 \\ -0.0652 \end{bmatrix}, \\ R &= \begin{bmatrix} 26.0314 & 14.4695 \\ -2.1661 & -1.0749 \end{bmatrix}, \\ H &= \begin{bmatrix} 1.2058 & 0.0001 \\ 0.0033 & 0.5007 \end{bmatrix}. \end{aligned}$$

The other control parameters are given as $\varepsilon_{s_1} = 0.3079$ and $\sigma_1 = 2$. For comparison, the DC motor system is also simulated under the nominal control $u = -K_x x + u_{n_1}$ without compensating the actuator and sensor faults. Initial conditions for the system and observer are set as $x(0) = [0.5\ 0.5]^\top$, $\xi_s(0) = 0$ and $\hat{\eta}_{s_1}(0) = 0$.

The fault estimation errors and the closed-loop system outputs are depicted in Figs. 5.4 and 5.5, respectively. It is seen from Fig. 5.4 that the actuator and sensor faults are estimated with small errors by using the proposed reduced-order ASUIO. The results in Fig. 5.5 show that the outputs of the armature current and angular velocity are regulated to zero by using the proposed FTC design. However, as shown in Fig. 5.6, the system outputs are unstable under the nominal control. The above results demonstrate effectiveness of the proposed state feedback integrated FE/FTC design.

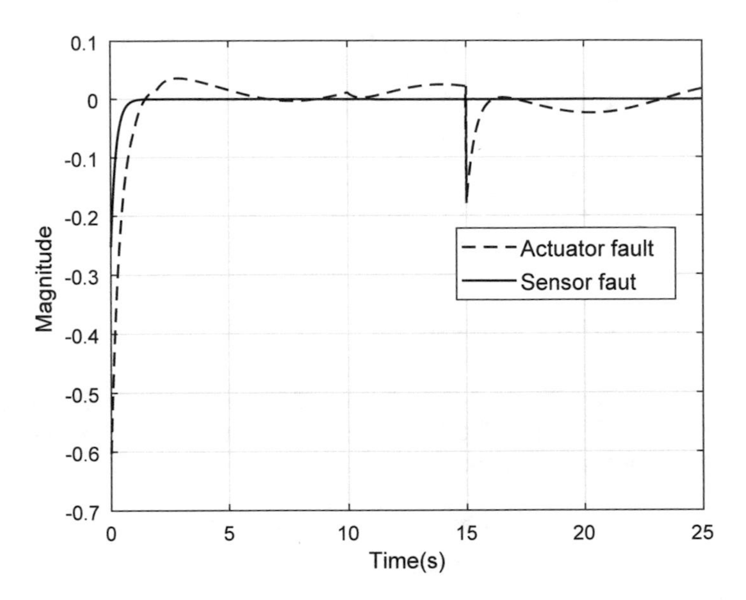

Fig. 5.4 FE performance: Additive fault, Case 1

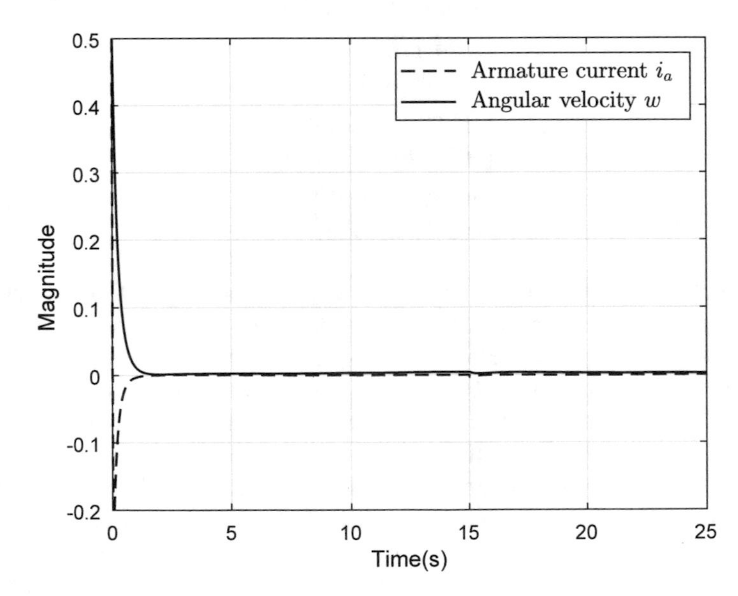

Fig. 5.5 FTC performance: Additive fault, Case 1

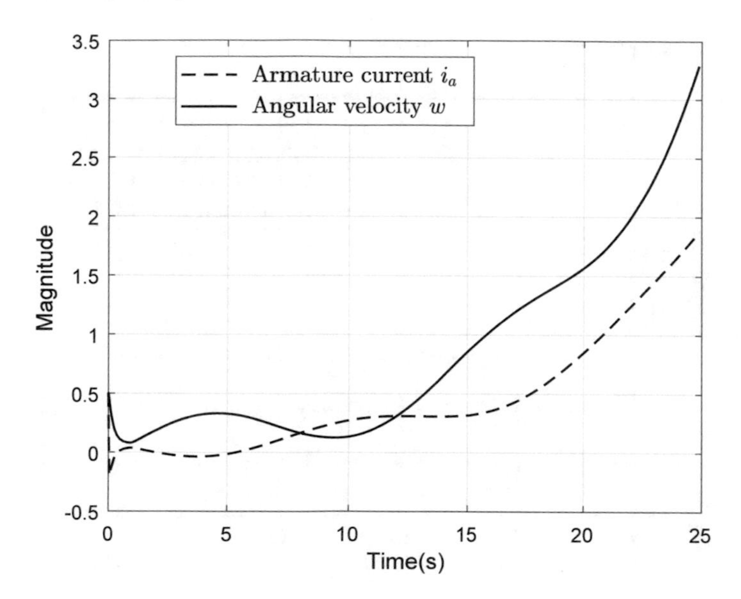

Fig. 5.6 Nominal control performance: Additive fault, Case 1

5.6.2 *Case 2: Output Feedback*

It is verified that Assumptions 5.1–5.4 and 5.5 are satisfied for the system (5.86). Given $Y_2 = [0.5\,0.5]$, $C_{ox} = I_2$, $C_{oe} = I_4$ and $\alpha_o = 10$, then solving the optimization problem (5.80) gives

$$\gamma_o = 0.5808,\ \beta_o = 1.3025 \times 10^{-8},\ K_o = [11.7376\ 6.7797],$$
$$M_o = \begin{bmatrix} -2.2724 & 0.4896 & 0.0483 & -1.0528 \\ 1.104 & -3.3947 & 0.0165 & 1.0407 \\ -1.2674 & 1.4827 & -4.8937 & -0.9071 \\ -0.2560 & 1.385 & -0.0062 & -1.4006 \end{bmatrix},\ G_o = \begin{bmatrix} 0.4835 \\ 0.1648 \\ -48.9369 \\ -0.0625 \end{bmatrix},$$
$$L_o = \begin{bmatrix} -2.5665 & -1.3059 \\ 4.3142 & 2.1486 \\ 44.1958 & 25.2536 \\ -2.1632 & -1.0762 \end{bmatrix},\ H_o = \begin{bmatrix} 0.9758 & 0.4850 \\ -0.0082 & -0.0066 \\ 2.4468 & 0.5791 \\ 0.0031 & 0.4987 \end{bmatrix}.$$

Simulations are performed with $\varepsilon_{s_2} = 0.0364$, $\sigma_2 = 0.1$, $x(0) = [0.5\ \ 0.5]^\top$, $\xi_o(0) = 0$ and $\hat{\eta}_{s_2}(0) = 0$. As seen from the results in Figs. 5.7 and 5.8, the proposed FE-based FTC design obtains accurate state and fault estimation and stable closed-loop state response. However, as shown in Fig. 5.9, the nominal control cannot stabilize the system.

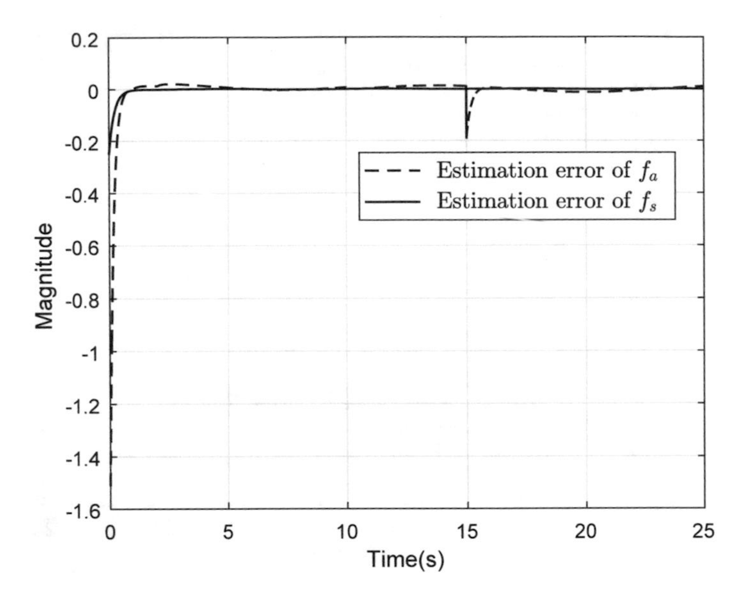

Fig. 5.7 FE performance: Additive fault, Case 2

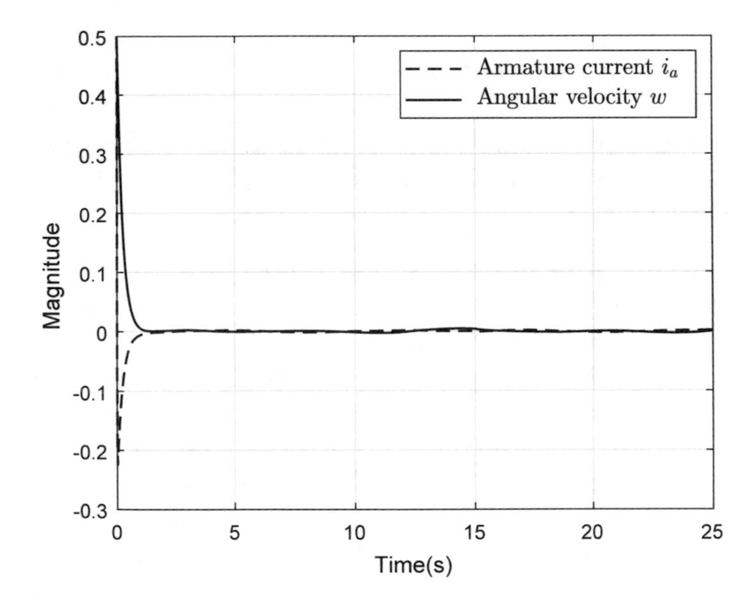

Fig. 5.8 FTC performance: Additive fault, Case 2

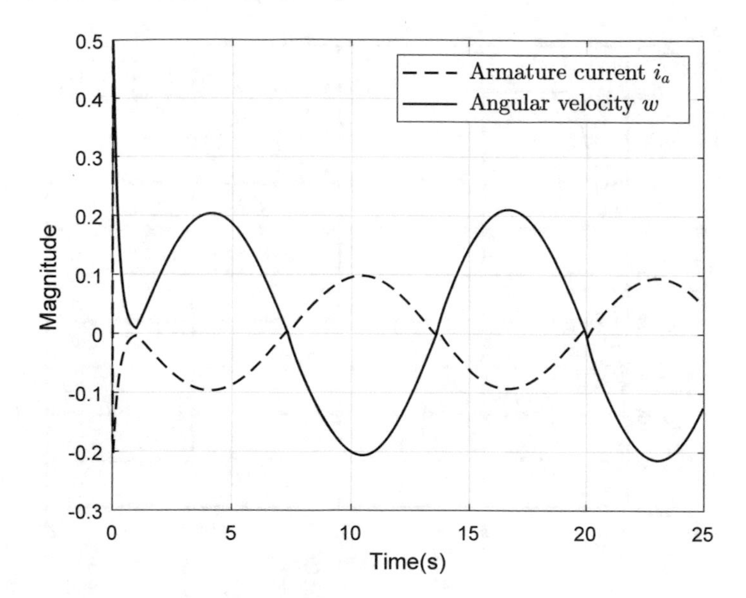

Fig. 5.9 Nominal control performance: Additive fault, Case 2

5.7 Tutorial Example 2: System with Multiplicative Fault

Consider the linear system

$$\begin{aligned} \dot{x} &= (A + \Delta A)x + B(1 - \theta(t))u + Dd \\ y &= Cx \end{aligned} \tag{5.87}$$

with

$$A = \begin{bmatrix} 24 & 12 \\ 4.4379 & -2.2189 \end{bmatrix}, \ \Delta A = \begin{bmatrix} 0 & 0.01 \\ 0.01 & 0 \end{bmatrix}, \ B = \begin{bmatrix} 20 \\ 0 \end{bmatrix}, \ D = \begin{bmatrix} 0 \\ 1 \end{bmatrix}, \ C = I_2,$$

where $\theta(t)$ is a time-varying scalar function representing the loss of actuator effectiveness and it takes values within $[0,\ 1)$.

According to (5.83), the partial loss of actuator effectiveness can be represented by a fictitious fault as below: $B(-\theta(t))u = F_m f_m$, $F_m = B$, and $f_m = -\theta(t)u$.

By setting $f_s = 0$ and replacing $F_a f_a$ with $F_m f_m$, the FE-based FTC design for system (5.87) is similar to that of (5.86). Therefore, the design for additive fault cases are easily amenable to cover the multiplicative case. Without loss of generality and to consider a more practically realizable situation, only the output feedback case is studied below.

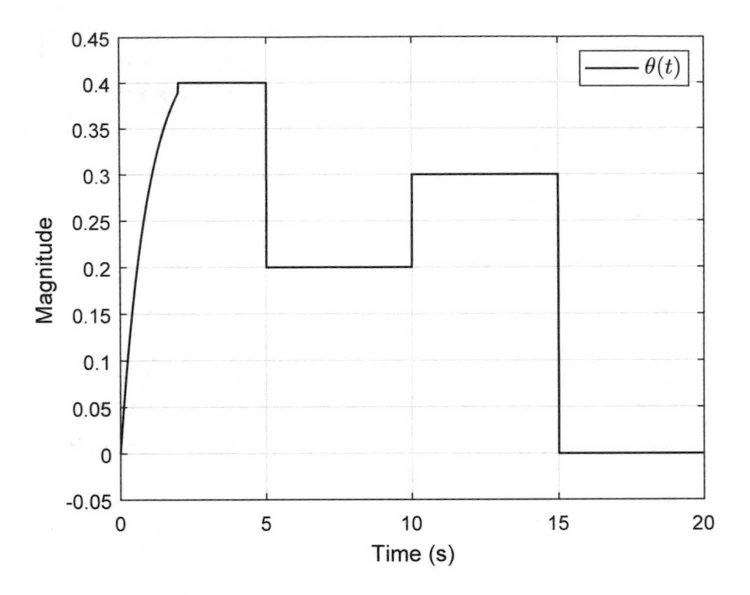

Fig. 5.10 Profile of $\theta(t)$: Multiplicative fault, Case 1

5.7.1 Case 1: FE-Based FTC

Given $Y_2 = [0.05\ 0.05]$, $C_{ox} = I_2$, $C_{oe} = I_4$ and $\alpha_o = 10$, then solving the optimization problem (5.80) gives $\gamma_o = 0.5049$, $\beta_o = 3.7091 \times 10^{-8}$, and

$$
\begin{aligned}
K_o &= [1.5748\ 1.3122], \\
M_o &= \begin{bmatrix} -1.2084 & 0.0041 & -0.0041 \\ -0.7604 & -1.0447 & 0.0054 \\ -2.7342 & -2.5218 & -5.1762 \end{bmatrix}, \quad G_o = \begin{bmatrix} -0.0041 \\ 0.0054 \\ -5.1762 \end{bmatrix}, \\
L_o &= \begin{bmatrix} -0.0062 & -0.0025 \\ 0.0081 & 0.0032 \\ -7.5510 & -3.1057 \end{bmatrix}, \quad H_o = \begin{bmatrix} 1.0002 & 0 \\ -0.0003 & 1 \\ 0.2588 & 0 \end{bmatrix}.
\end{aligned}
\tag{5.88}
$$

Simulations are performed with the profile of $\theta(t)$ depicted in Fig. 5.10 and initial conditions: $\varepsilon_{s_2} = 0.0115$, $\sigma_2 = 1$, $x(0) = [0.5\ 0.5]^\top$, $\xi_o(0) = 0$ and $\hat{\eta}_{s_2}(0) = 0$. It is seen from Fig. 5.11 that the fictitious fault f_m is estimated accurately. Moreover, it is observed from Figs. 5.12 and 5.13 that the proposed FTC design maintains system stability in the presence of partial actuator effectiveness loss, but the nominal control alone leads to system instability.

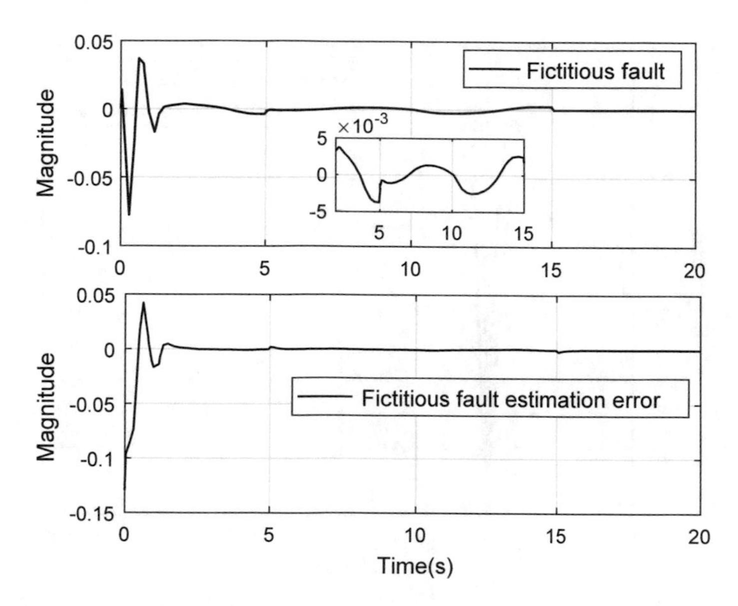

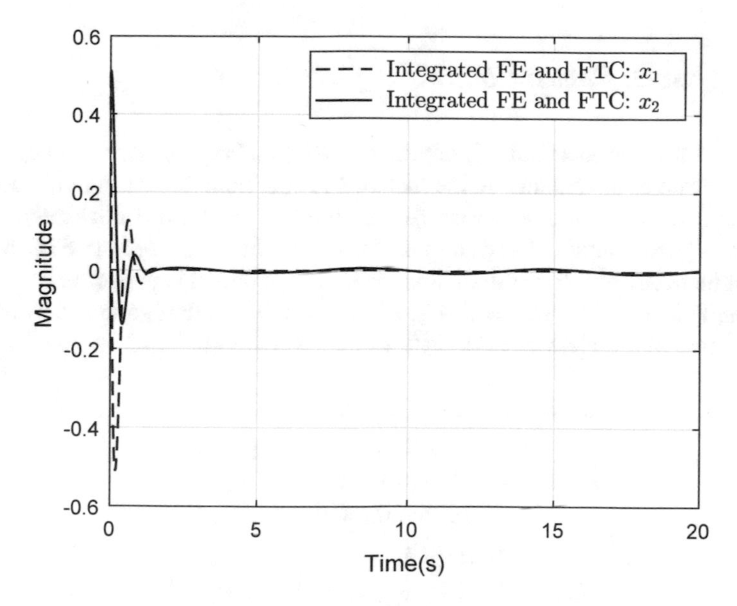

Fig. 5.11 Fictitious fault estimation: Multiplicative fault, Case 1

Fig. 5.12 FTC performance: Multiplicative fault, Case 1

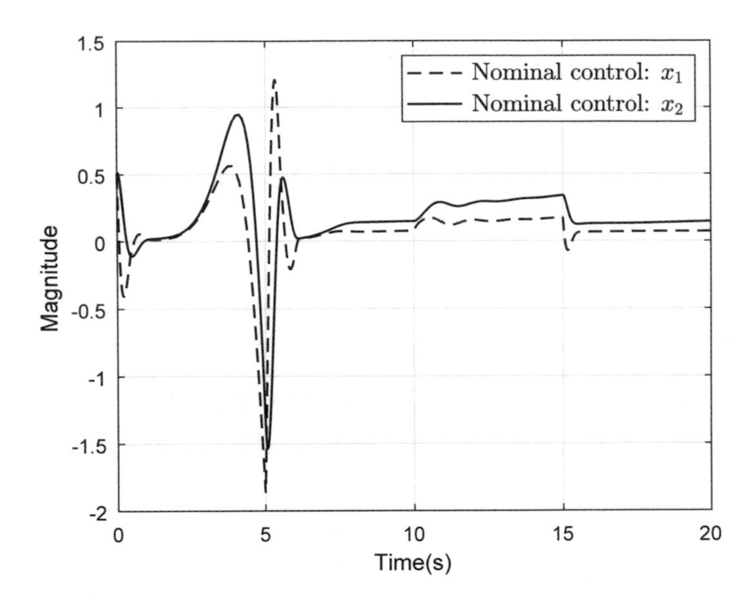

Fig. 5.13 Nominal control performance: Multiplicative fault, Case 1

5.7.2 *Case 2: Robust Control*

Notice that the fictitious fault f_m depends on $\theta(t)$ and u. Hence, it can be, to some extent, viewed as uncertainty on the matrix B. If the nominal controller is designed robust enough against the worst case f_m, then the nominal control alone might be able to achieve good enough closed-loop performance. In such case, the FTC function may not be necessary. To illustrate this, another simulation is performed.

Given $Y_2 = [0.5\,0.5]$, $C_{ox} = I_2$, $C_{oe} = I_4$ and $\alpha_o = 10$, then solving the optimization problem (5.80) gives $\gamma_o = 0.5049$, $\beta_o = 5.8116 \times 10^{-8}$ and

$$\begin{aligned}
K_o &= [10.3038\ 8.2552],\\
M_o &= \begin{bmatrix} -1.2196 & -0.0012 & 0.0155 \\ -0.7811 & -1.1336 & 0.0198 \\ -0.2935 & -0.1416 & -4.3482 \end{bmatrix},\\
G_o &= \begin{bmatrix} 0.0155 \\ 0.0198 \\ -4.3482 \end{bmatrix}, \ H_o = \begin{bmatrix} 0.9992 & 0 \\ -0.001 & 1 \\ 0.2174 & 0 \end{bmatrix},\\
L_o &= \begin{bmatrix} 0.0229 & 0.0093 \\ 0.0297 & 0.0119 \\ -6.1626 & -2.6088 \end{bmatrix}.
\end{aligned} \tag{5.89}$$

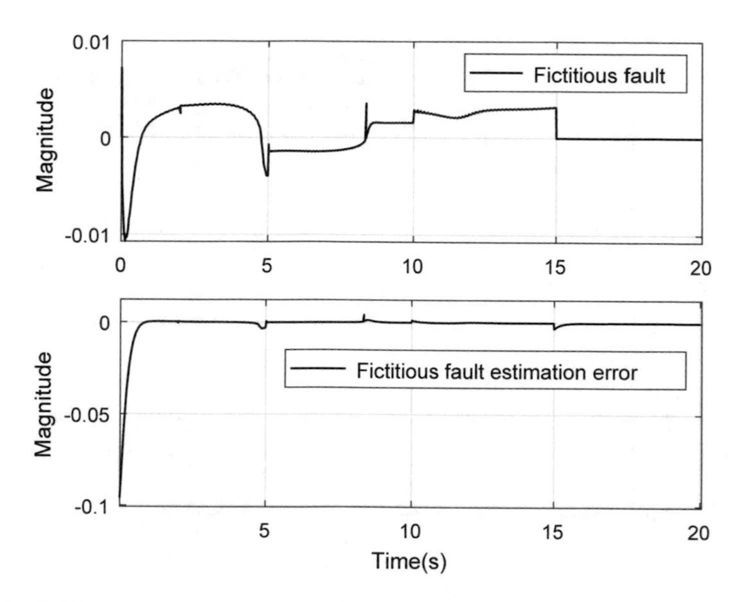

Fig. 5.14 Fictitious fault estimation: Multiplicative fault, Case 2

The eigenvalues of $(A + BK_x)$ with the control gains given in (5.88) and (5.89) are $(-19.9145, -9.357)$ and $(-156.3037, -7.6133)$, respectively. Hence, the nominal closed-loop systems under both gains have similar robustness. Simulations are performed with the same $\theta(t)$ and same initial conditions as in Sect. 5.7.1. It is shown in Fig. 5.14 that the fictitious fault is estimated accurately. As seen in Figs. 5.15 and 5.16, both the FTC and nominal designs are able to maintain system stability under partial actuator effectiveness loss.

In fact, the same phenomena have been observed in the simulation experiments for the DC motor system (5.85) with partial loss of actuator effectiveness. These phenomena happen because the matrix A in (5.85) is Hurwitz stable and a robust nominal controller is able to stabilize the system subject to partial loss of actuator effectiveness (or input uncertainty).

The above results may lead to a conclusion that the FTC function is unnecessary for systems subject to partial loss of actuator effectiveness faults. However, in order to attain a nominal controller that is robust enough to any possible f_m (combination of $\theta(t)$ and $u(t)$), the worst scenarios of f_m have to be considered. This will result in a rather conservative control design. In contrast, it is advantageous for using FE-based FTC design to actively estimate and compensate the faults without knowledge of their bounds at hand. Moreover, it will also result in less conservative designs.

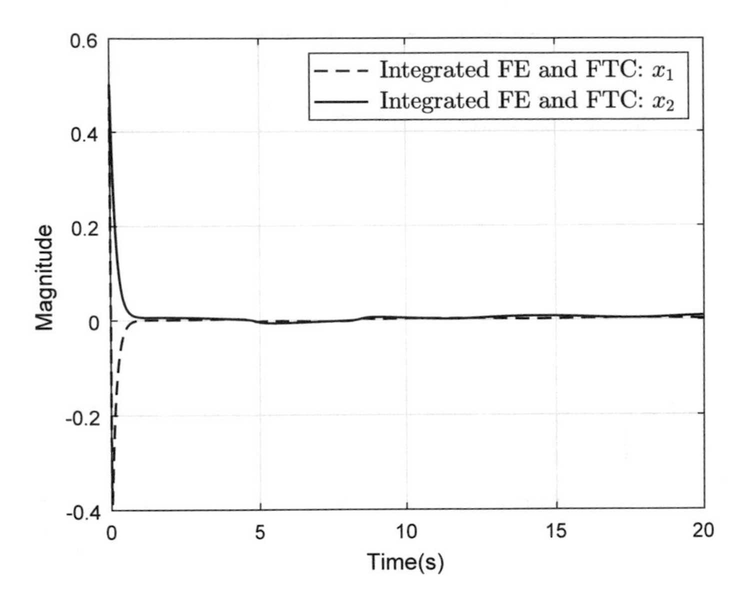

Fig. 5.15 FTC performance: Multiplicative fault, Case 2

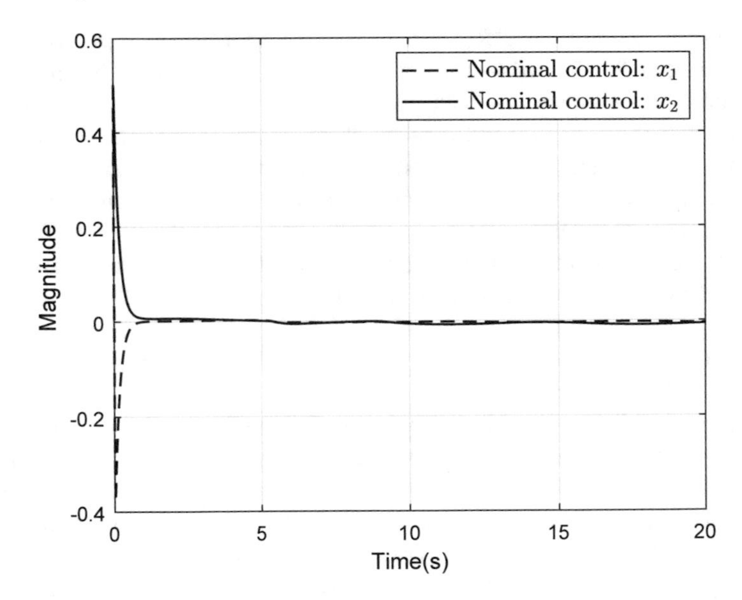

Fig. 5.16 Nominal control performance: Multiplicative fault, Case 2

5.8 Notes

In this chapter, the simultaneous integration strategy is developed for linear systems, but it can be extended to more complex systems, e.g. Lipschitz nonlinear system (Lan and Patton 2017) and T-S fuzzy system (Lan and Patton 2016). It will be extended and applied to nonlinear helicopter system in Chap. 8 and large-scale interconnected system in Chap. 9.

In order to obtain the single-step LMI formulation, the existing nonlinear constraints are converted into linear ones by introducing an extra equality constraints. Although this facilitates solving the considered optimization problems, the equality constraints impose restrictions on the controlled system models. As discussed in Lien (2004) and Kheloufi et al. (2013), necessary conditions to ensure feasibility of the obtained LMIs are (1) The system (5.1) is stabilizable and detectable and (2) the matrix B is full-column rank. These conditions are satisfied for most controlled systems. However, more conservativeness might be imposed on the optimization problem in some special cases, e.g. for the DC motor model studied in Sect. 5.6, the s.p.d. matrices P and P_0 are required to be diagonal as the matrix B is of the form $B = [B_1^\top \ 0]^\top$, where B_1 is a non-null matrix of appropriate dimension.

With the help of Young inequality (see Sect. 1.7.1), the use of equality constraint can be avoided (Kheloufi et al. 2013). The simultaneous integration of FE and FTC without equality constraint will be described in Chap. 8. However, as commented in Wang and Jiang (2014), the observer-based robust control design using Young inequality has no superiority over the one with equality constraint, although a less conservative design may be attained.

References

Bélanger PR (1995) Control engineering: a modern approach. Oxford University Press, Inc, Oxford

Chen J, Patton RJ (1999) Robust model-based fault diagnosis for dynamic systems. Kluwer Academic Publishers, London

Corless M, Tu J (1998) State and input estimation for a class of uncertain systems. Automatica 34(6):757–764

Ding SX (2008) Model-based fault diagnosis techniques: design schemes, algorithms, and tools. Springer Science & Business Media, Berlin

Edwards C, Spurgeon S (1998) Sliding mode control: theory and applications. CRC Press, Boca Raton

Gahinet P, Nemirovski A, Laub AJ, Chilali M (1995) LMI control toolbox for use with MATLAB. The MathWorks Inc, Natick, MA

Gao C, Duan G (2012) Robust adaptive fault estimation for a class of nonlinear systems subject to multiplicative faults. Circuits, Syst Signal Process 31(6):2035–2046

Isermann R (2011) Fault-diagnosis applications: model-based condition monitoring: actuators, drives, machinery, plants, sensors, and fault-tolerant systems. Springer Science & Business Media, Berlin

Kheloufi H, Zemouche A, Bedouhene F, Boutayeb M (2013) On LMI conditions to design observer-based controllers for linear systems with parameter uncertainties. Automatica 49(12):3700–3704

Lan J, Patton RJ (2016) Integrated design of fault-tolerant control for nonlinear systems based on fault estimation and T-S fuzzy modeling. IEEE Trans Fuzzy Syst 25(5):1141–1154
Lan J, Patton RJ (2017) Integrated fault estimation and fault-tolerant control for uncertain lipschitz nonlinear systems. Int J Robust Nonlinear Control 27(5):761–780
Li L, Ding S, Luo H, Peng K, Yang Y (2020) Performance-based fault-tolerant control approaches for industrial processes with multiplicative faults. IEEE Trans Ind Inf 16(7):4759–4768
Lien CH (2004) Robust observer-based control of systems with state perturbations via LMI approach. IEEE Trans Autom Control 49(8):1365–1370
Mosek A (2018) The MOSEK optimization software, version 8.1. http://www.mosek.com
Odgaard PF, Stoustrup J (2012) Fault tolerant control of wind turbines using unknown input observers. In: Proceedings of the IFAC Symposium on Fault Detection, Supervision and Safety of Technical Processes, pp 313–318
Rotondo D, López-Estrada FR, Nejjari F, Ponsart JC, Theilliol D, Puig V (2016) Actuator multiplicative fault estimation in discrete-time LPV systems using switched observers. J Frankl Inst 353(13):3176–3191
Tan CP, Edwards C (2004) Multiplicative fault reconstruction using sliding mode observers. In: Proceedings of the 5th Asian control conference, pp 957–962
Wang H, Daley S (1996) Actuator fault diagnosis: an adaptive observer-based technique. IEEE Trans Autom Control 41(7):1073–1078
Wang S, Jiang Y (2014) Comment on "on LMI conditions to design observer-based controllers for linear systems with parameter uncertainties [automatica 49 (2013) 3700–3704]". Automatica 50(10):2732–2733
Xiong Y, Saif M (2003) Unknown disturbance inputs estimation based on a state functional observer design. Automatica 39(8):1389–1398

Chapter 6
Robust Decoupling Integration of FE and FTC

6.1 Introduction

In Chap. 5, the simultaneous robust integration of FE and FTC (see Fig. 6.1(1)) is recast as the well-known robust observer-based control problem. Its solution relies on solving H_∞ optimization problem with BMI constraints. The BMI constraints are linearized by imposing an extra equality constraint, which needs to be further approximated by an LMI constraint. The above linearization imposes design conservativeness. Moreover, the simultaneous integration formulation has a lack of design freedom, where the FE and FTC performances cannot be tuned separately. These drawbacks become more obvious when applying the simultaneous strategy to complex nonlinear systems (Lan and Patton 2016).

The above background motivates the development of an integration strategy for FE and FTC which can avoid the BMI issue and have more design freedom. The strategy proposed in this chapter is called robust decoupling integration (see Fig. 6.1(2)). The keyword "*decoupling*" means that the FE and FTC designs are carried out separately. The keyword "*robust*" means that the *bidirectional robustness interactions* between the FE and FTC are attenuated in the spirit of the Small Gain Theorem (Zames 1966). For simplicity and clarity, the strategy will be illustrated using the ALSO FE observer and state feedback FTC controller developed in Chap. 3.

A two-phase synthesis strategy is first developed. In the first phase, the FTC controller is designed such that the closed-loop control system is robustly stable against the external disturbance and estimation error. In the second phase, the FE observer is determined based on the obtained controller such that the estimation error dynamics are robustly stable against the external disturbance, fault modelling error and system uncertainty. The separately designed FTC controller and FE observer can be assembled as a stable closed-loop system, by ensuring that the pure feedback interconnection of the FTC system and estimation error system is "small-gain" stable.

© The Author(s), under exclusive license to Springer Nature Switzerland AG 2021

J. Lan and R. J. Patton, *Robust Integration of Model-Based Fault Estimation and Fault-Tolerant Control*, Advances in Industrial Control,
https://doi.org/10.1007/978-3-030-58760-4_6

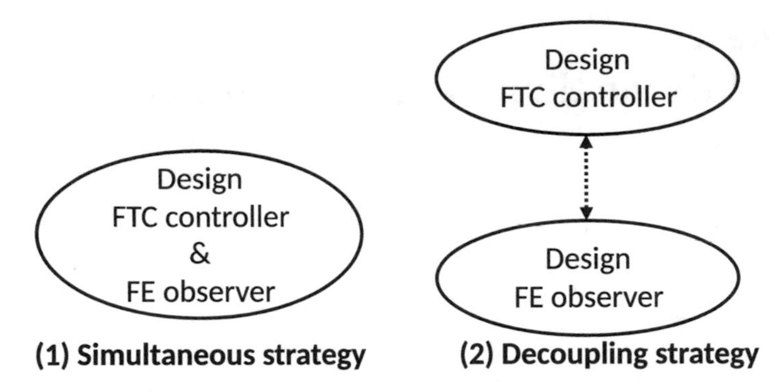

Fig. 6.1 Simultaneous and decoupling integration of FE and FTC

However, the above two-phase synthesis is restrictive in the sense that the FE observer can only be "passively" designed based on the FTC controller. In the first design phase, one has no idea of how the FE observer will perform after implementing the obtained FTC controller. That is, there is no "feedback" from the second phase to the first phase. Therefore, trial and error is generally required to achieve acceptable FTC and FE performances and their balance. It is desirable to overcome this restriction and attain acceptable FE and FTC performances in a more systematic way. To this end, an iterative algorithm is proposed to commutatively design the FTC controller and FE observer gains to simultaneously minimize the coupling effects. The iteration will terminate in finite steps and converge to a (local) minimum under prescribed stopping criteria.

6.2 Problem Description

Consider a class of uncertain linear systems described by

$$\begin{aligned} \dot{x} &= (A + \Delta A)x + Bu + Ff + Dd \\ y &= Cx \end{aligned} \tag{6.1}$$

where $x \in \mathbb{R}^n$, $u \in \mathbb{R}^m$, $y \in \mathbb{R}^p$, $f \in \mathbb{R}^q$ and $d \in \mathbb{R}^l$ are the state, control input, measured output, actuator fault and external disturbance, respectively. The constant matrices A, B, F, D and C are known and of compatible dimensions. The matrix ΔA represents the unknown system uncertainty. The system satisfies the following assumptions:

Assumption 6.1 The pair (A, B) is controllable and the triple (A, F, C) has no invariant zeros in the closed right-half complex plane. The actuator fault is matched, i.e. $\text{rank}[B\ F] = \text{rank}(B) = m$.

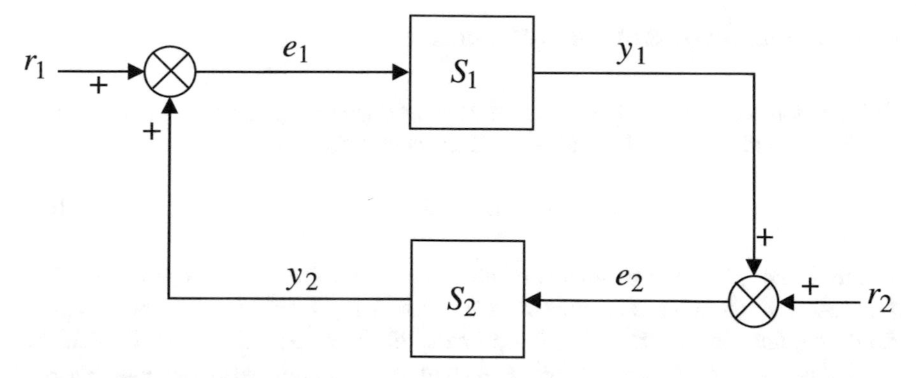

Fig. 6.2 A closed-loop system with two elements

Assumption 6.2 The uncertainty matrix ΔA satisfies the norm-bounded (energy-bounded) form $\Delta A = \mathcal{M}\mathcal{F}(t)\mathcal{N}$, where $\mathcal{M}$ and $\mathcal{N}$ are known matrices with appropriate dimensions, and $\mathcal{F}(t)$ is an unknown matrix satisfying $\mathcal{F}^\top(t)\mathcal{F}(t) \preceq I$.

Assumption 6.3 The fault f and disturbance d are norm-bounded. f has bounded first-order and second-order differentials.

This chapter aims to address the following problem:

Problem 6.1 For the system (6.1), design a control law (FTC) to compensate the fault effects and stabilize the system state, by using the estimates of state and fault obtained simultaneously from an observer (FE). Moreover, the FTC controller and FE observer are designed separately but with their bidirectional interactions robustly attenuated.

Before proceeding, the Small Gain Theorem (Zames 1966; Glad and Ljung 2000) is introduced below.

Lemma 6.1 (Small Gain Theorem) *Consider a feedback loop depicted in Fig. 6.2 composed of two stable systems S_1 and S_2, with inputs r_1 and r_2 and outputs e_1, e_2, y_1 and y_2. The closed-loop system is input-to-output stable if the gain product is less than 1, i.e.* $\|S_2\| \cdot \|S_1\| < 1$.

Remark 6.1 In the special case when $r_1 = 0$ and $r_2 = 0$, then $e_1 = y_2$ and $e_2 = y_1$. The loop is a pure feedback interconnection of S_1 and S_2. Consider the ℓ_2 gain properties $\|y_1\| < \gamma_1\|e_1\|$ and $\|y_2\| < \gamma_2\|e_2\|$, where γ_1 and γ_2 are two positive scalars. It follows from the Small Gain Theorem that the closed-loop system is asymptotic stable if $\gamma_1\gamma_2 < 1$.

6.3 Principle of Robust Decoupling

This section describes the key idea of the robust decoupling strategy. To compensate the fault f and stabilize the state x, the FTC controller is designed as

$$u = K\hat{x} + K_f\hat{f} \tag{6.2}$$

where $\hat{x}$ and $\hat{f}$ are the estimates of the state x and fault f to be obtained from the observer. The constant gains $K \in \mathbb{R}^{m\times n}$ and $K_f \in \mathbb{R}^{m\times q}$ are to be designed. Since the fault is matched, K_f is optimally designed as $K_f = -B^\dagger F$ such that $BK_f\hat{f} + Ff = F(f - \hat{f})$. Hence, if the fault is accurately estimated, then it can be completely compensated.

Substituting the controller (6.2) into (6.1) gives the closed-loop FTC system

$$\dot{x} = (A + BK)x + \Delta Ax + BB_e e + Dd \tag{6.3}$$

where $B_e = [-K\ B^\dagger F\ 0]$.

To obtain simultaneous estimation of state and fault, the system (6.1) is augmented with auxiliary state variables f and $\dot{f}$ and given as

$$\begin{aligned} \dot{\bar{x}} &= \bar{A}\bar{x} + \Delta\bar{A}x + \bar{B}u + \bar{D}\bar{d} \\ y &= \bar{C}\bar{x} \end{aligned} \tag{6.4}$$

where

$$\bar{x} = \begin{bmatrix} x \\ f \\ \dot{f} \end{bmatrix},\ \bar{d} = \begin{bmatrix} d \\ \ddot{f} \end{bmatrix},\ \bar{A} = \begin{bmatrix} A & F & 0 \\ 0 & 0 & I_q \\ 0 & 0 & 0 \end{bmatrix},\ \Delta\bar{A} = \begin{bmatrix} \Delta A \\ 0 \\ 0 \end{bmatrix},$$
$$\bar{B} = \begin{bmatrix} B \\ 0 \\ 0 \end{bmatrix},\ \bar{D} = \begin{bmatrix} D & 0 \\ 0 & 0 \\ 0 & I_q \end{bmatrix},\ \bar{C} = [C\ 0\ 0].$$

It can be verified that the pair $(\bar{A}, \bar{C})$ is observable under Assumption 6.1. Hence, the augmented state $\bar{x}$ can be estimated by the following ALSO:

$$\begin{aligned} \dot{\hat{\bar{x}}} &= \bar{A}\hat{\bar{x}} + \bar{B}u + L(y - \hat{y}) \\ \hat{y} &= \bar{C}\hat{\bar{x}}, \end{aligned} \tag{6.5}$$

where $\hat{\bar{x}} \in \mathbb{R}^{n+2q}$ and $\hat{y} \in \mathbb{R}^p$ are the estimates of $\bar{x}$ and y, respectively. The observer gain $L \in \mathbb{R}^{(n+2q)\times p}$ is to be determined. The estimates of x and f can be calculated by $\hat{x} = [I_n\ 0_{n\times 2q}]\hat{\bar{x}}$ and $\hat{f} = [0_{q\times n}\ I_q\ 0_{q\times q}]\hat{\bar{x}}$, respectively.

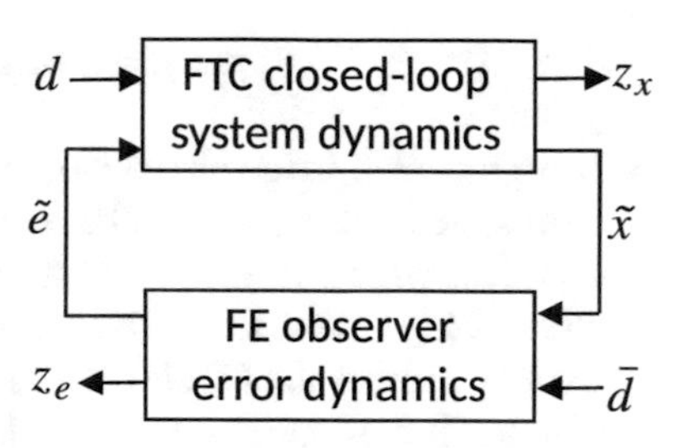

Fig. 6.3 Conceptual diagram of the FE-based FTC closed-loop system

Remark 6.2 This chapter focuses on illustrating the basic idea of the robust decoupling integration strategy. Hence, the simple FTC controller (6.2) and FE observer (6.5) are used. The reader will find that other forms of FTC controller and FE observer developed in Chaps. 4 and 5 can also be used in the robust decoupling strategy.

Define the estimation error as $e = \bar{x} - \hat{\bar{x}}$. Subtracting (6.5) from (6.4) yields the estimation error dynamics

$$\dot{e} = (\bar{A} - L\bar{C})e + \Delta\bar{A}x + \bar{D}\bar{d}. \tag{6.6}$$

Combining (6.3) and (6.6) gives the FE-based FTC closed-loop system

$$\begin{bmatrix} \dot{x} \\ \dot{e} \end{bmatrix} = \underbrace{\begin{bmatrix} A + BK + \Delta A & BB_e \\ \Delta\bar{A} & \bar{A} - L\bar{C} \end{bmatrix}}_{\mathcal{A}_c} \begin{bmatrix} x \\ e \end{bmatrix} + \begin{bmatrix} Dd \\ \bar{D}\bar{d} \end{bmatrix}. \tag{6.7}$$

The appearance of the off-diagonal elements in the above system matrix $\mathcal{A}_c$ clearly evidences the existence of *bidirectional robustness interactions* between the FE and FTC functions. Hence, the ideal Separation Principle (see Sect. 1.7.2) does not hold in this case. Stability of the FE-based FTC closed-loop system can no longer be guaranteed by independent choices of stabilizing controller and observer gains. Note that if the coupling effects $BB_e e$ and $\Delta\bar{A}x$ are minimized, then the Separation Principle is approximately recovered. In such case, the FTC controller and FE observer can be designed separately. This inspires the development of the robust decoupling strategy for minimizing the coupling effects and guaranteeing closed-loop stability when assembling the separately designed FTC controller and FE observer together.

For design purpose, the FE-based FTC closed-loop system (6.7) is conceptually shown in Fig. 6.3. In the diagram, the signals $z_x = C_x x$, $\tilde{x} = Wx$, $z_e = C_e e$ and $\tilde{e} = B_e e$ are performance outputs with the weights C_x, W, C_e and B_e.

By using Fig. 6.3, the goals of the robust decoupling strategy are illustrated below:

- *Design goal 1*: Ensuring that the closed-loop system is robustly stable against the disturbances d and $\bar{d}$ (including d and the fault modelling error $\ddot{f}$). To this end, the gains K and L are designed such that the system is stable and satisfies the ℓ_2 gain properties $\|z_x\| < \gamma_1\|d\|$ and $\|z_e\| < \gamma_2\|\bar{d}\|$, where $z_x = C_x x$ and $z_e = C_e e$ are the performance outputs associated with d and $\bar{d}$, respectively. $C_x \in \mathbb{R}^{n\times n}$ and

$C_e \in \mathbb{R}^{(n+2q)\times(n+2q)}$ are given constant weights. $\gamma_1, \gamma_2 > 0$ are the performance gains.

- *Design goal 2*: In the absence of disturbances d and $\bar{d}$, the FE-based FTC closed-loop system has a pure feedback structure with the inputs (outputs) $\tilde{x} = Wx$ and $\tilde{e} = B_e e$, where $W \in \mathbb{R}^{n\times n}$ is a positive definite matrix to be determined. The closed-loop system should be robustly stable against the coupling effects between the FTC system and FE observer. To this end, the gains K and L are designed such that the closed-loop system is stable and satisfies the ℓ_2 gain properties $\|\tilde{x}\| < \|\tilde{e}\|$ and $\|\tilde{e}\| < \gamma_3\|\tilde{x}\|$. If $\gamma_3 \leq 1$, then it follows directly from Remark 6.1 that the closed-loop system is asymptotic stable.

To achieve the above goals, the proposed robust decoupling strategy has the following two design phases:

- *Phase 1*: Design the controller gain K and the weight W such that the control system (6.3) is stable and satisfies the ℓ_2 gain properties $\|z_x\| < \gamma_1\|d\|$ and $\|\tilde{x}\| < \|\tilde{e}\|$.
- *Phase 2*: Based on the matrices K and W obtained in Phase 1, design the observer gain L such that the estimation error dynamics (6.6) are stable and satisfy the ℓ_2 gain properties $\|z_e\| < \gamma_2\|\bar{d}\|$ and $\|\tilde{e}\| < \gamma_3\|\tilde{x}\|$.

Remark 6.3 Note that two different performance outputs z_e and $\tilde{e}$ are used in Phase 2 to characterize the estimation performance against the disturbances $\bar{d}$ and $\tilde{x}$, respectively. The reason for this is to reduce the design conservativeness. Since only the affine mapping $B_e e$ of the estimation error e intrudes the FTC closed-loop system, it is desirable to directly reduce $B_e e$ rather than e.

6.4 Phase 1: FTC Controller Design

The FTC closed-loop system perturbed by the disturbance d is represented by

$$\begin{aligned} \dot{x} &= (A + BK)x + \Delta A x + Dd \\ z_x &= C_x x \end{aligned} \tag{6.8}$$

where $z_x \in \mathbb{R}^n$ is the performance output and C_x is a given constant matrix. The controller gain K is designed to ensure that the system (6.8) is stable with the ℓ_2 gain property $\|z_x\| < \gamma_1\|d\|$ by using the following lemma.

Lemma 6.2 *Under Assumptions 6.1–6.3, the FTC closed-loop system (6.8) is stable and satisfies the ℓ_2 gain property $\|z_x\| < \gamma_1\|d\|$, if the following optimization problem is feasible:*

$$\min_{X,P,\gamma_1} \gamma_1 \tag{6.9}$$

$$\text{s.t.} \begin{bmatrix} \text{He}(AP+BX) & D & \mathcal{M} & P\mathcal{N}^\top & PC_x^\top \\ \star & -\gamma_1 I & 0 & 0 & 0 \\ \star & \star & -I & 0 & 0 \\ \star & \star & \star & -I & 0 \\ \star & \star & \star & \star & -\gamma_1 I \end{bmatrix} \prec 0 \tag{6.10}$$

$$P = P^\top \succ 0, \ \ \gamma_1 > 0. \tag{6.11}$$

Then the controller gain is obtained as $K = XP^{-1}$.

Proof Consider the Lyapunov function $V_1 = x^\top P_1 x$ with a s.p.d. matrix $P_1 \in \mathbb{R}^{n\times n}$. Under Assumption 6.2, the following inequality holds

$$\text{He}(x^\top P_1 \Delta A x) \leq x^\top P_1 \mathcal{M}(P_1\mathcal{M})^\top x + x^\top \mathcal{N}^\top \mathcal{N} x.$$

Hence, the time derivative of V_1 along the system (6.8) is derived as

$$\dot{V}_1 \leq x^\top[\text{He}(P_1(A+BK)) + P_1\mathcal{M}(P_1\mathcal{M})^\top + \mathcal{N}^\top\mathcal{N}]x + \text{He}(x^\top P_1 Dd) \tag{6.12}$$

According to Isidori (2017), the ℓ_2 gain property $\|z_x\| < \gamma_1\|d\|$ is equivalent to the H_∞ performance $\|G_{z_x d}\|_\infty < \gamma_1$. Hence, it can be equivalently quantified by

$$J_1 = \int_0^\infty \Big(\frac{1}{\gamma_1} z_x(t)^\top z_x(t) - \gamma_1 d(t)^\top d(t)\Big)\mathrm{d}t < 0. \tag{6.13}$$

Under zero initial condition $x(0) = 0$, it holds that

$$\begin{aligned} J_1 &= \int_0^\infty \Big(\frac{1}{\gamma_1} z_x(t)^\top z_x(t) - \gamma_1 d(t)^\top d(t) + \dot{V}_1(t)\Big)\mathrm{d}t - \int_0^\infty \dot{V}_1(t)\mathrm{d}t \\ &= \int_0^\infty \Big(\frac{1}{\gamma_1} z_x(t)^\top z_x(t) - \gamma_1 d(t)^\top d(t) + \dot{V}_1(t)\Big)\mathrm{d}t - V_1(\infty) + V_1(0) \\ &\leq \int_0^\infty \Big(\frac{1}{\gamma_1} z_x(t)^\top z_x(t) - \gamma_1 d(t)^\top d(t) + \dot{V}_1(t)\Big)\mathrm{d}t. \end{aligned} \tag{6.14}$$

A sufficient condition for (6.14) is given by

$$\frac{1}{\gamma_1} z_x^\top z_x - \gamma_1 d^\top d + \dot{V}_1 < 0. \tag{6.15}$$

Notice that $z_x^\top z_x = x^\top C_x^\top C_x x$. Substituting (6.12) into (6.15) gives

$$\begin{bmatrix} x \\ d \end{bmatrix}^\top \begin{bmatrix} \Phi_{1,1} & \Phi_{1,2} \\ \star & \Phi_{2,2} \end{bmatrix} \begin{bmatrix} x \\ d \end{bmatrix} < 0 \tag{6.16}$$

and equivalently,

$$\begin{bmatrix} \Phi_{1,1} & \Phi_{1,2} \\ \star & \Phi_{2,2} \end{bmatrix} \prec 0 \tag{6.17}$$

where

$$\Phi_{1,1} = \text{He}(P_1(A + BK)) + P_1\mathcal{M}(P_1\mathcal{M})^\top + \mathcal{N}^\top\mathcal{N} + \frac{1}{\gamma_1}C_x^\top C_x,$$
$$\Phi_{1,2} = P_1 D,\ \Phi_{2,2} = -\gamma_1 I.$$

Applying Schur Complement (see Sect. 1.7.1) repeatedly to (6.17) and rearranging some entries, it then yields

$$\begin{bmatrix} \text{He}(P_1(A + BK)) & P_1 D & P_1\mathcal{M} & \mathcal{N}^\top & C_x^\top \\ \star & -\gamma_1 I & 0 & 0 & 0 \\ \star & \star & -I & 0 & 0 \\ \star & \star & \star & -I & 0 \\ \star & \star & \star & \star & -\gamma_1 I \end{bmatrix} \prec 0. \tag{6.18}$$

Let $P = P_1^{-1}$. Pre- and post-multiplying both sides of (6.18) with $\text{diag}(P, I, I, I, I)$ and its transpose, respectively, then it gives

$$\begin{bmatrix} \text{He}(AP + BKP) & D & \mathcal{M} & P\mathcal{N}^\top & PC_x^\top \\ \star & -\gamma_1 I & 0 & 0 & 0 \\ \star & \star & -I & 0 & 0 \\ \star & \star & \star & -I & 0 \\ \star & \star & \star & \star & -\gamma_1 I \end{bmatrix} \prec 0. \tag{6.19}$$

Further define $X = KP$, then the design of K can be formulated as the optimization problem (6.9). □

The FTC closed-loop system perturbed by estimation error e is represented by

$$\begin{aligned} \dot{x} &= (A + BK)x + \Delta Ax + B\tilde{e} \\ \tilde{x} &= Wx \end{aligned} \tag{6.20}$$

where $\tilde{e} = B_e e$ and $W \in \mathbb{R}^{n\times n}$ is a positive definite matrix to be determined. Therefore, the controller gain K should also be designed such that the system (6.20) is stable with $\|\tilde{x}\| < \|\tilde{e}\|$.

To have better attenuation of the “disturbance” $\tilde{e}$, it is desirable to maximize W. To illustrate this point, assume that the two matrices W_1 and W_2 satisfying $W_1 \prec W_2$ are both solutions to the design problem under the same $\tilde{e}$. Then one gets the inequalities $\|W_1 x\| \leq \|W_2 x\| \leq \|\tilde{e}\|$. It is clear that the matrix W_2 provides tighter information of the effect of $\tilde{e}$ on the state x. To characterize the worst-case effect of disturbance $\tilde{e}$ on the state trajectories, it is hence natural to maximize W. To this end, the design of K and W is described below.

Lemma 6.3 *Under Assumptions 6.1–6.3, the closed-loop system (6.20) is stable and satisfies the ℓ_2 gain property $\|\tilde{x}\| < \|\tilde{e}\|$, if the following optimization problem is feasible:*

$$\min_{X,P,S} \operatorname{trace}(S) \tag{6.21}$$

$$\text{s.t.} \begin{bmatrix} \operatorname{He}(AP+BX) & B & \mathcal{M} & P\mathcal{N}^\top & P \\ \star & -I & 0 & 0 & 0 \\ \star & \star & -I & 0 & 0 \\ \star & \star & \star & -I & 0 \\ \star & \star & \star & \star & -S \end{bmatrix} \prec 0 \tag{6.22}$$

$$P = P^\top \succ 0, \ S = S^\top \succ 0. \tag{6.23}$$

Then the design matrices are obtained as $K = XP^{-1}$ and $W = \sqrt{S^{-1}}$.

Proof Consider the Lyapunov function $V_2 = x^\top P_2 x$ with a s.p.d. matrix $P_2 \in \mathbb{R}^{n\times n}$. Under Assumption 6.2, it can be derived that

$$\operatorname{He}(x^\top P_2 \Delta A x) \leq x^\top P_2 \mathcal{M}(P_2\mathcal{M})^\top x + x^\top \mathcal{N}^\top \mathcal{N} x.$$

Hence, the time derivative of V_2 along the system (6.20) is derived as

$$\dot{V}_2 \leq x^\top[\operatorname{He}(P_2(A+BK)) + P_2\mathcal{M}(P_2\mathcal{M})^\top + \mathcal{N}^\top\mathcal{N}]x + \operatorname{He}(x^\top P_2 B\tilde{e}). \tag{6.24}$$

The ℓ_2 gain property $\|\tilde{x}\| < \|\tilde{e}\|$ is equivalent to the H_∞ performance $\|G_{\tilde{x}\tilde{e}}\|_\infty < 1$ and can be equivalently quantified by

$$J_2 = \int_0^\infty \left(\tilde{x}(t)^\top \tilde{x}(t) - \tilde{e}(t)^\top \tilde{e}(t)\right) \mathrm{d}t < 0. \tag{6.25}$$

Under zero initial condition $x(0) = 0$, it holds that

$$\begin{aligned} J_2 &= \int_0^\infty \left(\tilde{x}(t)^\top \tilde{x}(t) - \tilde{e}(t)^\top \tilde{e}(t) + \dot{V}_2(t)\right) \mathrm{d}t - \int_0^\infty \dot{V}_2(t)\mathrm{d}t \\ &= \int_0^\infty \left(\tilde{x}(t)^\top \tilde{x}(t) - \tilde{e}(t)^\top \tilde{e}(t) + \dot{V}_2(t)\right) \mathrm{d}t - V_2(\infty) + V_2(0) \\ &\leq \int_0^\infty \left(\tilde{x}(t)^\top \tilde{x}(t) - \tilde{e}(t)^\top \tilde{e}(t) + \dot{V}_2(t)\right) \mathrm{d}t. \end{aligned} \tag{6.26}$$

A sufficient condition for (6.26) is given by

$$\tilde{x}^\top \tilde{x} - \tilde{e}^\top \tilde{e} + \dot{V}_2 < 0. \tag{6.27}$$

Substituting (6.24) into (6.27) gives

$$\begin{bmatrix} x \\ \tilde{e} \end{bmatrix}^\top \begin{bmatrix} \Phi_{1,1} & \Phi_{1,2} \\ \star & \Phi_{2,2} \end{bmatrix} \begin{bmatrix} x \\ \tilde{e} \end{bmatrix} < 0 \tag{6.28}$$

and equivalently,

$$\begin{bmatrix} \Phi_{1,1} & \Phi_{1,2} \\ \star & \Phi_{2,2} \end{bmatrix} \prec 0 \tag{6.29}$$

where

$$\Phi_{1,1} = \text{He}(P_2(A + BK)) + P_2\mathcal{M}(P_2\mathcal{M})^\top + \mathcal{N}^\top \mathcal{N} + W^\top W,$$
$$\Phi_{1,2} = P_2 B, \ \ \Phi_{2,2} = -I.$$

Let $P = P_2^{-1}$, $X = KP$ and $S = (W^\top W)^{-1}$. Pre- and post-multiplying both sides of (6.29) with $\text{diag}(P, I)$ and its transpose, respectively, and further applying Schur Complement, then it yields

$$\begin{bmatrix} \text{He}(AP + BX) & B & \mathcal{M} & P\mathcal{N}^\top & P \\ \star & -I & 0 & 0 & 0 \\ \star & \star & -I & 0 & 0 \\ \star & \star & \star & -I & 0 \\ \star & \star & \star & \star & -S \end{bmatrix} \prec 0. \tag{6.30}$$

Therefore, the design of K and W is formulated as the optimization problem (6.21). □

Based on Lemmas 6.2 and 6.3, the design of K and W is described below.

Theorem 6.1 *Under Assumptions 6.1–6.3, the FTC closed-loop system (6.3) is stable and satisfies the ℓ_2 gain properties $\|z_x\| < \gamma_1\|d\|$ and $\|\tilde{x}\| < \|\tilde{e}\|$, if the following optimization problem is feasible:*

$$\min_{X,P,S,\gamma_1} \alpha_1\gamma_1 + \alpha_2\text{trace}(S) \tag{6.31}$$

$$\text{s.t.} \begin{bmatrix} \text{He}(AP+BX) & D & \mathcal{M} & P\mathcal{N}^\top & PC_x^\top \\ \star & -\gamma_1 I & 0 & 0 & 0 \\ \star & \star & -I & 0 & 0 \\ \star & \star & \star & -I & 0 \\ \star & \star & \star & \star & -\gamma_1 I \end{bmatrix} \prec 0 \tag{6.32}$$

$$\begin{bmatrix} \text{He}(AP+BX) & B & \mathcal{M} & P\mathcal{N}^\top & P \\ \star & -I & 0 & 0 & 0 \\ \star & \star & -I & 0 & 0 \\ \star & \star & \star & -I & 0 \\ \star & \star & \star & \star & -S \end{bmatrix} \prec 0 \tag{6.33}$$

$$P = P^\top \succ 0,\ S = S^\top \succ 0,\ \gamma_1 > 0 \tag{6.34}$$

where α_1 and α_2 are prescribed positive scalars. Then the design matrices are obtained as $K = XP^{-1}$ and $W = \sqrt{S^{-1}}$.

6.5 Phase 2: FE Observer Design

The estimation error system perturbed only by the disturbance $\bar{d}$ is represented by

$$\begin{aligned} \dot{e} &= (\bar{A} - L\bar{C})e + \bar{D}\bar{d} \\ z_e &= C_e e \end{aligned} \tag{6.35}$$

where $z_e \in \mathbb{R}^{n+2q}$ is the performance output and C_e is a given constant matrix. The design of L to make the above system stable and satisfy the ℓ_2 gain property $\|z_e\| < \gamma_2\|\bar{d}\|$ is described below.

Lemma 6.4 *Under Assumptions 6.1 and 6.3, the error system (6.35) is stable and satisfies the ℓ_2 gain property $\|z_e\| < \gamma_2\|\bar{d}\|$, if the following optimization problem is feasible:*

$$\min_{Y,Q,\gamma_2} \gamma_2 \tag{6.36}$$

$$\text{s.t.} \begin{bmatrix} \text{He}(Q\bar{A} - Y\bar{C}) & Q\bar{D} & C_e^\top \\ \star & -\gamma_2 I & 0 \\ \star & \star & -\gamma_2 I \end{bmatrix} \prec 0 \tag{6.37}$$

$$Q = Q^\top \succ 0,\ \gamma_2 > 0. \tag{6.38}$$

Then the observer gain is obtained as $L = Q^{-1}Y$.

Proof Consider the Lyapunov function $V_3 = e^\top Q e$ with a s.p.d. matrix $Q \in \mathbb{R}^{(n+2q)\times(n+2q)}$. The time derivative of V_3 is obtained as

$$\dot{V}_3 \leq e^\top \mathrm{He}(Q(\bar{A} - L\bar{C}))e + \mathrm{He}(e^\top Q\bar{D}\bar{d}). \tag{6.39}$$

The ℓ_2 gain property $\|z_e\| < \gamma_2 \|\bar{d}\|$ is equivalent to the H_∞ performance $\|G_{z_e\bar{d}}\|_\infty < \gamma_2$ and can be equivalently quantified by

$$J_3 = \int_0^\infty \Big(\frac{1}{\gamma_2} z_e(t)^\top z_e(t) - \gamma_2 \bar{d}(t)^\top \bar{d}(t) \Big) \mathrm{d}t < 0. \tag{6.40}$$

Under zero initial condition $e(0) = 0$, it holds that

$$\begin{aligned} J_3 &= \int_0^\infty \Big(\frac{1}{\gamma_2} z_e(t)^\top z_e(t) - \gamma_2 \bar{d}(t)^\top \bar{d}(t) + \dot{V}_3(t) \Big) \mathrm{d}t - \int_0^\infty \dot{V}_3(t) \mathrm{d}t \\ &= \int_0^\infty \Big(\frac{1}{\gamma_2} z_e(t)^\top z_e(t) - \gamma_2 \bar{d}(t)^\top \bar{d}(t) + \dot{V}_3(t) \Big) \mathrm{d}t - V_3(\infty) + V_3(0) \\ &\leq \int_0^\infty \Big(\frac{1}{\gamma_2} z_e(t)^\top z_e(t) - \gamma_2 \bar{d}(t)^\top \bar{d}(t) + \dot{V}_3(t) \Big) \mathrm{d}t. \end{aligned} \tag{6.41}$$

A sufficient condition for (6.41) is given as

$$\frac{1}{\gamma_2} z_e^\top z_e - \gamma_2 \bar{d}^\top \bar{d} + \dot{V}_3 < 0. \tag{6.42}$$

Substituting (6.39) into (6.42) yields

$$\begin{bmatrix} e \\ \bar{d} \end{bmatrix}^\top \begin{bmatrix} \Phi_{1,1} & \Phi_{1,2} \\ \star & \Phi_{2,2} \end{bmatrix} \begin{bmatrix} e \\ \bar{d} \end{bmatrix} < 0 \tag{6.43}$$

and equivalently,

$$\begin{bmatrix} \Phi_{1,1} & \Phi_{1,2} \\ \star & \Phi_{2,2} \end{bmatrix} \prec 0 \tag{6.44}$$

where

$$\Phi_{1,1} = \mathrm{He}(Q(\bar{A} - L\bar{C})) + \frac{1}{\gamma_2} C_e^\top C_e, \ \Phi_{1,2} = Q\bar{D}, \ \Phi_{2,2} = -\gamma_2 I.$$

Define $Y = QL$. Applying Schur Complement to (6.44) yields

$$\begin{bmatrix} \mathrm{He}(Q\bar{A} - Y\bar{C}) & Q\bar{D} & C_e^\top \\ \star & -\gamma_2 I & 0 \\ \star & \star & -\gamma_2 I \end{bmatrix} \prec 0. \tag{6.45}$$

Therefore, the design of L can be formulated as the optimization problem (6.36). □

The estimation error system perturbed only by the coupling term $\Delta\bar{A}x$ is represented by

$$\begin{aligned}\dot{e} &= (\bar{A} - L\bar{C})e + \Delta\bar{A}W^{-1}\tilde{x} \\ \tilde{e} &= B_e e\end{aligned} \tag{6.46}$$

where $\tilde{x} = Wx$. Hence, the observer gain L should also be designed such that the above error system is stable and satisfies $\|\tilde{e}\| < \gamma_3\|\tilde{x}\|$ with $0 < \gamma_3 \le 1$, which is described below.

Lemma 6.5 *Under Assumptions 6.1–6.3, the error system (6.46) is stable and satisfies the ℓ_2 gain property $\|\tilde{e}\| < \gamma_3\|\tilde{x}\|$, if the following optimization problem is feasible:*

$$\min_{Y,Q,\gamma_3} \gamma_3 \tag{6.47}$$

$$\text{s.t.} \begin{bmatrix} \mathrm{He}(Q\bar{A} - Y\bar{C}) & Q\bar{\mathcal{M}} & B_e^\top & 0 \\ \star & -I & 0 & 0 \\ \star & \star & -\gamma_3 I & 0 \\ \star & \star & \star & \mathcal{N}^\top\mathcal{N} - \gamma_3 S^{-1} \end{bmatrix} \prec 0 \tag{6.48}$$

$$Q = Q^\top \succ 0,\ 0 < \gamma_3 \le 1. \tag{6.49}$$

Then the observer gain is obtained as $L = Q^{-1}Y$.

Proof Consider the Lyapunov function $V_4 = e^\top Qe$. Under Assumption 6.2, it can be derived that

$$\mathrm{He}(e^\top Q\Delta\bar{A}W^{-1}\tilde{x}) \le e^\top Q\bar{\mathcal{M}}(Q\bar{\mathcal{M}})^\top e + \tilde{x}^\top(W^{-1})^\top\mathcal{N}^\top\mathcal{N}W^{-1}\tilde{x},$$

where $\bar{\mathcal{M}} = [\mathcal{M}^\top\ 0]^\top$. Hence, the time derivative of V_4 is obtained as

$$\dot{V}_4 \le e^\top\left[\mathrm{He}(Q(\bar{A} - L\bar{C})) + Q\bar{\mathcal{M}}(Q\bar{\mathcal{M}})^\top\right]e + \tilde{x}^\top(W^{-1})^\top\mathcal{N}^\top\mathcal{N}W^{-1}\tilde{x}. \tag{6.50}$$

The ℓ_2 gain property $\|\tilde{e}\| < \gamma_3\|\tilde{x}\|$ is equivalent to the H_∞ performance $\|G_{\tilde{e}\tilde{x}}\|_\infty < \gamma_3$ and can be equivalently quantified by

$$J_4 = \int_0^\infty \Big(\frac{1}{\gamma_3}\tilde{e}(t)^\top\tilde{e}(t) - \gamma_3\tilde{x}(t)^\top\tilde{x}(t)\Big)\mathrm{d}t < 0. \tag{6.51}$$

Under zero initial condition $e(0) = 0$, it holds that

$$\begin{aligned}J_4 &= \int_0^\infty \Big(\frac{1}{\gamma_3}\tilde{e}(t)^\top\tilde{e}(t) - \gamma_3\tilde{x}(t)^\top\tilde{x}(t) + \dot{V}_4(t)\Big)\mathrm{d}t - \int_0^\infty \dot{V}_4(t)\mathrm{d}t \\ &\le \int_0^\infty \Big(\frac{1}{\gamma_3}\tilde{e}(t)^\top\tilde{e}(t) - \gamma_3\tilde{x}(t)^\top\tilde{x}(t) + \dot{V}_4(t)\Big)\mathrm{d}t.\end{aligned} \tag{6.52}$$

A sufficient condition for (6.52) is given as

$$\frac{1}{\gamma_3}\tilde{e}^\top \tilde{e} - \gamma_3 \tilde{x}(t)^\top \tilde{x}(t) + \dot{V}_4 < 0. \tag{6.53}$$

Notice that $\tilde{e}^\top \tilde{e} = e^\top B_e^\top B_e e$. Substituting (6.50) into (6.53), then it yields

$$\begin{bmatrix} e \\ x \end{bmatrix}^\top \begin{bmatrix} \Phi_{1,1} & 0 \\ \star & \Phi_{2,2} \end{bmatrix} \begin{bmatrix} e \\ x \end{bmatrix} < 0 \tag{6.54}$$

and equivalently,

$$\begin{bmatrix} \Phi_{1,1} & 0 \\ \star & \Phi_{2,2} \end{bmatrix} \prec 0 \tag{6.55}$$

where

$$\Phi_{1,1} = \text{He}(Q(\bar{A} - L\bar{C})) + Q\bar{\mathcal{M}}(Q\bar{\mathcal{M}})^\top + \frac{1}{\gamma_3} B_e^\top B_e,$$
$$\Phi_{2,2} = \mathcal{N}^\top \mathcal{N} - \gamma_3 S^{-1}, \ S = (W^\top W)^{-1}.$$

Further applying Schur Complement to (6.55) and defining $Y = QL$, then the design of L can be formulated as the optimization problem (6.47). □

A necessary condition for feasibility of the LMI (6.55) (also the optimization problem (6.47)) is

$$\mathcal{N}^\top \mathcal{N} - \gamma_3 S^{-1} \prec 0 \tag{6.56}$$

under $0 < \gamma_3 \leq 1$. This inequality is always satisfied provided that

$$\lambda_{\max}(\mathcal{N}^\top \mathcal{N}) < \lambda_{\min}(S^{-1}). \tag{6.57}$$

Hence, an intuitive method to ensure the satisfaction of (6.56) is to include (6.57) as a constraint for the optimization problem (6.31) when solving K and W. However, this is difficult for the present setting because the optimization is performed over the decision variable S rather than S^{-1}. Alternatively, it may be easier to impose the following constraint that is equivalent to (6.57):

$$\lambda_{\max}(S) < \lambda_{\min}((\mathcal{N}^\top \mathcal{N})^{-1}) \tag{6.58}$$

provided that $(\mathcal{N}^\top \mathcal{N})$ is non-singular.

To ease the solvability of the optimization problem, this chapter adopts another method to ensure (6.56) without imposing any extra constraint. Note that given an uncertainty matrix ΔA, the matrices $\mathcal{N}$ and $\mathcal{M}$ can be arbitrarily specified, as long as Assumption 6.2 is satisfied. By observing this, the method adopted in this chapter is described below.

First, when solving the optimization problem (6.31), the matrix $\mathcal{N}$ is given as

$$\mathcal{N} = \delta_0 \times I_n$$

with a scalar δ_0 taking value within (0, 1]. The matrices $\mathcal{M}$ and $\mathcal{F}$ are subsequently chosen to satisfy Assumption 6.2.

Second, when solving the optimization problem (6.47), the matrices $\mathcal{N}$ and $\mathcal{M}$ are re-scaled to be

$$\hat{\mathcal{N}} = \delta \mathcal{N}, \ \hat{\mathcal{M}} = \delta^{-1} \mathcal{M} \tag{6.59}$$

where δ is a positive scaling factor defined as $\delta = \sqrt{\lambda_{\min}(S^{-1})} - \epsilon_0$ with a given small positive scalar ϵ_0. It is obvious that the new matrices $\hat{\mathcal{N}}$ and $\hat{\mathcal{M}}$ always satisfy Assumption 6.2. Moreover, they automatically lead to the satisfaction of the constraint (6.57) and furthermore (6.56).

In summary, the observer design can be stated below.

Theorem 6.2 *Under Assumptions 6.1–6.3, the estimation error system (6.6) is stable and satisfies the ℓ_2 gain properties $\|z_e\| < \gamma_2 \|\bar{d}\|$ and $\|\tilde{e}\| < \gamma_3 \|\tilde{x}\|$, if the following optimization problem is feasible:*

$$\min_{Y, Q, \gamma_2, \gamma_3} \ \beta_1 \gamma_2 + \beta_2 \gamma_3 \tag{6.60}$$

$$\text{s.t.} \quad \begin{bmatrix} \mathrm{He}(Q\bar{A} - Y\bar{C}) & Q\bar{D} & C_e^\top \\ \star & -\gamma_2 I & 0 \\ \star & \star & -\gamma_2 I \end{bmatrix} \prec 0 \tag{6.61}$$

$$\begin{bmatrix} \mathrm{He}(Q\bar{A} - Y\bar{C}) & Q\hat{\hat{\mathcal{M}}} & B_e^\top & 0 \\ \star & -I & 0 & 0 \\ \star & \star & -\gamma_3 I & 0 \\ \star & \star & \star & \hat{\mathcal{N}}^\top \hat{\mathcal{N}} - \gamma_3 S^{-1} \end{bmatrix} \prec 0 \tag{6.62}$$

$$Q = Q^\top \succ 0, \ \gamma_2 > 0, \ 0 < \gamma_3 \leq 1 \tag{6.63}$$

where β_1 and β_2 are prescribed positive scalars, $\hat{\hat{\mathcal{M}}} = [\hat{\mathcal{M}}^\top \ 0]^\top$, $\hat{\mathcal{N}} = \delta \mathcal{N}$, $\hat{\mathcal{M}} = \delta^{-1} \mathcal{M}$ and $\delta = \sqrt{\lambda_{\min}(S^{-1})} - \epsilon_0$ with a given small positive scalar ϵ_0. Then the observer gain is obtained as $L = Q^{-1} Y$.

6.6 Robustness Enhancement via Iteration

The two-phase synthesis strategy described in the previous sections is virtually a sequential strategy similar to that in Chap. 3, where there is no "feedback" from the second phase to the first phase. Hence, acceptable FTC and FE performances and their balance should be achieved through trial and error. To overcome this, an iterative algorithm is proposed in this section to commutatively design the FTC controller and FE observer. This allows the *bidirectional robustness interactions* between the FTC system and FE observer to be taken into account in a systematic way.

To achieve this, the output $\tilde{e}$ in Fig. 6.3 is redefined as $\tilde{e} = VB_e e$, where $V \in \mathbb{R}^{m\times m}$ is another positive definite weighting matrix to be determined in the FE observer design phase. The proposed iterative algorithm is based on the following two-phase synthesis strategy modified from the previous sections.

6.6.1 Phase 1: FTC Controller Design

Without consideration of the estimation error e, the controller gain K is designed such that the control system (6.8) is stable with $\|z_x\| < \gamma_1\|d\|$. The design is formulated as the optimization problem (6.9) in Lemma 6.2.

The FTC closed-loop system perturbed only by estimation error is represented by

$$\begin{aligned} \dot{x} &= (A + BK)x + \Delta Ax + \hat{B}\tilde{e} \\ \tilde{x} &= Wx \end{aligned} \tag{6.64}$$

where $\hat{B} = BV^{-1}, \tilde{e} = VB_e e$ and $W \in \mathbb{R}^{n\times n}$ is a positive definite matrix to be determined. Hence, the controller gain K should also be designed such that the system (6.64) is stable with $\|\tilde{x}\| < \|\tilde{e}\|$. The design is stated below.

Lemma 6.6 *Under Assumptions 6.1–6.3, the FTC closed-loop system (6.64) is stable and satisfies the ℓ_2 gain property $\|\tilde{x}\| < \|\tilde{e}\|$, if the following optimization problem is feasible:*

$$\min_{X,P,S} \operatorname{trace}(S) \tag{6.65}$$

$$\text{s.t.} \begin{bmatrix} \mathrm{He}(AP + BX) & B & \mathcal{M} & P\mathcal{N}^\top & P \\ \star & -R^{-1} & 0 & 0 & 0 \\ \star & \star & -I & 0 & 0 \\ \star & \star & \star & -I & 0 \\ \star & \star & \star & \star & -S \end{bmatrix} \prec 0 \tag{6.66}$$

$$P = P^\top \succ 0, \; S = S^\top \succ 0. \tag{6.67}$$

Then the design matrices are obtained as $K = XP^{-1}$ and $W = \sqrt{S^{-1}}$.

Proof The proof follows directly from that of Lemma 6.3 by defining $R = (V^\top V)^{-1}$. □

By using Lemmas 6.2 and 6.6, the design of K and W is described below.

Theorem 6.3 *Under Assumptions 6.1–6.3, the FTC closed-loop system (6.3) is stable and satisfies the ℓ_2 gain properties $\|z_x\| < \gamma_1\|d\|$ and $\|\tilde{x}\| < \|\tilde{e}\|$, if the following optimization problem is feasible:*

$$\min_{X,P,S,\gamma_1} \alpha_1\gamma_1 + \alpha_2 \text{trace}(S) \tag{6.68}$$

$$\text{s.t.} \begin{bmatrix} \text{He}(AP+BX) & D & \mathcal{M} & P\mathcal{N}^\top & PC_x^\top \\ \star & -\gamma_1 I & 0 & 0 & 0 \\ \star & \star & -I & 0 & 0 \\ \star & \star & \star & -I & 0 \\ \star & \star & \star & \star & -\gamma_1 I \end{bmatrix} \prec 0 \tag{6.69}$$

$$\begin{bmatrix} \text{He}(AP+BX) & B & \mathcal{M} & P\mathcal{N}^\top & P \\ \star & -R^{-1} & 0 & 0 & 0 \\ \star & \star & -I & 0 & 0 \\ \star & \star & \star & -I & 0 \\ \star & \star & \star & \star & -S \end{bmatrix} \prec 0 \tag{6.70}$$

$$P = P^\top \succ 0, \ S = S^\top \succ 0, \ \gamma_1 > 0 \tag{6.71}$$

where α_1 and α_2 are prescribed positive scalars. Then the design matrices are obtained as $K = XP^{-1}$ and $W = \sqrt{S^{-1}}$.

6.6.2 Phase 2: FE Observer Design

When the error system (6.35) is perturbed only by the disturbance $\bar{d}$, the gain L is designed such that this system is stable and satisfies the ℓ_2 gain property $\|z_e\| < \gamma_2\|\bar{d}\|$. The design is formulated as the optimization problem (6.36) in Lemma 6.4.

The estimation error system perturbed only by $\Delta\bar{A}x$ is represented by

$$\begin{aligned} \dot{e} &= (\bar{A} - L\bar{C})e + \Delta\bar{A}W^{-1}\tilde{x} \\ \tilde{e} &= VB_e e \end{aligned} \tag{6.72}$$

where $\tilde{x} = Wx$ and $V \in \mathbb{R}^{m\times m}$ is a positive definite matrix to be determined. Similarly to that of W, it is desirable to maximize V to have better attenuation of the "disturbance" $\tilde{x}$. The design is summarized below.

Lemma 6.7 *Under Assumptions 6.1–6.3, the error system (6.72) is stable with the ℓ_2 gain property $\|\tilde{e}\| < \|\tilde{x}\|$, if the following optimization problem is feasible:*

$$\min_{Y,Q,R} \operatorname{trace}(R) \tag{6.73}$$

$$\text{s.t.} \begin{bmatrix} \mathrm{He}(Q\bar{A} - Y\bar{C}) & Q\bar{\mathcal{M}} & B_e^\top & 0 \\ \star & -I & 0 & 0 \\ \star & \star & -R & 0 \\ \star & \star & \star & \mathcal{N}^\top \mathcal{N} - S^{-1} \end{bmatrix} \prec 0 \tag{6.74}$$

$$Q = Q^\top \succ 0, \ R = R^\top \succ 0. \tag{6.75}$$

Then the design matrices are obtained as $L = Q^{-1}Y$ *and* $V = \sqrt{R^{-1}}$.

Proof Consider the Lyapunov function $V_5 = e^\top Q e$. Under Assumption 6.2, it can be derived that

$$\mathrm{He}(e^\top Q \Delta \bar{A} W^{-1} \tilde{x}) \leq e^\top Q \bar{\mathcal{M}} (Q\bar{\mathcal{M}})^\top e + \tilde{x}^\top (W^{-1})^\top \mathcal{N}^\top \mathcal{N} W^{-1} \tilde{x},$$

where $\bar{\mathcal{M}} = [\mathcal{M}^\top \ 0]^\top$. Hence, the derivative of V_5 is obtained as

$$\begin{aligned} \dot{V}_5 \leq\ & e^\top \left[\mathrm{He}(Q(\bar{A} - L\bar{C})) + Q\bar{\mathcal{M}}(Q\bar{\mathcal{M}})^\top\right] e \\ & + \tilde{x}^\top (W^{-1})^\top \mathcal{N}^\top \mathcal{N} W^{-1} \tilde{x}. \end{aligned} \tag{6.76}$$

The ℓ_2 gain property $\|\tilde{e}\| < \|\tilde{x}\|$ is equivalent to the H_∞ performance $\|G_{\tilde{e}\tilde{x}}\|_\infty < 1$. Hence, it can be quantified by

$$J_5 = \int_0^\infty \Big(\tilde{e}(t)^\top \tilde{e}(t) - \tilde{x}(t)^\top \tilde{x}(t)\Big) \mathrm{d}t < 0. \tag{6.77}$$

Under zero initial condition $e(0) = 0$, it holds that

$$\begin{aligned} J_5 &= \int_0^\infty \Big(\tilde{e}(t)^\top \tilde{e}(t) - \tilde{x}(t)^\top \tilde{x}(t) + \dot{V}_5(t)\Big) \mathrm{d}t - \int_0^\infty \dot{V}_5(t) \mathrm{d}t \\ &= \int_0^\infty \Big(\tilde{e}(t)^\top \tilde{e}(t) - \tilde{x}(t)^\top \tilde{x}(t) + \dot{V}_5(t)\Big) \mathrm{d}t - V_5(\infty) + V_5(0) \\ &\leq \int_0^\infty \Big(\tilde{e}(t)^\top \tilde{e}(t) - \tilde{x}(t)^\top \tilde{x}(t) + \dot{V}_5(t)\Big) \mathrm{d}t. \end{aligned} \tag{6.78}$$

A sufficient condition for (6.78) is given by

$$\tilde{e}^\top \tilde{e} - \tilde{x}^\top \tilde{x} + \dot{V}_5 < 0. \tag{6.79}$$

Notice that $\tilde{e}^\top \tilde{e} = e^\top B_e^\top V^\top V B_e e$. Substituting (6.76) into (6.79) gives

$$\begin{bmatrix} e \\ x \end{bmatrix}^\top \begin{bmatrix} \Phi_{1,1} & 0 \\ \star & \Phi_{2,2} \end{bmatrix} \begin{bmatrix} e \\ x \end{bmatrix} < 0 \tag{6.80}$$

and equivalently,

$$\begin{bmatrix} \Phi_{1,1} & 0 \\ \star & \Phi_{2,2} \end{bmatrix} \prec 0 \tag{6.81}$$

where

$$\Phi_{1,1} = \mathrm{He}(Q(\bar{A} - L\bar{C})) + Q\bar{\mathcal{M}}(Q\bar{\mathcal{M}})^\top + B_e^\top V^\top V B_e,$$
$$\Phi_{2,2} = \mathcal{N}^\top \mathcal{N} - W^\top W.$$

Define $R = (V^\top V)^{-1}$ and $Y = QL$. Applying Schur Complement to (6.81) and using $S = (W^\top W)^{-1}$ gives

$$\begin{bmatrix} \mathrm{He}(Q\bar{A} - Y\bar{C}) & Q\bar{\mathcal{M}} & B_e^\top & 0 \\ \star & -I & 0 & 0 \\ \star & \star & -R & 0 \\ \star & \star & \star & \mathcal{N}^\top \mathcal{N} - S^{-1} \end{bmatrix} \prec 0. \tag{6.82}$$

Therefore, the design of L can be formulated as the optimization problem (6.73). □

A necessary condition for feasibility of the LMI (6.82) is

$$\mathcal{N}^\top \mathcal{N} - S^{-1} \prec 0. \tag{6.83}$$

According to the discussions in Sect. 6.5, (6.83) can be guaranteed by rescaling the matrices $\mathcal{N}$ and $\mathcal{M}$ as

$$\hat{\mathcal{N}} = \delta \mathcal{N}, \ \hat{\mathcal{M}} = \delta^{-1} \mathcal{M} \tag{6.84}$$

with $\delta = \sqrt{\lambda_{\min}(S^{-1})} - \epsilon_0$ and a small positive scalar ϵ_0.

Based on Lemmas 6.4 and 6.7, the FE observer design is summarized below.

Theorem 6.4 *Under Assumptions 6.1–6.3, the estimation error system (6.6) is stable and satisfies the ℓ_2 gain properties $\|z_e\| < \gamma_2 \|\bar{d}\|$ and $\|\tilde{e}\| < \|\tilde{x}\|$, if the following optimization problem is feasible:*

$$\min_{Y,Q,R,\gamma_2} \ \beta_1 \gamma_2 + \beta_2 \mathrm{trace}(R) \tag{6.85}$$

$$\text{s.t.} \quad \begin{bmatrix} \mathrm{He}(Q\bar{A} - Y\bar{C}) & Q\bar{D} & C_e^\top \\ \star & -\gamma_2 I & 0 \\ \star & \star & -\gamma_2 I \end{bmatrix} \prec 0 \tag{6.86}$$

$$\begin{bmatrix} \mathrm{He}(Q\bar{A} - Y\bar{C}) & Q\bar{\mathcal{M}} & B_e^\top & 0 \\ \star & -I & 0 & 0 \\ \star & \star & -R & 0 \\ \star & \star & \star & \mathcal{N}^\top \mathcal{N} - S^{-1} \end{bmatrix} \prec 0 \tag{6.87}$$

$$Q = Q^\top \succ 0, \ R = R^\top \succ 0, \ \gamma_2 > 0 \tag{6.88}$$

where β_1 and β_2 are prescribed positive scalars, $\hat{\hat{\mathcal{M}}} = [\hat{\mathcal{M}}^\top \ 0]^\top$, $\hat{\mathcal{N}} = \delta\mathcal{N}$, $\hat{\mathcal{M}} = \delta^{-1}\mathcal{M}$ and $\delta = \sqrt{\lambda_{\min}(S^{-1})} - \epsilon_0$ with a given small positive scalar ϵ_0. Then the design matrices are obtained as $L = Q^{-1}Y$ and $V = \sqrt{R^{-1}}$.

Feasibility of the optimization problem (6.85) is analysed below.

Proposition 6.1 *Feasibility of the optimization problem (6.85) is guaranteed and independent of the matrices K and S obtained in the FTC controller design.*

Proof By using Schur Complement, (6.86) and (6.87) are compactly represented by

$$\Upsilon_{o1}(Q, Y) + [Q\bar{D} \ C_e^\top]\gamma_2^{-1}[Q\bar{D} \ C_e^\top]^\top \prec 0 \tag{6.89}$$

$$\Upsilon_{o2}(Q, Y) + Q\hat{\mathcal{M}}(Q\hat{\mathcal{M}})^\top + B_e^\top R^{-1} B_e \prec 0 \tag{6.90}$$

$$\hat{\mathcal{N}}^\top \hat{\mathcal{N}} - S^{-1} \prec 0, \tag{6.91}$$

where (6.89) equals to (6.86), and the combination of (6.90) and (6.91) equals to (6.87).

Since the pair $(\bar{A}, \bar{C})$ is observable, there is a large enough γ_2 satisfying (6.89). Similarly, there is a small enough Q and a big enough R such that $(Q\hat{\mathcal{M}}(Q\hat{\mathcal{M}})^\top + B_e^\top R^{-1} B_e)$ is small enough and subsequently the inequality (6.90) is feasible. The inequality (6.91) is always satisfied by defining $\hat{\mathcal{N}}$ as in (6.83). □

6.6.3 Iterative Algorithm

An iterative algorithm is presented in this subsection based on the optimization problems (6.68) and (6.85). The relevant definitions are given below.

Define $\bar{V} = V^\top V$. At iteration j, the matrix $\bar{V}$ is redesigned as

$$\bar{V}^{(j)} = \left(\bar{V}_0 + \sum_{i=1}^{j} \Delta\bar{V}^{(i)}\right) \tag{6.92}$$

where $\bar{V}_0$ is the initial value of $\bar{V}$. The extra components $\Delta\bar{V}^{(i)}$ are used to enhance the FE observer robustness against the perturbation from the FTC system. They are determined at iteration i, $i = 1, 2, \ldots, j$ with the integer $j \geq 1$.

In correspondence to (6.92), at iteration j, the following matrices are defined:

$$R_{\text{known}}^{(j-1)} = \left(\bar{V}_0 + \sum_{i=1}^{j-1} \Delta\bar{V}^{(i)}\right)^{-1}, \ \Delta R^{(j)} = \left(\Delta\bar{V}^{(j)}\right)^{-1} \tag{6.93}$$

where $R_{\text{known}}^{(j-1)}$ is determined at iteration $j-1$ and treated known at iteration j, and the matrix $\Delta R^{(j)}$ is the decision variable at iteration j.

Under the above definitions, at iteration j, the inequality constraint (6.87) becomes (6.97). All the other constraints of the optimization problem (6.68) and those of the optimization problem (6.85) can be equivalently represented with the superscript j and used in the iterative procedure, as described in Algorithm 6.1.

To start the iteration, the initial gains K_0 and W_0 are solved offline from the optimization problem (6.68) with a given positive definite matrix V_0.

The stopping criterion of the iterative algorithm is defined as

$$\sigma^{(j)} = \frac{|J_c^{(j)} - J_c^{(j-1)}|}{J_c^{(j-1)}} < \epsilon \tag{6.94}$$

where $\sigma^{(j)}$ represents the relative change of the FTC cost function $J_c^{(j)} = \alpha_1\gamma_1^{(j)} + \alpha_2\text{trace}(S^{(j)})$ and ϵ is a prescribed positive scalar.

6.6.4 Convergence Analysis

This subsection provides convergence analysis of the Algorithm 6.1. It starts with the analysis of the optimization problem $\mathcal{P}_2$ given below.

Proposition 6.2 *The solution to the optimization problem $\mathcal{P}_2$ at iteration j is always one feasible solution to this problem at iteration $j-1$.*

Proof By using Schur Complement, the two LMI constraints of the optimization problem $\mathcal{P}_2$ can be compactly represented as

$$\mathcal{L}_{c1}(P^{(j)}, X^{(j)}, \gamma_1^{(j)}) \prec 0 \tag{6.103}$$

$$\mathcal{L}_{c2}(P^{(j)}, X^{(j)}, S^{(j)}) + BR_{\text{known}}^{(j)}B^\top \prec 0. \tag{6.104}$$

Since B has full column rank (see Assumption 6.1), the inequality (6.104) is equivalent to

$$B^\dagger\mathcal{L}_{c2}(P^{(j)}, X^{(j)}, S^{(j)})(B^\dagger)^\top + R_{\text{known}}^{(j)} \prec 0. \tag{6.105}$$

By construction, $0 \prec R_{\text{known}}^{(j)} \prec R_{\text{known}}^{(j-1)}$. Hence, by induction, the solution to problem $\mathcal{P}_2$ at iteration j is always one feasible solution to this problem at iteration $j-1$, by setting $X^{(j)} = X^{(j-1)}$, $P^{(j)} = P^{(j-1)}$, $S^{(j)} = S^{(j-1)}$ and $\gamma_1^{(j)} = \gamma_1^{(j-1)}$. □

Based on Proposition 6.2, the following statement is given.

Proposition 6.3 *The FTC cost function sequence $\{J_c^{(j)}\}_{j=0}^\infty$ converges to a (local) minimum J_c^*.*

Proof By construction, Algorithm 6.1 creates a series of positive scalars $J_c^{(j)}$. Moreover, according to Proposition 6.2, it holds that $J_c^{(j+1)} \leq J_c^{(j)}$. Therefore, the

Algorithm 6.1 Iterative algorithm for robust decoupling integration of FE and FTC

Input: $A, B, F, D, C, \mathcal{M}, \mathcal{N}, \bar{A}, \bar{C}, \bar{D}, C_e, C_x, K_0, W_0, V_0, \gamma_{1,0}, \epsilon, \epsilon_0, \alpha_1, \alpha_2, \beta_1, \beta_2$.

Initialization: Set $j = 1$, $K^{(0)} = K_0$, $S^{(0)} = (W_0^\top W_0)^{-1}$, $\gamma_1^{(0)} = \gamma_{1,0}$, $J_c^{(0)} = \alpha_1\gamma_1^{(0)} + \alpha_2\text{trace}(S^{(0)})$, $\bar{V}^{(0)} = V_0^\top V_0$, $R_{\text{known}}^{(0)} = (V_0^\top V_0)^{-1}$.

while ($j \geq 1$) **do**

Step 1: FE observer design

(1) Calculate $\delta = \sqrt{\lambda_{\min}((S^{(j-1)})^{-1})} - \epsilon_0$, $\hat{\mathcal{M}} = \delta^{-1}\mathcal{M}$, $\hat{\mathcal{N}} = \delta\mathcal{N}$, $\hat{\hat{\mathcal{M}}} = [\hat{\mathcal{M}}^\top\ 0]^\top$, and $B_e = [-K^{(j-1)}\ B^\dagger F\ 0]$.

(2) Solve the optimization problem $\mathcal{P}_1$:

$$\min_{Y^{(j)},\Delta R^{(j)},Q^{(j)},\gamma_2^{(j)}} \quad \beta_1\gamma_2^{(j)} + \beta_2\text{trace}(\Delta R^{(j)}) \tag{6.95}$$

$$\text{s.t.} \quad \begin{bmatrix} \text{He}(Q^{(j)}\bar{A} - Y^{(j)}\bar{C}) & Q^{(j)}\bar{D} & C_e^\top \\ \star & -\gamma_2^{(j)}I & 0 \\ \star & \star & -\gamma_2^{(j)}I \end{bmatrix} \prec 0 \tag{6.96}$$

$$\begin{bmatrix} \text{He}(Q^{(j)}\bar{A} - Y^{(j)}\bar{C}) & Q^{(j)}\hat{\hat{\mathcal{M}}} & B_e^\top & B_e^\top & 0 \\ \star & -I & 0 & 0 & 0 \\ \star & \star & -R_{\text{known}}^{(j-1)} & 0 & 0 \\ \star & \star & \star & -\Delta R^{(j)} & 0 \\ \star & \star & \star & \star & \hat{\mathcal{N}}^\top\hat{\mathcal{N}} - (S^{(j-1)})^{-1} \end{bmatrix} \prec 0 \tag{6.97}$$

$$Q^{(j)} = (Q^{(j)})^\top \succ 0,\ \Delta R^{(j)} = (\Delta R^{(j)})^\top \succ 0,\ \gamma_2^{(j)} > 0. \tag{6.98}$$

(3) Calculate $L^{(j)} = (Q^{(j)})^{-1}Y^{(j)}$, $\Delta\bar{V}^{(j)} = (\Delta R^{(j)})^{-1}$ and $R_{\text{known}}^{(j)} = (\bar{V}^{(j-1)} + \Delta\bar{V}^{(j)})^{-1}$.

Step 2: FTC controller design

(1) Solve the optimization problem $\mathcal{P}_2$:

$$\min_{X^{(j)},P^{(j)},S^{(j)},\gamma_1^{(j)}} \quad \alpha_1\gamma_1^{(j)} + \alpha_2\text{trace}(S^{(j)}) \tag{6.99}$$

$$\text{s.t.} \quad \begin{bmatrix} \text{He}(AP^{(j)} + BX^{(j)}) & D & \mathcal{M} & P^{(j)}\mathcal{N}^\top & P^{(j)}C_x^\top \\ \star & -\gamma_1^{(j)}I & 0 & 0 & 0 \\ \star & \star & -I & 0 & 0 \\ \star & \star & \star & -I & 0 \\ \star & \star & \star & \star & -\gamma_1^{(j)}I \end{bmatrix} \prec 0 \tag{6.100}$$

$$\begin{bmatrix} \text{He}(AP^{(j)} + BX^{(j)}) & B & \mathcal{M} & P^{(j)}\mathcal{N}^\top & P^{(j)} \\ \star & -(R_{\text{known}}^{(j)})^{-1} & 0 & 0 & 0 \\ \star & \star & -I & 0 & 0 \\ \star & \star & \star & -I & 0 \\ \star & \star & \star & \star & -S^{(j)} \end{bmatrix} \prec 0 \tag{6.101}$$

$$P^{(j)} = (P^{(j)})^\top \succ 0,\ S^{(j)} = (S^{(j)})^\top \succ 0,\ \gamma_1^{(j)} > 0. \tag{6.102}$$

(2) Calculate $K^{(j)} = X^{(j)}(P^{(j)})^{-1}$ and $W^{(j)} = \sqrt{(S^{(j)})^{-1}}$.

if $\sigma^{(j)} < \epsilon$ **then**

Set $j^* = j$ and stop.

else

Set $j = j + 1$.

end if

end while

Output: $j^*, \gamma_1^{(j^*)}, \gamma_2^{(j^*)}, K = K^{(j^*)}, L = L^{(j^*)}, W = W^{(j^*)}, V = \sqrt{\bar{V}_0 + \sum_{i=1}^{j^*}\Delta\bar{V}^{(i)}}$.

sequence $\{J_c^{(j)}\}_{j=0}^{\infty}$ is non-increasing and bounded below by zero. Define J_c^* as the greatest lower bound of the sequence, then it holds that

$$J_c^{(j)} \geq J_c^*, \ \forall j = 0, 1, 2, \ldots, \infty.$$

However, for every $\varepsilon > 0$, there is an integer N such that $J_c^{(N)} < J_c^* + \varepsilon$, otherwise J_c^* is not the greatest lower bound of $\{J_c^{(j)}\}_{j=0}^{\infty}$. Since $\{J_c^{(j)}\}_{j=0}^{\infty}$ is non-increasing, for all $j \geq N$, the following relation is true

$$J_c^* - \varepsilon \leq J_c^{(j)} \leq J_c^* + \varepsilon.$$

This means that $\{J_c^{(j)}\}_{j=0}^{\infty}$ converges to J_c^* and it is a Cauchy sequence (Rudin 1964). Therefore, for any $\varepsilon > 0$, there is an integer N_0 such that $|J_c^{(j+1)} - J_c^{(j)}| < \varepsilon, \ \forall j > N_0$. This implies that Algorithm 6.1 will terminate in finite iterations and an arbitrarily close approximation to the true local minimum J_c^* is found. □

The above results imply that the sequence $\{J_c^{(j)}\}_{j=0}^{\infty}$ will converge in finite iterations given feasible initial FTC design matrices K_0 and W_0. However, it generally converges to local minima because of the nonlinear nature of iteration. Therefore, the sequence $\{\sigma^{(j)}\}_{j=1}^{\infty}$ defined in (6.94) also converges to local minima.

The convergence of the optimization problem $\mathcal{P}_1$ is also analysed below. By using Schur Complement, the two LMI constraints of the optimization problem $\mathcal{P}_1$ can be compactly represented as

$$\mathcal{L}_{o1}(Q^{(j)}, Y^{(j)}, \gamma_2^{(j)}) \prec 0 \tag{6.106}$$

$$\mathcal{L}_{o2}(Q^{(j)}, Y^{(j)}, K^{(j-1)}, R_{\text{known}}^{(j-1)}, \Delta R^{(j)}) \prec 0 \tag{6.107}$$

$$\hat{\mathcal{N}}^{\top} \hat{\mathcal{N}} - (S^{(j-1)})^{-1} \prec 0 \tag{6.108}$$

where (6.106) is equivalent to (6.96), and the combination of (6.107) and (6.108) is equivalent to (6.97).

By setting $Q^{(j)} = Q^{(j-1)}, Y^{(j)} = Y^{(j-1)}, \gamma_2^{(j)} = \gamma_2^{(j-1)}$, and $\Delta R^{(j)} = 0$, the solution to (6.106) at iteration j is always one feasible solution to it at iteration $j - 1$. However, the same result does not apply for (6.107) and (6.108). As seen from (6.97), the inequality (6.107) includes a nonlinear term of the controller gain $K^{(j-1)}$ (appears at B_e). Hence, the solution to (6.107) at iteration j is not necessarily a solution to it at iteration $j - 1$. By using the definition of $\hat{\mathcal{N}}$ in (6.84), the inequality (6.108) can be reformulated as

$$\mathcal{N}^{\top} \mathcal{N} \prec (S^{(j-1)})^{-1} / \delta^2 \tag{6.109}$$

where $\delta = \sqrt{\lambda_{\min}((S^{(j-1)})^{-1})} - \epsilon_0$. Since the right-hand side term $(S^{(j-1)})^{-1}/\delta^2$ is not necessarily decreasing, the solution to (6.109) at iteration j is not necessarily a solution to it at iteration $j - 1$.

The above analysis implies that convergence of the optimization problem $\mathcal{P}_1$ cannot be guaranteed, neither do that of the FE cost function and γ_2.

Summarizing the above analysis, the Algorithm 6.1 will terminate in finite iterations by using the stopping criterion defined in (6.94) given feasible initial FTC design matrices K_0 and W_0. The iteration can improve the robustness of the FE observer against the FTC system perturbation by gradually increasing the weight $\bar{V}$ and equivalently V. Since the FTC cost function J_c converges, robustness of the FTC system against external disturbance and estimation error is also gradually improved. By specifying appropriate weights α_1 and α_2, a balance between the robustness against external disturbance and estimation error is also realized.

Remark 6.4 In the special case when there is no external disturbance, the FTC cost function $J_c^{(j)}$ merely depends on trace$(S^{(j)})$. In such case, convergence of $J_c^{(j)}$ means convergence of trace$(S^{(j)})$, which leads to increase in the weight W. Therefore, the robustness of the FTC system against the estimation error and vice versa can both be gradually enhanced by using the proposed iterative algorithm.

6.7 Tutorial Example

Consider an aircraft system modified from the example in Sect. 5.7.1 of the book Edwards and Spurgeon (1998). The system is given in the form of (6.1) with

$$A = \begin{bmatrix} 0 & 0 & 1 & 0 & 0 \\ 0 & -0.154 & -0.0042 & 1.54 & 0 \\ 0 & 0.249 & -1 & -5.2 & 0 \\ 0.0386 & -0.996 & -0.0003 & -0.117 & 0 \\ 0 & 0.5 & 0 & 0 & -0.5 \end{bmatrix}, \quad B = \begin{bmatrix} 0 & 0 \\ -3.72 & -0.16 \\ 1.685 & -5.6 \\ 0.1 & 0 \\ 0 & 0 \end{bmatrix},$$

$$F = \begin{bmatrix} 0 \\ -3.72 \\ 1.685 \\ 0.1 \\ 0 \end{bmatrix}, \quad D = \begin{bmatrix} 0 & 0 \\ 1 & 1 \\ 0 & 1 \\ 1 & 0 \\ 0 & 0 \end{bmatrix}, \quad C = \begin{bmatrix} 0 & 1 & 0 & 0 & 0 \\ 0 & 0 & 0 & 1 & 1 \\ 1 & 1 & 1 & 0 & 0 \end{bmatrix},$$

$$\Delta A = 0.05 \sin(0.1\pi t) \times A_p,$$

$$A_p = \begin{bmatrix} 0 & 0 & 0 & 0 & 0 \\ 0 & -0.154 & -0.0042 & 1.54 & 0 \\ 0 & 0.249 & -1 & -5.2 & 0 \\ 0.0386 & -0.996 & -0.0003 & -0.117 & 0 \\ 0 & 0.5 & 0 & 0 & -0.5 \end{bmatrix}.$$

It can be verified that the above system satisfies Assumptions 6.1–6.3. The matrices $\mathcal{M}$, $\mathcal{F}(t)$ and $\mathcal{N}$ are given as

$$\mathcal{M} = \frac{10}{9} \times A_p,\ \mathcal{F}(t) = \sin(0.1\pi t) \times I_5,\ \mathcal{N} = \frac{9}{10} I_5.$$

The disturbance vector is represented by $d(t) = [d_1(t)\ d_2(t)]^\top$, where $d_1(t)$ is a normally distributed random signal taking value within the interval $[-0.05, 0.05]$ and $d_2(t) = 0.01\cos(0.5t)$. The actuator fault f is characterized by

$$f(t) = \begin{cases} 0, & 0\text{ s} \leq t \leq 10\text{ s} \\ 1, & 10\text{ s} < t \leq 20\text{ s} \\ 2, & 20\text{ s} < t \leq 30\text{ s} \\ 1, & 30\text{ s} < t \leq 40\text{ s} \\ 0, & 40\text{ s} < t \leq 50\text{ s} \end{cases}.$$

In order to ease the iteration, the optimization problem in Theorem 6.1 is solved by replacing $\min(\alpha_1\gamma_1 + \alpha_2\text{trace}(S))$ with $\text{Find}(\gamma_1, S)$. Without loss of generality, in this example, the value $V_0 = I_2$ is used. The following initial gains are obtained:

$$\begin{aligned} \gamma_{1,0} &= 5.2211, \\ K_0 &= \begin{bmatrix} -6.0102 & 3.2491 & -5.9696 & -3.6257 & 0.0442 \\ 3.2655 & 3.3046 & 3.0024 & -4.3765 & 0.1594 \end{bmatrix}, \end{aligned}$$

$$W_0 = \begin{bmatrix} 0.4618 & 0.0518i & 0.1592 & 0.0436i & 0.0367 \\ 0.0518i & 0.4605 & 0.0535 & 0.1178i & 0.0488 \\ 0.1592 & 0.0535 & 0.4507 & 0.0498 & 0.0282i \\ 0.0436i & 0.1178i & 0.0498 & 0.4736 & 0.0146i \\ 0.0367 & 0.0488i & 0.0282i & 0.0146 & 0.4537 \end{bmatrix}.$$

The Algorithm 6.1 is run with the following parameters:

$$\begin{aligned} &\alpha_1 = 100,\ \alpha_2 = 10,\ \beta_1 = 50, \\ &\beta_2 = 1,\ \epsilon = 1.0 \times 10^{-5},\ \epsilon_0 = \text{eps}. \end{aligned}$$

The scalar eps $= 2.2204 \times 10^{-16}$ is defined in MATLAB as the distance from 1.0 to the next larger double-precision number.

The algorithm terminates in 5 steps with the trajectories of σ, γ_1, $\text{trace}(W^\top W)$, γ_2, and $\text{trace}(V^\top V)$ depicted in Figs. 6.4, 6.5, 6.6, 6.7 and 6.8. It can be seen from Fig. 6.4 that the value of σ monotonically decreases and converges to the local minimum 1.5884×10^{-6}. As shown in Fig. 6.5, the value of γ_1 decreases to 2.5249, which is much smaller than the initial value 5.2211. As expected, the value of γ_2 does not necessarily monotonically decrease and converge, which can be observed from Fig. 6.7. The values of $\text{trace}(W^\top W)$ and $\text{trace}(V^\top V)$ increase, as shown in Figs. 6.6 and 6.8, respectively. This implies that the coupling effects between FTC system and FE observer are gradually minimized.

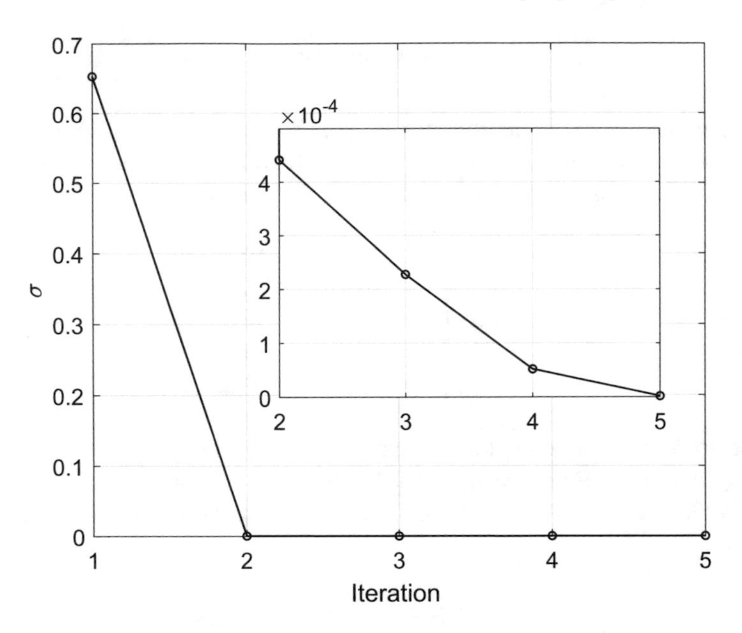

Fig. 6.4 Evolution of the relative change σ

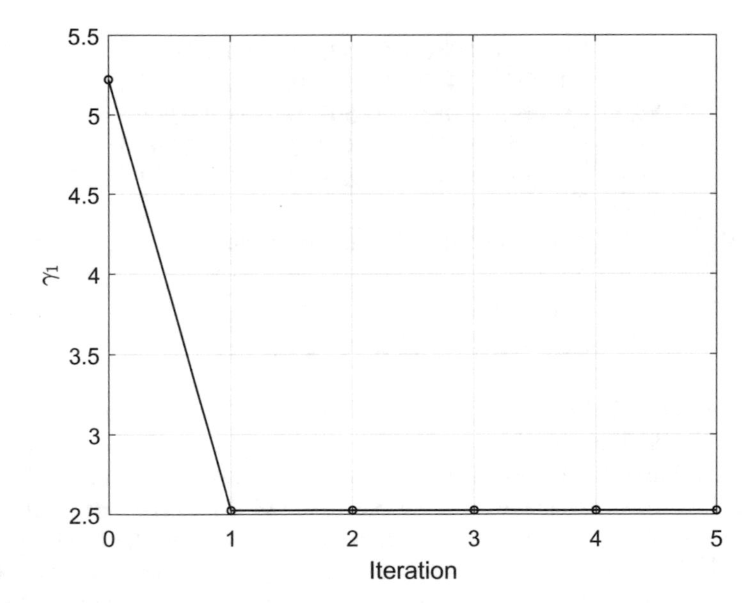

Fig. 6.5 Evolution of the gain γ_1

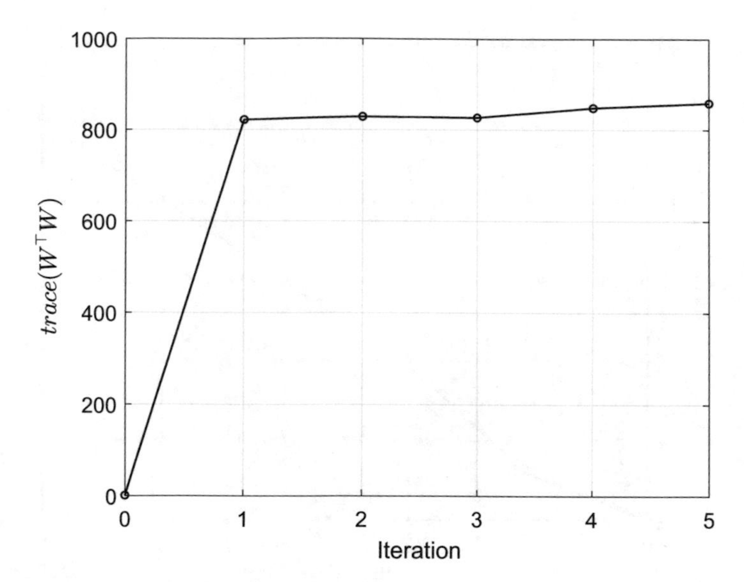

Fig. 6.6 Evolution of trace$(W^{\top}W)$

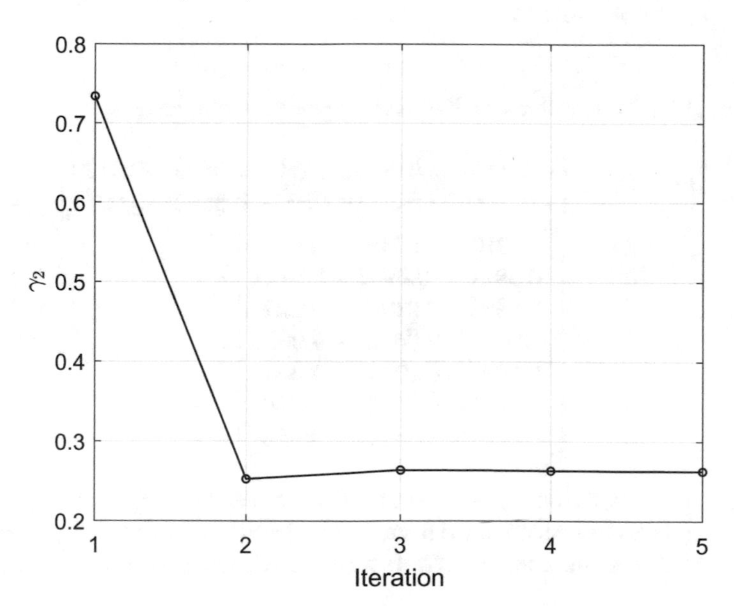

Fig. 6.7 Evolution of the gain γ_2

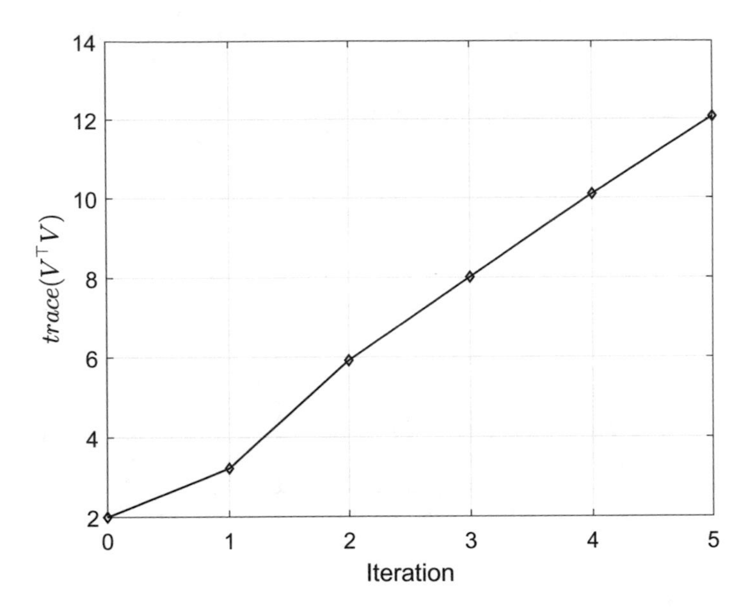

Fig. 6.8 Evolution of trace$(V^\top V)$

The final FTC controller and FE observer gains are obtained as

$$K = 10^3 \times \begin{bmatrix} -0.0958 & 0.2011 & -0.0054 & -1.0505 & -0.0129 \\ -0.0121 & 0.0642 & 0.0069 & -0.3342 & -0.0031 \end{bmatrix},$$

$$L = 10^3 \times \begin{bmatrix} -0.0107 & 0.0049 & 0.0070 \\ 0.3921 & -0.0901 & -0.2161 \\ 0.0490 & 0.0178 & 0.0281 \\ -0.0669 & 0.0618 & -0.0026 \\ 0.0097 & -0.0035 & -0.0045 \\ -0.7446 & 0.3182 & 0.4350 \\ -3.2234 & 1.3817 & 1.8958 \end{bmatrix}.$$

For comparison, the example system is also simulated by implementing the FTC controller and FE observer obtained from the separated integration strategy described in Sect. 2.4.2, where the coupling effects are ignored. The gains are obtained as

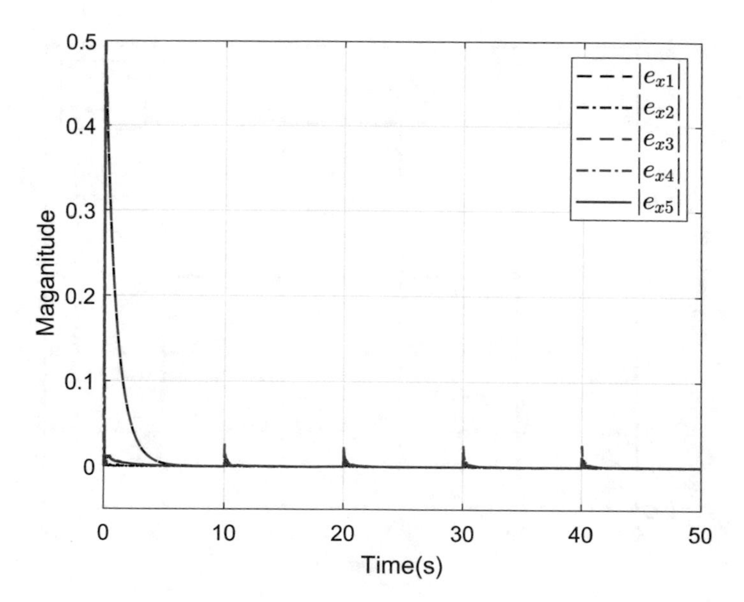

Fig. 6.9 State estimation performance: robust decoupling strategy

$$\gamma_c = 2.8255,$$
$$K = \begin{bmatrix} 26.0391 & 3.7336 & 18.5609 & -7.3562 & -0.3730 \\ 11.9322 & -10.7148 & 7.3144 & 30.0809 & 7.2964 \end{bmatrix},$$
$$\gamma_e = 0.6859,$$
$$L = \begin{bmatrix} -0.6835 & 0.0961 & 0.6881 \\ 21.3456 & -11.7039 & -7.8628 \\ -8.3541 & 1.2083 & 6.0522 \\ 21.5868 & -4.1322 & -11.6232 \\ 1.5118 & -1.6413 & -0.1572 \\ -19.3612 & 16.6467 & 8.0990 \\ -31.0858 & 27.5841 & 13.3844 \end{bmatrix}.$$

The systems under the proposed robust decoupling strategy and the separated strategy are simulated with initial conditions $x(0) = [0.5\ 0\ 0.5\ 0\ 0]^\top$ and $\hat{\bar{x}}(0) = 0_{7\times 1}$.

Compared the results in Figs. 6.9 and 6.10 with those in Figs. 6.12 and 6.13, it is seen that the proposed FE observer achieves much more robust and accurate estimation of state and fault than the separated strategy. It is observed from the closed-loop system state responses in Figs. 6.11 and 6.14 that the robust decoupling strategy achieves much more robust and stable state response than the separated strategy.

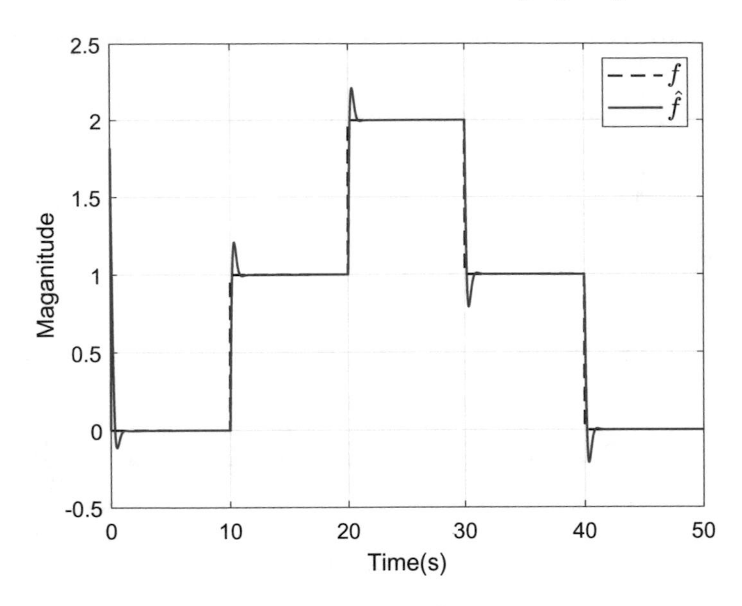

Fig. 6.10 Fault estimation performance: robust decoupling strategy

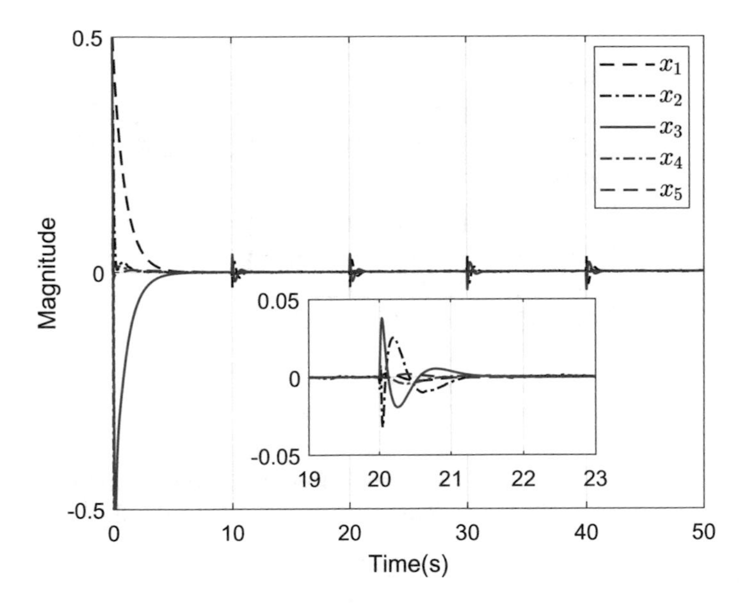

Fig. 6.11 State response: robust decoupling strategy

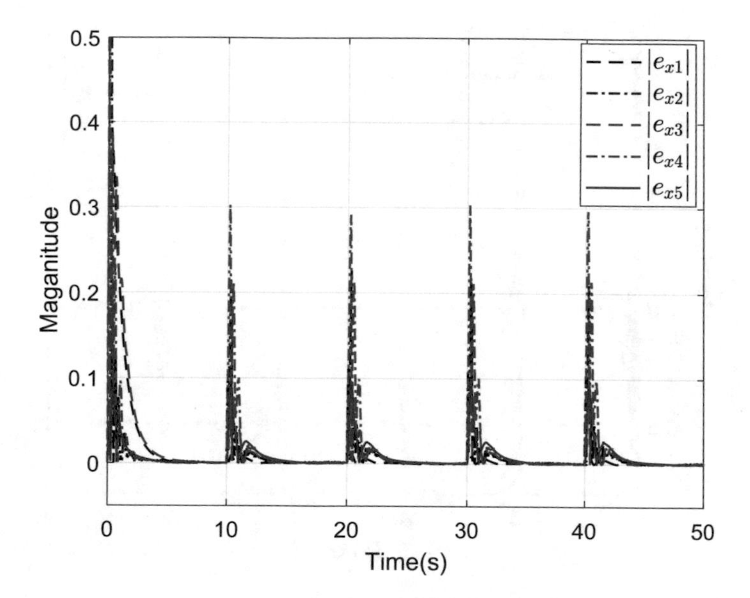

Fig. 6.12 State estimation performance: separated strategy

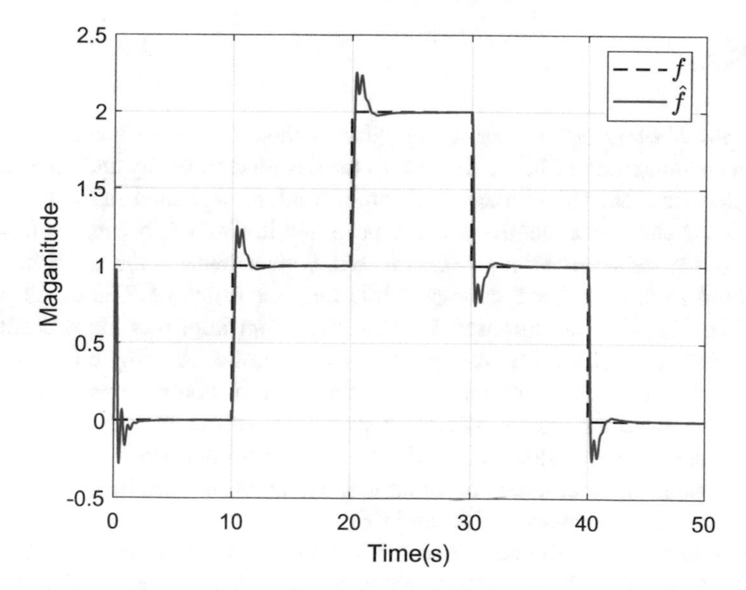

Fig. 6.13 Fault estimation performance: separated strategy

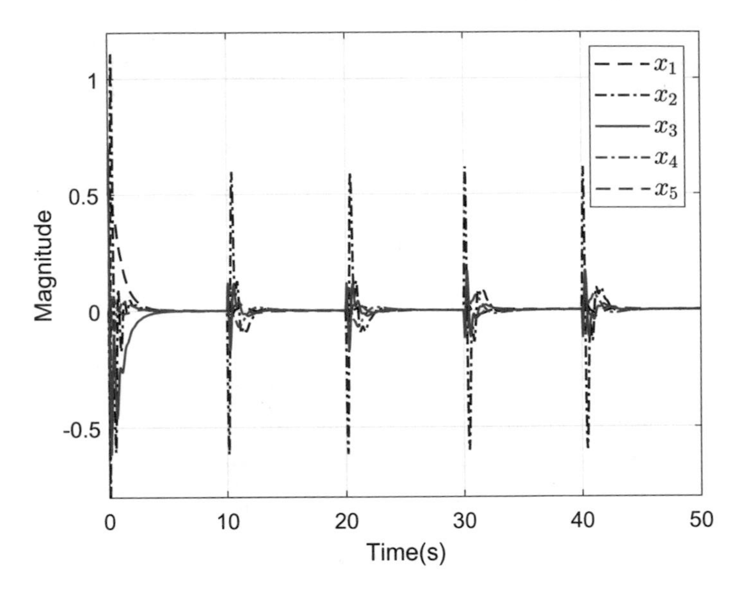

Fig. 6.14 State response: separated strategy

6.8 Notes

The proposed robust decoupling strategy avoids the difficulty of handling the BMI problem encountered in Chap. 5. It reduces the design complexity and facilitates the real implementation. This strategy is different from the separated strategy described in Sect. 2.4.2 and the sequential strategy proposed in Chap. 3, because it takes into account of the *bidirectional robustness interactions*. However, the coupling effects are ignored in the separated strategy, while only the effect of FTC on FE can be handled in Chap. 3. The proposed iterative algorithm improves the robustness of both the FTC system and FE observer by attenuating the coupling effects between them. This is realized via commutatively tuning the FTC controller and FE observer gains. In this sense, the proposed iterative algorithm is different from the one used in Chap. 4, where only robustness of the FTC controller is improved by tuning the FTC controller gain. Therefore, the presented iterative algorithm has better capability in balancing the performances of FTC and FE.

Robust observer-based control for discrete-time systems with uncertainties is studied in Peaucelle et al. (2017) by using a two-phase design strategy similar to the one described in Sects. 6.4 and 6.5. However, faults and external disturbances are not considered in their work. Moreover, the observer and controller designs are carried out following a heuristic procedure and no iterative algorithm is used.

References

Edwards C, Spurgeon S (1998) Sliding mode control: theory and applications. CRC Press, Boca Raton

Glad T, Ljung L (2000) Control theory. CRC Press, Boca Raton

Isidori A (2017) Lectures in feedback design for multivariable systems, vol 3. Springer, Berlin

Lan J, Patton RJ (2016) Integrated design of fault-tolerant control for nonlinear systems based on fault estimation and T-S fuzzy modeling. IEEE Trans Fuzzy Syst 25(5):1141–1154

Peaucelle D, Ebihara Y, Hosoe Y (2017) Robust observed-state feedback design for discrete-time systems rational in the uncertainties. Automatica 76:96–102

Rudin W (1964) Principles of mathematical analysis, vol 3. McGraw-hill, New York

Zames G (1966) On the input-output stability of time-varying nonlinear feedback systems part one: Conditions derived using concepts of loop gain, conicity, and positivity. IEEE Trans Autom Control 11(2):228–238

Chapter 7
Adaptive Decoupling Integration of FE and FTC

7.1 Introduction

It has been shown in Chaps. 3–6 that the robust integration of FE and FTC can be cast as a robust observer-based control problem. The solution to this problem can be obtained by solving H_∞ optimization problems with BMI constraints. To circumvent the BMI problem, sequential and iterative strategies are proposed in Chaps. 3 and 4, respectively. However, they can only obtain a suboptimal solution to the overall system design. A simultaneous strategy is proposed in Chap. 5 aiming to obtain an optimal solution in one shot. However, it uses an equality constraint to linearize the BMI, which imposes conservativeness. Moreover, the single-step LMI formulation in Chap. 5 leads to a loss of design freedom. A robust decoupling strategy is proposed in Chap. 6 to circumvent the BMI problem and have more design freedom. It approximately recovers the Separation Principle through minimizing the coupling effects between the FTC system and FE observer, based on the Small Gain Theorem (Zames 1966). The robust decoupling strategy can ensure closed-loop system stability by assembling the separately designed FTC controller and FE observer. However, the coupling effects are "passively" attenuated using H_∞ control theory. All the above strategies are based on H_∞ control theory to attenuate the perturbations (including external disturbance and/or system uncertainty), resulting in conservative designs.

This chapter will develop an adaptive decoupling strategy to realize the robust integration of FE and FTC. The decoupling is realized via "actively" estimating and compensating the coupling effects based on the use of an adaptive control technique and a new observer formulation. After decoupling, the FE observer and FTC controller can be designed independently, i.e. their designs are not dependent on each other. This is different from the strategies in Chaps. 3–6, where the FE observer design depends on the FTC controller gain. Moreover, the adaptive decoupling strategy is not based on H_∞ control theory and thus avoids the design conservativeness of Chap. 3–6.

© The Author(s), under exclusive license to Springer Nature Switzerland AG 2021

J. Lan and R. J. Patton, *Robust Integration of Model-Based Fault Estimation and Fault-Tolerant Control*, Advances in Industrial Control,
https://doi.org/10.1007/978-3-030-58760-4_7

The FE design is effective unless there is a guarantee of good robustness against system perturbations that act on the state dynamics and/or output measurements. The following two approaches can be employed to improve the FE robustness:

- *Attenuation approach.* It focuses on suppressing the perturbation effects on the estimation errors. This is a well-known way to enhance robustness and is realized using techniques such as H_∞ optimization, as described in Chaps. 3–6.
- *Partial decoupling approach.* It aims to eliminate from estimation error dynamics the perturbation whose distribution matrix satisfies the matching condition $\text{rank}(CD) = \text{rank}(D)$ (see Sect. 1.7.3), where C and D are the distribution matrices of output and perturbation, respectively. In Gao et al. (2016) the state perturbations satisfying the matching condition are decoupled through observer gain design. Sliding mode observer (SMO) also has a similar decoupling function because they are insensitive to perturbations matched to the control input (Edwards et al. 2000). However, the decoupling approach has limited applicability because in practice usually only partial perturbations can meet the matching condition. For the unmatched part, H_∞ optimization is incorporated to attenuate their effects (Gao et al. 2016).
  Moreover, the faults considered in Chaps. 3–6 and most existing literature are assumed to be continuously differentiable and matched with respect to the control input. This limits the applicability of the FE-based FTC designs. Therefore, this chapter will design a new FE observer with enhanced robustness and the capability of estimating a wider class of faults.
  The above background inspires the work of this chapter with a focus on developing a new FE observer and an adaptive decoupling integration strategy. The proposed strategy has the following salient features:
- *The proposed FE observer and FTC controller can estimate and compensate a more general class of faults.* This chapter considers a more general class of actuator faults, which can be (1) differentiable or non-differentiable and (2) matched or unmatched. An adaptive sliding mode augmented state unknown input observer (ASUIO) is proposed to estimate the system state, actuator fault and perturbation without using H_∞ control theory. The faults, either matched or unmatched, can be fully compensated by the adaptive backstepping FTC controller developed in this chapter.
- *The decoupling integration offers more design freedom.* By using a descriptor approach, the system perturbation is augmented as auxiliary state and estimated, which ensures that the proposed observer is unaffected by FTC system perturbations. Moreover, with an appropriately designed switching function, the effect of actuator fault on the estimation error dynamics is removed. By combining the above techniques, the FE observer is decoupled from the FTC system, which recovers the Separation Principle (see Sect. 1.7.2) and allows more freedom for FE-based FTC design. It should be noted that the proposed decoupling strategy is different from the separated strategy described in Sect. 2.4.2, since it takes into account of the *bidirectional robustness interactions*.

- *Active perturbation cancellation contributes to a more robust FTC system.* Instead of suppressing the perturbation and estimation error as in Chap. 6, an adaptive backstepping FTC controller is designed in this chapter to actively compensate them. This results in a more robust FTC system. Moreover, the use of an adaptive control technique makes the proposed design an "active" strategy, in contrast to the "passive" strategy in Chap. 6.

7.2 Problem Description

Consider a class of linear systems described by

$$\begin{aligned} \dot{x} &= Ax + Bu + Ff + D_1 d \\ y &= Cx + D_2 d \end{aligned} \tag{7.1}$$

where $x \in \mathbb{R}^n$, $u \in \mathbb{R}^m$ and $y \in \mathbb{R}^p$ are the state, control input and measured output, respectively. $f \in \mathbb{R}^l$ is the actuator fault vector. $d \in \mathbb{R}^q$ denotes the perturbation including external disturbance and/or system uncertainty. The constant matrices A, B, F, D_1, C and D_2 are known and of compatible dimensions. The system satisfies the following assumptions:

Assumption 7.1 The pair (A, B) is controllable. None of the invariant zeros of $(A, [F\ D_1], C, [0\ D_2])$ are in the closed right-half complex plane.

Assumption 7.2 There exists an unknown positive constant f_0 such that $\|f\| \leq f_0$. The perturbation d is norm-bounded with bounded first-order time derivative.

It is rational to assume that the perturbation d, including system uncertainty and/or external disturbance, is differentiable, because: (1) The system uncertainty is a function of state variables and continuously differentiable; (2) According to the output regulation theory (Isidori 1995), a differentiable exogenous system can be used to represent many types of disturbances in engineering, e.g. constant and harmonics. Although normally the distribution matrix D_1 of the perturbation cannot be obtained directly, an approximate modelling of it can be determined through several ways described in Chen and Patton (1999).

This chapter aims to propose an FE-based FTC design for the system (7.1) such that the output can track its reference in the presence of actuator faults and perturbation. The FTC system design includes (1) an adaptive sliding mode ASUIO for estimating the system state, fault and perturbation and (2) an adaptive backstepping FTC controller for compensating the fault and perturbation to achieve satisfactory output tracking performance.

A decoupling strategy will be used to achieve robust integration of the observer and controller. Specifically, the adaptive sliding mode ASUIO will be designed to decouple from the control system and recover the Separation Principle, which then enables separated design of estimation and control.

7.3 Principle of Adaptive Decoupling

To illustrate the basic idea of adaptive decoupling, the ASUIO FE observer proposed in Chaps. 4 and 5 is used to estimate the state and fault of the system (7.1).

Define f as auxiliary system state, then the system (7.1) can be augmented as

$$\begin{aligned} \dot{\bar{x}}_o &= \bar{A}_o\bar{x}_o + \bar{B}_o u + \bar{D}_1\bar{d} \\ y &= \bar{C}_o\bar{x}_o + D_2 d \end{aligned} \tag{7.2}$$

where

$$\bar{x}_o = \begin{bmatrix} x \\ f \end{bmatrix},\ \bar{d} = \begin{bmatrix} d \\ \dot{f} \end{bmatrix},\ \bar{A}_o = \begin{bmatrix} A & F \\ 0 & 0 \end{bmatrix},\ \bar{B}_o = \begin{bmatrix} B \\ 0 \end{bmatrix},\ \bar{D}_1 = \begin{bmatrix} D_1 & 0 \\ 0 & I_l \end{bmatrix},\ \bar{C}_o = [C\ 0].$$

The augmented state $\bar{x}_o$ is estimated by the following ASUIO:

$$\begin{aligned} \dot{\xi}_o &= M_o\xi_o + G_o u + L_o y \\ \hat{\bar{x}}_o &= \xi_o + H_o y \end{aligned} \tag{7.3}$$

where ξ_o, $\hat{\bar{x}}_o \in \mathbb{R}^{n+l}$ are the observer state and estimate of $\bar{x}_o$, respectively. The constant matrix M_o, G_o, L_o and H_o are to be determined.

By defining the estimation error as $e_o = \bar{x}_o - \hat{\bar{x}}_o$, then the error dynamics are obtained as

$$\dot{e}_o = \Xi_2 e_o + \Xi_3\xi_o + \Xi_4 u + \Xi_5\bar{C}_o\bar{x}_o + \chi_o \tag{7.4}$$

where

$$\begin{aligned} &\Xi_1 = I_{n+l} - H_o\bar{C}_o,\ L_o = L_{o1} + L_{o2},\ \Xi_2 = \Xi_1\bar{A}_o - L_{o1}\bar{C}, \\ &\Xi_3 = \Xi_1\bar{A}_o - L_{o1}\bar{C}_o - M_o,\ \Xi_4 = \Xi_1\bar{B}_o - G_o, \\ &\Xi_5 = (\Xi_1\bar{A}_o - L_{o1}\bar{C}_o)H_o - L_{o2},\ \chi_o = \Xi_1\bar{D}_1\bar{d} - L_o D_2 d - H_o D_2\dot{d}. \end{aligned}$$

By designing $\Xi_i = 0$, $i = 3, 4, 5$, the error dynamics (7.4) becomes

$$\dot{e}_o = \Xi_2 e_o + \chi_o. \tag{7.5}$$

The ASUIO (7.3) is restrictive in two aspects: (1) The actuator fault f is required to be differentiable so that it can be augmented as a new system state. (2) The error dynamics (7.5) are perturbed by the uncertain term χ_0 that is a function of the system perturbations (d and $\dot{d}$) and fault modelling errors $\dot{f}$. The estimation errors in turn affect the FTC system performance because the controller uses the state and fault estimates. Therefore, there exist *bidirectional robustness interactions* between the FE observer (7.3) and the FTC system. In other words, the FE and FTC functions

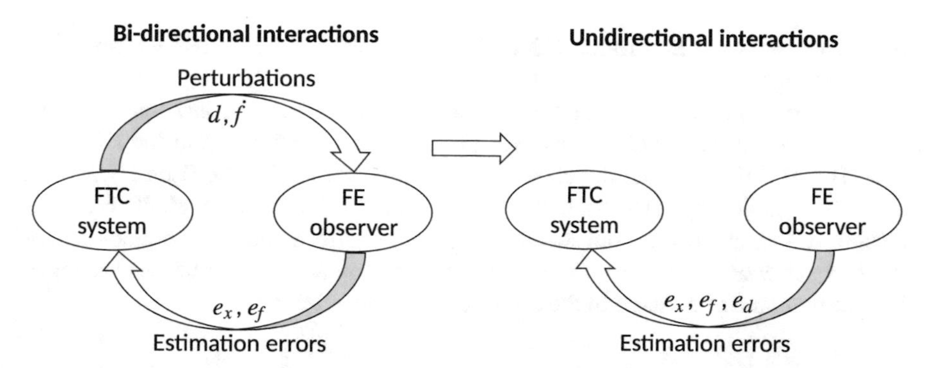

Fig. 7.1 Interactions within (**a**) simultaneous and (**b**) adaptive decoupling integration

are coupled with each other, as shown in Fig. 7.1a. The above results also apply for the FE-based FTC designs in Chaps. 3 and 6.

The second aspect implies that if there is neither system perturbations nor fault modelling errors acting on the error dynamics (7.5), then the estimation performance is not affected by the system uncertainty and the FE observer is decoupled from the control system. Therefore, one way to achieve the decoupling is to design an observer with error dynamics free from the perturbations (d and $\dot{d}$) and fault modelling error $\dot{f}$, using a combination of the following two methods:

- *Descriptor augmentation*. The system (7.1) can be augmented into a descriptor form (7.6) in Sect. 7.4 with d as auxiliary system state. In the augmented system, the only unknown input signal is the actuator fault f. Therefore, there will be no perturbation acting on the state estimation error dynamics, as shown in (7.13).
- *SMO*. It can be seen from the ASUIO (7.3) that modelling f as the auxiliary system state requires the differentiability assumption and leads to existence of the fault modelling errors $\dot{f}$. They can be removed by the SMO method (Edwards et al. 2000) for FE, where the actuator fault is reconstructed through the equivalent output injection signal corresponding to an ideal sliding motion, without a need of modelling the fault as the auxiliary system state.

Such an SMO exists provided that (1) no system perturbation in the state estimation error dynamics and (2) the fault appearing in the error dynamics is matched with respect to the switching function. The former is fulfilled by the descriptor augmentation described above, while the latter can be met using appropriate matrix design (i.e. the design of T and W in the proposed observer (7.7)), details of which is given in Remark 7.1.

Therefore, the proposed adaptive sliding mode ASUIO (7.7) will combine the descriptor augmentation method with SMO. It can then remove the effects of perturbations and fault modelling errors. This decouples the FE observer from the FTC system and results in the existence of just a *unidirectional robustness interaction* (Fig. 7.1b), rather than the bidirectional interactions described in Chap. 4 (Fig. 7.1a).

It then recovers the Separation Principle for the proposed FE observer and FTC system designs.

It is worth mentioning that this chapter follows a new Separation Principle achieved by using a novel FE observer (7.7) that is decoupled from the FTC system. This is different from the classical Separation Principle (see Sect. 1.7.2) used extensively in the separated strategies (e.g. Jiang et al. 2006; Gao et al. 2016). Their designs cannot achieve overall robust FTC system performance due to ignorance of the *bidirectional robustness interactions*. In this chapter, the interactions are taken into account and eliminated in the observer and controller designs.

7.4 Adaptive Sliding Mode ASUIO Design: Special Case

Following the basic idea of adaptive decoupling described in Sect. 7.3, this section presents the observer design for the special case when the system (7.1) satisfies the extra assumption given below:

Assumption 7.3 $\mathrm{rank}(D_2) = q$.

The above assumption means that all the perturbations affect the measured output through different channels.

7.4.1 *Observer Design*

By defining the perturbation d as auxiliary state, then the system (7.1) can be reformulated into a descriptor form

$$\begin{aligned} E\dot{\bar{x}} &= \bar{A}\bar{x} + Bu + Ff \\ y &= \bar{C}\bar{x} \end{aligned} \tag{7.6}$$

with

$$\bar{x} = [x^\top \ d^\top]^\top, \ E = [I_n \ 0_{n\times q}], \ \bar{A} = [A \ D_1], \ \bar{C} = [C \ D_2].$$

It can be verified that the augmented system (7.6) is infinitely observable (Dai 1989) under Assumptions 7.1 and 7.3. Hence, there exists an asymptotic state observer for the augmented system. Considering this, the following observer is proposed to estimate the augmented state $\bar{x}$:

$$\begin{aligned} \dot{z} &= Nz + Ju + Ly + Wv \\ \hat{\bar{x}} &= z + Hy \\ \hat{y} &= \bar{C}\hat{\bar{x}} \end{aligned} \tag{7.7}$$

where $z \in \mathbb{R}^{n+q}$ is the observer state and $\hat{\bar{x}} \in \mathbb{R}^{n+q}$ is the estimate of $\bar{x}$. The matrices N, J, L, W and H are to be designed. The switching function v is defined as

$$v = \rho_v \text{sign}(e_y),$$

where $e_y = y - \hat{y}$ and ρ_v is a design scalar.

Define $\varepsilon = TE\bar{x} - z$ with a design matrix T. By using (7.6) and (7.7), it can be derived that

$$\begin{aligned}
\dot{\varepsilon} &= TE\dot{\bar{x}} - \dot{z} \\
&= T\bar{A}\bar{x} + TBu + TFf - Nz - Ju - Ly - Wv \\
&= T\bar{A}\bar{x} + TBu + TFf - N(TE\bar{x} - \varepsilon) - Ju - L\bar{C}\bar{x} - Wv \\
&= N\varepsilon + (T\bar{A} - NTE - L\bar{C})\bar{x} + (TB - J)u + TFf - Wv. \qquad (7.8)
\end{aligned}$$

Define the estimation error of $\bar{x}$ as $e = \bar{x} - \hat{\bar{x}}$. According to (7.8), the error system is derived as

$$\begin{aligned}
\dot{\varepsilon} &= N\varepsilon + (T\bar{A} - NTE - L\bar{C})\bar{x} + (TB - J)u + TFf - Wv \\
e &= \varepsilon + (I_{n+q} - H\bar{C} - TE)\bar{x}.
\end{aligned} \qquad (7.9)$$

Define the following matrix equations:

$$T\bar{A} - NTE - L\bar{C} = 0 \qquad (7.10)$$

$$TB - J = 0 \qquad (7.11)$$

$$I_{n+q} - H\bar{C} - TE = 0. \qquad (7.12)$$

By substituting the above equations into (7.9), it is derived that $e = \varepsilon$ and

$$\dot{e} = Ne + TFf - Wv. \qquad (7.13)$$

Remark 7.1 If the matrices T and W are designed such that $TF = W\overline{W}$, where $\overline{W}$ is some matrix with compatible dimension, then the error system (7.13) becomes

$$\dot{e} = Ne + W(\overline{W}f - v).$$

Since the fault function $\overline{W}f$ is matched with respect to the switching function v, its effect can be totally cancelled by an appropriately designed v in Sect. 7.4.2. In such case, the error system becomes

$$\dot{e} = Ne,$$

which is an autonomous system and is asymptotically stable by designing a Hurwitz stable N. Therefore, the observer (7.7) is decoupled from the FTC system.

7.4.2 Estimation Performance Analysis

This section provides performance analysis of the observer (7.7) with the main results stated below.

Theorem 7.1 *Under Assumptions 7.1–7.3, the observer (7.7) achieves asymptotic estimation of the augmented state $\bar{x}$ and actuator fault f, if there exists a s.p.d. matrix $P \in \mathbb{R}^{(n+q)\times(n+q)}$, a matrix $Q \in \mathbb{R}^{p\times l}$ and a positive constant ξ, such that*

$$PN + N^{\top}P \prec -\xi I_{n+q} \tag{7.14}$$

$$PTF = \bar{C}^{\top}Q. \tag{7.15}$$

Then the fault estimate is obtained as

$$\hat{f} = (Q^{\top}\bar{C}P^{-1}\bar{C}^{\top}Q)^{-1}Q^{\top}\bar{C}P^{-1}\bar{C}^{\top}v_{eq},$$

where v_{eq} is the equivalent signal of the switching function v.

Proof **(1) Augmented system state estimation**

To analyse the stability of the error system (7.13), the following Lyapunov function is used:

$$V_{e_0} = e^{\top}Pe.$$

The time derivative of V_{e_0} along the error system is obtained as

$$\dot{V}_{e_0} = e^{\top}(PN + N^{\top}P)e + 2(e^{\top}PTFf - e^{\top}PWv). \tag{7.16}$$

Design $W = P^{-1}\bar{C}^{\top}$ and $PTF = \bar{C}^{\top}Q$. Since $TF = WQ$, the matching condition described in Remark 7.1 is satisfied. By further using $e_y^{\top}v = \rho_v\|e_y\|$, then (7.16) becomes

$$\begin{aligned} \dot{V}_{e_0} &= e^{\top}(PN + N^{\top}P)e + 2(e_y^{\top}Qf - e_y^{\top}v) \\ &\leq e^{\top}(PN + N^{\top}P)e + 2\|e_y\|(\rho - \rho_v) \end{aligned} \tag{7.17}$$

where $\rho = \|Q\|f_0$.

In order to compensate the unknown scalar ρ, the parameter ρ_v is designed as $\rho_v = \hat{\rho} + \epsilon$, where ϵ is a positive constant and $\hat{\rho}$ is the estimate of ρ generated by

$$\dot{\hat{\rho}} = \sigma_0\|e_y\|,\ \hat{\rho}(0) = 0 \tag{7.18}$$

with a positive design constant σ_0.

Define the estimation error of ρ as $\tilde{\rho} = \rho - \hat{\rho}$. Consider the Lyapunov function

$$V_e = V_{e_0} + \frac{1}{\sigma_0}\tilde{\rho}^2.$$

By using (7.14), (7.17), (7.18) and the fact that $\dot{\rho} = 0$, it can be derived that

$$\begin{aligned}\dot{V}_e &= \dot{V}_{e_0} + \frac{2}{\sigma_0}\tilde{\rho}\left(-\sigma_0\|e_y\|\right) \\ &\leq e^\top(PN + N^\top P)e + 2\|e_y\|\left(\rho - \hat{\rho} - \epsilon - \tilde{\rho}\right) \\ &\leq -\xi\|e\| - 2\epsilon\|e_y\| \\ &\leq 0. \end{aligned} \tag{7.19}$$

It follows from (7.19) and the Barbalat's Lemma (see Sect. 1.7.1) that $\lim_{t\to\infty} V_e(t) = 0$. This implies that $V_e(t) \leq V_e(0)$ and the boundedness of e and $\tilde{\rho}$. It further implies that $|e(t)| \leq \sqrt{2V_e(0)}$. Under zero initial condition $V_e(0) = 0$ (i.e. $e(0) = \tilde{\rho}(0) = 0$), then it holds that $\lim_{t\to\infty} e(t) = 0$ and $\lim_{t\to\infty} \tilde{\rho}(t) = 0$. Therefore, under the zero initial condition, the sliding surface $Q^\top e_y = 0$ is reachable and the observer (7.7) estimates the augmented system state $\bar{x}$ asymptotically.

(2) Actuator fault estimation

By using $e_y = \bar{C}e$, $W = P^{-1}\bar{C}^\top$ and $TF = WQ$, it follows from (7.13) that

$$Q^\top \dot{e}_y = Q^\top \bar{C}Ne + Zf - Q^\top \bar{C}P^{-1}\bar{C}^\top v \tag{7.20}$$

where $Z = Q^\top \bar{C}P^{-1}\bar{C}^\top Q$.

Note that during the sliding motion, $\dot{e}_y = 0$ and thus $Q^\top \dot{e}_y = 0$. This implies that

$$0 = Q^\top \bar{C}Ne + Zf - Q^\top \bar{C}P^{-1}\bar{C}^\top v_{eq} \tag{7.21}$$

where v_{eq} is the equivalent control input signal representing the average behaviour of the switching function v and the effort necessary to maintain the sliding motion.

It can be derived from (7.21) that

$$f = Z^{-1}(-Q^\top \bar{C}Ne + Q^\top \bar{C}P^{-1}\bar{C}^\top v_{eq}). \tag{7.22}$$

Design the actuator fault estimate as

$$\hat{f} = Z^{-1}Q^\top \bar{C}P^{-1}\bar{C}^\top v_{eq}. \tag{7.23}$$

Define the fault estimation error as $e_f = f - \hat{f}$, then it is derived that

$$e_f = -Z^{-1}Q^\top \bar{C}Ne. \tag{7.24}$$

Since $\lim_{t\to\infty} e(t) = 0$, it holds that $\lim_{t\to\infty} e_f(t) = 0$. This means that accurate fault estimation is obtained asymptotically. □

Remark 7.2 In practice, the equivalent control input signal v_{eq} can be obtained by passing the switching function v through an appropriately designed low-pass filter $v_{eq} \cong \frac{1}{\tau s+1}v$ with a time constant τ. The error between v_{eq} and v can be made arbitrarily small by choosing a small enough τ, details of which are referred to Utkin (1992). The signal v_{eq} can also be obtained using a continuous approximation of v using the method in Edwards et al. (2000).

7.4.3 Observer Gain Parametrization

In Sects. 7.4.1–7.4.2, the FE observer (7.7) is described and its estimation performance is analysed. This section proposes a way to determine the observer gains using parametrization based on a theorem equivalent to Theorem 7.1.

The matrix equation (7.12) can be rearranged as

$$[T \ H]\begin{bmatrix} E \\ \bar{C} \end{bmatrix} = I_{n+q}. \tag{7.25}$$

Denote $\Omega_1 = \begin{bmatrix} E \\ \bar{C} \end{bmatrix}$ and $\Sigma_1 = I_{n+q}$. Since $\text{rank}(\Omega_1) = \text{rank}\begin{bmatrix} \Omega_1 \\ \Sigma_1 \end{bmatrix} = n + q$, the matrix equation (7.25) is solvable and its general solution is

$$[T \ H] = \Sigma_1\Omega_1^{\dagger} - Y_1(I_{n+p} - \Omega_1\Omega_1^{\dagger}),$$

where Y_1 is a design matrix with the dimension $(n + q) \times (n + p)$. Therefore, the matrices T and H can be parametrized as

$$T = T_1 - Y_1T_2, \ H = H_1 - Y_1H_2 \tag{7.26}$$

with

$$T_1 = \Sigma_1\Omega_1^{\dagger}\begin{bmatrix} I_n \\ 0 \end{bmatrix}, \ T_2 = (I_{n+p} - \Omega_1\Omega_1^{\dagger})\begin{bmatrix} I_n \\ 0 \end{bmatrix},$$
$$H_1 = \Sigma_1\Omega_1^{\dagger}\begin{bmatrix} 0 \\ I_p \end{bmatrix}, \ H_2 = (I_{n+p} - \Omega_1\Omega_1^{\dagger})\begin{bmatrix} 0 \\ I_p \end{bmatrix}.$$

From (7.12), one has $TE = I_{n+q} - H\bar{C}$. Substituting it into (7.10) gives

$$[N \ \bar{L}]\begin{bmatrix} I_{n+q} \\ \bar{C} \end{bmatrix} = T\bar{A} \tag{7.27}$$

where $\bar{L} = L - NH$.

Denote $\Omega_2 = \begin{bmatrix} I_{n+q} \\ \bar{C} \end{bmatrix}$ and $\Sigma_2 = T\bar{A}$. Since $\text{rank}(\Omega_2) = \text{rank}\begin{bmatrix} \Omega_2 \\ \Sigma_2 \end{bmatrix} = n + q$, the matrix equation (7.27) is solvable and its general solution is

$$[N\ \bar{L}] = \Sigma_2 \Omega_2^{\dagger} - Y_2(I_{n+q+p} - \Omega_2 \Omega_2^{\dagger}),$$

where Y_2 is a design matrix with the dimension $(n+q) \times (n+q+p)$. Therefore, the matrices N and $\bar{L}$ are parametrized as

$$N = N_1 - Y_2 N_2,\ \bar{L} = \bar{L}_1 - Y_2 \bar{L}_2 \tag{7.28}$$

with

$$N_1 = \Sigma_2 \Omega_2^{\dagger} \begin{bmatrix} I_{n+q} \\ 0 \end{bmatrix},\ N_2 = (I_{n+q+p} - \Omega_2 \Omega_2^{\dagger}) \begin{bmatrix} I_{n+q} \\ 0 \end{bmatrix},$$
$$\bar{L}_1 = \Sigma_2 \Omega_2^{\dagger} \begin{bmatrix} 0 \\ I_p \end{bmatrix},\ \bar{L}_2 = (I_{n+q+p} - \Omega_2 \Omega_2^{\dagger}) \begin{bmatrix} 0 \\ I_p \end{bmatrix}.$$

It can be seen that once the matrices Y_1 and Y_2 are determined, by using the parameterizations (7.26) and (7.28), the matrix equations (7.10)–(7.12) can be solved and all the observer gains can thus be obtained.

However, it is worth asking the question: Do such matrices Y_1 and Y_2 really exist? Their existence will be proved in Lemma 7.2 based on Lemma 7.1.

Lemma 7.1 *For all $s \in \mathbb{C}$, $\mathrm{Re}(s) \geq 0$, it holds that*

$$\mathrm{rank} \begin{bmatrix} sI_{n+q} & -\Omega_1^{\dagger} \\ 0 & I_{n+p} - \Omega_1 \Omega_1^{\dagger} \end{bmatrix} = n+q.$$

Proof Note that

$$\mathrm{rank} \begin{bmatrix} I_{n+q} & 0 \\ 0 & \Omega_1 \end{bmatrix} = \mathrm{rank} \begin{bmatrix} sI_{n+q} & 0 & I_{n+q} \\ 0 & I_{n+p} & \Omega_1 \end{bmatrix}. \tag{7.29}$$

The left-hand side of (7.29) is equivalent to

$$\mathrm{rank} \begin{bmatrix} I_{n+q} & 0 \\ 0 & \Omega_1 \end{bmatrix} = n+q+\mathrm{rank}(\Omega_1). \tag{7.30}$$

The right-hand side of (7.29) is equivalent to

$$\begin{aligned} &\mathrm{rank} \begin{bmatrix} sI_{n+q} & 0 & I_{n+q} \\ 0 & I_{n+p} & \Omega_1 \end{bmatrix} \\ &= \mathrm{rank} \left\{ \begin{bmatrix} I_{n+q} & -\Omega_1^{\dagger} \\ 0 & I_{n+p} - \Omega_1 \Omega_1^{\dagger} \\ 0 & \Omega_1 \Omega_1^{\dagger} \end{bmatrix} \begin{bmatrix} sI_{n+q} & 0 & I_{n+q} \\ 0 & I_{n+p} & \Omega_1 \end{bmatrix} \right\} \end{aligned}$$

$$= \text{rank} \begin{bmatrix} sI_{n+q} & -\Omega_1^\dagger & 0 \\ 0 & I_{n+p} - \Omega_1\Omega_1^\dagger & 0 \\ 0 & \Omega_1\Omega_1^\dagger & \Omega_1 \end{bmatrix}$$

$$= \text{rank} \begin{bmatrix} sI_{n+q} & -\Omega_1^\dagger \\ 0 & I_{n+p} - \Omega_1\Omega_1^\dagger \end{bmatrix} + \text{rank}(\Omega_1). \tag{7.31}$$

By comparing (7.30) with (7.31), it can be seen that

$$\text{rank} \begin{bmatrix} sI_{n+q} & -\Omega_1^\dagger \\ 0 & I_{n+p} - \Omega_1\Omega_1^\dagger \end{bmatrix} = n + q.$$

This completes the proof. □

Lemma 7.2 *There exist matrices Y_1 and Y_2 such that the matrix equations (7.10)–(7.12) are solvable.*

Proof It follows from (7.26) and (7.28) that

$$\begin{aligned} N &= (T_1 - Y_1 T_2)\bar{A}\Omega_2^\dagger \begin{bmatrix} I_{n+q} \\ 0 \end{bmatrix} - Y_2 N_2 \\ &= T_1\Phi - Y T_{2N}, \end{aligned}$$

where $\Phi = \bar{A}\Omega_2^\dagger \begin{bmatrix} I_{n+q} \\ 0 \end{bmatrix}$, $T_{2N} = \begin{bmatrix} T_2\Phi \\ N_2 \end{bmatrix}$ and $Y = [Y_1 \ Y_2]$.

Therefore, the matrix Y exists if the pair $(T_1\Phi, T_{2N})$ is observable, i.e.

$$\text{rank} \begin{bmatrix} sI_{n+q} - T_1\Phi \\ T_2\Phi \\ N_2 \end{bmatrix} = n + q. \tag{7.32}$$

A sufficient condition for (7.32) is

$$\text{rank} \begin{bmatrix} sI_{n+q} - T_1\Phi \\ T_2\Phi \end{bmatrix} = n + q. \tag{7.33}$$

Define $\check{\Phi} = \begin{bmatrix} I_n \\ 0 \end{bmatrix} \Phi$. Since

$$\Sigma_1 = I_{n+q}, \ T_1 = \Sigma_1\Omega_1^\dagger \begin{bmatrix} I_n \\ 0 \end{bmatrix}, \ T_2 = (I_{n+p} - \Omega_1\Omega_1^\dagger) \begin{bmatrix} I_n \\ 0 \end{bmatrix},$$

it holds that

$$\begin{bmatrix} sI_{n+q} - T_1\Phi \\ T_2\Phi \end{bmatrix} = \begin{bmatrix} sI_{n+q} & -\Omega_1^\dagger \\ 0 & I_{n+p} - \Omega_1\Omega_1^\dagger \end{bmatrix} \begin{bmatrix} I_{n+q} \\ \check{\Phi} \end{bmatrix}.$$

It is clear that rank $\begin{bmatrix} I_{n+q} \\ \check{\Phi} \end{bmatrix} = n + q$. Hence, the relation (7.33) holds if

$$\text{rank} \begin{bmatrix} sI_{n+q} & -\Omega_1^{\dagger} \\ 0 & I_{n+p} - \Omega_1\Omega_1^{\dagger} \end{bmatrix} = n + q. \tag{7.34}$$

Since (7.34) has already been proved in Lemma 7.1, the sufficient condition (7.33) holds. This proves that the pair $(T_1\Phi, T_{2N})$ is observable and the matrices Y_1 and Y_2 exist. □

7.4.4 Feasibility and Gain Determination

This subsection presents a way to determine the design gains using (7.14) and (7.15) in Theorem 7.1. Before preceding, it is necessary to prove the feasibility of these conditions by using Lemma 7.3.

Lemma 7.3 *Under Assumptions 7.1–7.3, the conditions (7.14) and (7.15) in Theorem 7.1 are always feasible.*

Proof Multiplying both sides of (7.6) with T and using (7.12), then (7.6) is equivalently represented by

$$\begin{aligned} \dot{\chi} &= T\bar{A}\chi + TBu + TFf + H\dot{y} \\ y &= \bar{C}\chi. \end{aligned} \tag{7.35}$$

According to Lemma 3 in Corless and Tu (1998), there exists a s.p.d. matrix $\check{P}$ and a matrix $\check{L}$ such that

$$\text{He}(\check{P}(T\bar{A} - \check{L}\bar{C})) \prec 0 \tag{7.36}$$

$$\check{P}TF = \bar{C}^{\top}Q \tag{7.37}$$

if and only if the following conditions hold:

- **C1**: $\text{rank}(\bar{C}TF) = \text{rank}(TF) = l + q$.
- **C2**: None of the invariant zeros of $(T\bar{A}, TF, \bar{C})$ are in $\mathbb{C}_+$.

The next step is to prove satisfaction of C1 and C2 under Assumptions 7.1–7.3.

Satisfaction of C1. By using the definitions of E and $\bar{C}$ given in (7.6), it can be derived that

$$\text{rank} \begin{bmatrix} E & F \\ \bar{C} & 0 \end{bmatrix} = \text{rank} \begin{bmatrix} I_n & 0 & F \\ C & D_2 & 0 \end{bmatrix} = n + q + l. \tag{7.38}$$

According to (7.12), the following equation holds:

$$[T\ H]\begin{bmatrix} E \\ \bar{C} \end{bmatrix} = I_{n+q}.$$

This implies that

$$\begin{bmatrix} T & H \\ 0 & I_p \end{bmatrix}\begin{bmatrix} E \\ \bar{C} \end{bmatrix} = \begin{bmatrix} I_{n+q} \\ [0\ \bar{C}] \end{bmatrix}. \tag{7.39}$$

Since $[E^\top\ \bar{C}^\top]^\top$ has full rank, it follows from (7.39) that

$$\begin{bmatrix} T & H \\ 0 & I_p \end{bmatrix} = \begin{bmatrix} I_{n+q} \\ [0\ \bar{C}] \end{bmatrix}\begin{bmatrix} E \\ \bar{C} \end{bmatrix}^{-1}.$$

The left-hand side term of the above equation has full column rank $n+q$. Hence, it follows from (7.38) that

$$\text{rank}\begin{bmatrix} E & F \\ \bar{C} & 0 \end{bmatrix} = \text{rank}\left\{\begin{bmatrix} T & H \\ 0 & I_p \end{bmatrix}\begin{bmatrix} E & F \\ \bar{C} & 0 \end{bmatrix}\right\} = \text{rank}\begin{bmatrix} I_{n+q} & TF \\ \bar{C} & 0 \end{bmatrix} = n+q+l. \tag{7.40}$$

According to Lemma 1 in Yu and Liu (2009), (7.40) is equivalent to condition C1.

Satisfaction of C2. Under Assumption 7.1, for $\forall s \in \mathbb{C}_+$, it holds that

$$\text{rank}\begin{bmatrix} sE-\bar{A} & F \\ \bar{C} & 0 \end{bmatrix} = \text{rank}\begin{bmatrix} sI_n - A & -D_1 & F \\ C & D_2 & 0 \end{bmatrix} = n+q+l. \tag{7.41}$$

It is also true that

$$\text{rank}\begin{bmatrix} sE-\bar{A} & F \\ \bar{C} & 0 \end{bmatrix} = \text{rank}\left\{\begin{bmatrix} T & sH \\ 0 & I_p \end{bmatrix}\begin{bmatrix} sE-\bar{A} & F \\ \bar{C} & 0 \end{bmatrix}\right\} = \text{rank}\begin{bmatrix} sI_{n+q} - T\bar{A} & TF \\ \bar{C} & 0 \end{bmatrix}. \tag{7.42}$$

Comparing (7.42) with (7.41) confirms the satisfaction of condition C2.

The above analysis shows that C1 and C2 are always satisfied under Assumptions 7.1–7.3. By using (7.10), (7.12) and $\bar{L} = L - NH$, it is derived that $N = T\bar{A} - \bar{L}\bar{C}$. Therefore, by letting $P = \check{P}$ and $\bar{L} = \check{L}$, the feasibility of (7.36) and (7.37) implies that of (7.14) and (7.15) in Theorem 7.1. □

According to Lemma 7.2, by substituting the parametrizations of N and T into (7.14) and (7.15) and solving Theorem 7.2, the matrices Y_1 and Y_2 can be obtained and so do all the observer gains.

Theorem 7.2 *Under Assumptions 7.1–7.3, the observer (7.7) can achieve asymptotic estimation of the augmented state $\bar{x}$ and actuator fault f, if there exists a s.p.d. matrix $P \in \mathbb{R}^{(n+q)\times(n+q)}$, matrices $Q \in \mathbb{R}^{p\times l}$ and $M \in \mathbb{R}^{(n+q)\times(2n+2p+q)}$, and positive constants ξ and β, such that*

$$\mathrm{He}\left(PT_1\Phi - M\begin{bmatrix} T_2\Phi \\ N_2 \end{bmatrix}\right) \prec -\xi I_{n+q} \tag{7.43}$$

$$\begin{bmatrix} \beta I & \left(PT_1 - M\hat{T}_2\right)F - \bar{C}^\top Q \\ \star & \beta I \end{bmatrix} \succ 0 \tag{7.44}$$

where $\hat{T}_2 = [T_2^\top \ 0]^\top$. *Then the matrices* Y_1 *and* Y_2 *are obtained as*

$$Y_1 = P^{-1}M\begin{bmatrix} I_{n+p} \\ 0 \end{bmatrix}, \ Y_2 = P^{-1}M\begin{bmatrix} 0 \\ I_{n+p+q} \end{bmatrix}.$$

Proof Substituting (7.26) and (7.28) into (7.14) and (7.15) and defining $M = PY$, then the inequality (7.43) is derived from (7.14) directly. Moreover, by using the method described in Corless and Tu (1998), the equality constraint (7.15) can be converted into the inequality (7.44). □

7.5 Adaptive Sliding Mode ASUIO Design: General Case

This section considers the observer design under a more general case when system (7.1) satisfies, instead of Assumption 7.3, the following assumptions:

Assumption 7.4 rank $\begin{bmatrix} D_1 \\ D_2 \end{bmatrix} = q$.

Assumption 7.5 rank$[D_2 \ CF \ CD_1] = l + q$.

It is seen that Assumption 7.4 contains the condition rank$(D_2) = q$ in Assumption 7.3 as a special case. This section focuses on the general case when rank$(D_2) = r < q$. In such case, a direct extension of the design in Sect. 7.4 is impossible. This is because following the same system augmentation strategy leads to an infinitely unobservable descriptor system and for which no asymptotic observer exists. To address this challenge, a system reformulation is performed in Sect. 7.5.1 under the Assumptions 7.4 and 7.5. The reformulation results in an infinitely observable descriptor system, which can be used for the observer design.

Remark 7.3 Notice that under Assumption 7.4, the perturbation d considered in this section is able to represent a wide class of unknown inputs acting on the system. For example, when $d = [w_a^\top \ w_s^\top]^\top$, it can represent a mixture of state perturbation w_a and output perturbation w_s; when $d = [w^\top \ f_s^\top]^\top$, it can represent co-existence of system perturbation w acting on state dynamics and/or output measurement, and sensor fault f_s.

7.5.1 System Reformulation

Since $\text{rank}(D_2) = r < q$, the singular value decomposition of D_2 is given by

$$D_2 = U_d \Sigma_d V_d^\top,$$

where $U_d \in \mathbb{R}^{p \times p}$ and $V_d \in \mathbb{R}^{q \times q}$ are orthogonal matrices, $\Sigma_d = \text{diag}(\Sigma_r, 0_{(p-r)\times(q-r)})$ and $\Sigma_r \in \mathbb{R}^{r \times r}$ is a diagonal matrix containing all the r singular values of D_2 on the diagonal.

Define the terms

$$V_d^\top d = \begin{bmatrix} d_1 \\ d_2 \end{bmatrix} \begin{matrix} \updownarrow r \\ \updownarrow q-r \end{matrix}, \ U_d^\top y = \bar{y}, \ D_1 V_d = [D_{11} \ D_{12}], \ U_d^\top C = \tilde{C}. \quad (7.45)$$

By using (7.45), the system (7.1) is rewritten as

$$\begin{aligned} \dot{x} &= Ax + Bu + \mathcal{F}\bar{f} + D_{11} d_1 \\ \bar{y} &= \tilde{C}x + \bar{D}_2 d_1 \end{aligned} \quad (7.46)$$

where $\bar{f} = [f^\top \ d_2^\top]^\top, \mathcal{F} = [F \ D_{12}]$ and $\bar{D}_2 = [\Sigma_r^\top \ 0]^\top$.

By defining the new state vector as $\chi = [x^\top \ d_1^\top]^\top$, the system (7.46) can be rearranged into the descriptor form

$$\begin{aligned} \mathcal{E}\dot{\chi} &= \mathcal{A}\chi + \mathcal{B}u + \mathcal{F}\bar{f} \\ \bar{y} &= C\chi \end{aligned} \quad (7.47)$$

where $\mathcal{E} = [I_n \ 0_{n\times r}]$, $\mathcal{A} = [A \ D_{11}]$, $\mathcal{B} = B$ and $C = [\tilde{C} \ \bar{D}_2]$. Here, d_1 is treated as auxiliary state and d_2 as auxiliary fault. By defining the new fault vector $\bar{f}$, d_2 will be estimated together with the real fault f.

Remark 7.4 If following Sect. 7.4 by regarding the entire vector d as auxiliary state, then $\mathcal{E} = [I_n \ 0_{n\times q}]$, $\mathcal{A} = [A \ D_1]$, $\mathcal{F} = F$ and $C = [C \ D_2]$. This will result in an infinitely unobservable descriptor system that violates Corollary 7.1.

Corollary 7.1 *Under Assumptions 7.1, 7.2, 7.4 and 7.5, the descriptor system (7.47) holds the following properties:*

$$\text{rank}\begin{bmatrix} \mathcal{E} \\ C \end{bmatrix} = n + r \quad (7.48)$$

$$\text{rank}\begin{bmatrix} \mathcal{E} & \mathcal{F} \\ C & 0 \end{bmatrix} = n + l + q \quad (7.49)$$

$$\text{rank}\begin{bmatrix} s\mathcal{E} - \mathcal{A} & \mathcal{F} \\ C & 0 \end{bmatrix} = n + l + q, \ \forall s \in \mathbb{C}_+. \quad (7.50)$$

Proof Since $\text{rank}(\bar{D}_2) = r$, it holds that

$$\text{rank}\begin{bmatrix}\mathcal{E}\\ \mathcal{C}\end{bmatrix} = \text{rank}\begin{bmatrix}I_n & 0_{n\times r}\\ \tilde{C} & \bar{D}_2\end{bmatrix} = n + r.$$

In view of the structure of $\bar{D}_2$ in (7.46), it follows that

$$\text{rank}\begin{bmatrix}D_{11} & D_{12}\\ \bar{D}_2 & 0\end{bmatrix} = r + \text{rank}(D_{12}). \tag{7.51}$$

According to Assumption 7.4 and (7.45), it holds that

$$\text{rank}\begin{bmatrix}D_{11} & D_{12}\\ \bar{D}_2 & 0\end{bmatrix} = \text{rank}\left\{\begin{bmatrix}I_n & 0\\ 0 & U_d^\top\end{bmatrix}\begin{bmatrix}D_1\\ D_2\end{bmatrix}\right\} = q. \tag{7.52}$$

Combining (7.51) and (7.52) yields $\text{rank}(D_{12}) = q - r$ and $\text{col}(\tilde{C}D_{11}) \subseteq \text{col}(\bar{D}_2)$, where $\text{col}(\cdot)$ represents the column space spanned by the matrix. This implies that

$$\text{rank}[\bar{D}_2\ \tilde{C}F\ \tilde{C}D_1V_d] = \text{rank}[\bar{D}_2\ \tilde{C}F\ \tilde{C}D_{12}]. \tag{7.53}$$

Since U_d and V_d are non-singular matrices, it can also be verified that

$$\text{rank}[\bar{D}_2\ \tilde{C}F\ \tilde{C}D_1V_d] = \text{rank}[D_2\ CF\ CD_1]. \tag{7.54}$$

Under Assumption 7.5, combining (7.53) and (7.54) yields $\text{rank}[\bar{D}_2\ \tilde{C}F\ \tilde{C}D_{12}] = l + q$, which further leads to

$$\begin{aligned}\text{rank}\begin{bmatrix}\mathcal{E} & \mathcal{F}\\ \mathcal{C} & 0\end{bmatrix} &= \text{rank}\left\{\begin{bmatrix}I_n & 0\\ -\tilde{C} & I_p\end{bmatrix}\begin{bmatrix}I_n & 0 & F & D_{12}\\ \tilde{C} & \bar{D}_2 & 0 & 0\end{bmatrix}\right\}\\ &= \text{rank}\begin{bmatrix}I_n & 0 & F & D_{12}\\ 0 & \bar{D}_2 & \tilde{C}F & \tilde{C}D_{12}\end{bmatrix}\\ &= n + l + q.\end{aligned}$$

Under Assumption 7.1, it is true that

$$\text{rank}\begin{bmatrix}s\mathcal{E} - \mathcal{A} & \mathcal{F}\\ \mathcal{C} & 0\end{bmatrix} = \text{rank}\left\{X_1\begin{bmatrix}sI_n - A & F & D_1\\ C & 0 & D_2\end{bmatrix}\right\} = n + l + q,$$

where $X_1 = \text{diag}(I_n, U_d^\top)$ and $X_2 = \text{diag}(I_n, I_l, V_d)$ are full rank because U_d and V_d are both non-singular matrices. Therefore, the property (7.50) holds. □

The properties (7.48) and (7.50) ensure that the descriptor system (7.47) is infinitely observable (Dai 1989). The property (7.49) further ensures that the vector $\bar{f}$ can be fully reconstructed. Therefore, there exists an observer for the descriptor system (7.47) to estimate accurately χ and $\bar{f}$, and subsequently the state x, fault f and perturbation d of the original system (7.1).

7.5.2 Observer Design

Since the descriptor system (7.47) is in the form of (7.6), the observer design is similar to that in Sect. 7.4. Therefore, the design details are left for the readers. Once estimates of the augmented state χ and auxiliary fault $\bar{f}$ are obtained, the estimates of x, f and d are given by

$$\hat{x} = [I_n \ 0]\hat{\bar{x}}, \ \hat{f} = [I_l \ 0]\hat{\bar{f}}, \ \hat{d} = V_d \begin{bmatrix} \hat{d}_1 \\ \hat{d}_2 \end{bmatrix}, \ \hat{d}_1 = [0 \ I_r]\hat{\bar{x}}, \ \hat{d}_2 = [0 \ I_{q-r}]\hat{\bar{f}}.$$

One advantage of the presented observer is in achieving simultaneous asymptotic estimation of the state, faults and perturbations. However, the estimation of perturbations cannot be obtained using the attenuation-based observer in Chaps. 4 and 5 or the partial decoupling-based observer in Gao et al. (2016). The existence conditions for these observers are different, as compared below:

- The attenuation-based observer requires: (1) None of the invariant zeros of (A, F, C) are in $\mathbb{C}_+$; (2) $\text{rank}(F) = l$; (3) $l \leq p$.
- Partition D_1 as $D_1 = [\check{D}_{11} \ \check{D}_{12}]$ corresponding to the partition $d = [\check{d}_1^\top \ \check{d}_2^\top]^\top$ with $\check{d}_1 \in \mathbb{R}^{q_1}$, $\check{d}_2 \in \mathbb{R}^{q_2}$ and $q_1 + q_2 = q$. The partial decoupling-based observer requires: (1) None of the invariant zeros of $(A, [F \ \check{D}_{11}], C)$ are in $\mathbb{C}_+$; (2) $\text{rank}(C\check{D}_{11}) = \text{rank}(\check{D}_{11}) = q_1$; (3) $\text{rank}(F) = l$; (4) $l + q_1 \leq p$.
- The proposed observer requires: (1) Assumptions 7.1, 7.2, 7.4 and 7.5; (2) $l + q \leq p$.

7.6 Adaptive Backstepping FTC Design

Since the fault f and perturbation d are unmatched, their effects on the system dynamics cannot be compensated through direct control actions. In this section, an adaptive backstepping FTC controller is proposed to compensate f and d, by using the estimates obtained from the observer. Before describing controller design in Sect. 7.6.2, the system (7.1) needs to be transferred into a strict-feedback form using the technique presented in Sect. 7.6.1.

7.6.1 System Reformulation

The system (7.1) can be represented in the following strict-feedback form

$$\begin{aligned}\dot{x}_1 &= A_1x_1 + B_1(x_2 + F_1 f + S_1 d)\\ \dot{x}_i &= A_i\bar{x}_i + B_i(x_{i+1} + F_i f + S_i d),\ i = 2, \ldots, r-1\\ \dot{x}_r &= A_r\bar{x}_r + B_r(u + F_r f + S_r d)\end{aligned} \tag{7.55}$$

where $x_i \in \mathbb{R}^{n_i}$ are the new system state, $x_{r+1} = u$ and x_1 is the system output. $\bar{x}_i = [x_1^\top, \ldots, x_i^\top]^\top$, $\text{rank}(B_i) = n_i$ and $\sum_{i=1}^{r} n_i = n$. The matrices A_i, F_i and S_i are of compatible dimensions. The original system state is $x = [x_1^\top, \ldots, x_r^\top]^\top$.

Remark 7.5 Many physical systems can be rearranged into a strict-feedback form required for backstepping control design (Krstic et al. 1995). Moreover, using the decomposition algorithm described in Polyakov (2012), a controllable system (7.1) can always be decomposed into the required block-controllable (triangular) form.

7.6.2 FTC Design

The adaptive backstepping FTC design aims to (1) compensate the actuator fault f and perturbation d and (2) ensure that the system output x_1 can track a given reference x_d, by using the system state estimate $\hat{x}_i$, fault estimate $\hat{f}$ and perturbation estimate $\hat{d}$.

Define the estimation errors as

$$\begin{aligned} e_{x_i} &= x_i - \hat{x}_i,\ e_{\bar{x}_i} = \bar{x}_i - \hat{\bar{x}}_i,\ i = 1, 2, \ldots, r,\\ e_f &= f - \hat{f},\ e_d = d - \hat{d}.\end{aligned}$$

Although it is shown in Theorem 7.1 that all these estimation errors are bounded and converge to zero in finite time, they still have side effects on the transient performance of the FTC system, which should be taken into account in the control design. Therefore, an adaptive method is incorporated with backstepping control to estimate and compensate effects of the estimation errors automatically.

- **Step i:** $1 \leq i \leq r-1$

Define $z_i = \hat{x}_i - \alpha_{i-1}$, where α_{i-1} is a virtual control input to be designed and $z_0 = 0$ and $\alpha_0 = x_d$. It follows from (7.55) that

$$\dot{z}_i = A_i\bar{x}_i + B_i\left(x_{i+1} + F_i f + S_i d\right) - \dot{e}_{x_i} - \dot{\alpha}_{i-1}.$$

To analyse stability of the above dynamics, the following Lyapunov function is considered:

$$V_{z_{i0}} = \frac{1}{2} z_i^\top z_i.$$

Its derivative is obtained as

$$\begin{aligned}\dot{V}_{z_{i0}} &= z_i^\top \left[A_i \bar{x}_i + B_i \left(\alpha_i + F_i f + S_i d\right)\right] + z_i^\top B_i z_{i+1} \\ &\quad + z_i^\top (B_i e_{x_{i+1}} - \dot{e}_{x_i} - \dot{\alpha}_{i-1}).\end{aligned} \tag{7.56}$$

To ensure satisfactory tracking, compensate the fault and perturbation and cancel the side effect from the estimation error system (7.13) (i.e. the term $z_1^\top (B_i e_{x_{i+1}} - \dot{e}_{x_i})$ in (7.56)), the virtual control input α_i is designed as

$$\alpha_i = -B_i^{-1} \left[c_i z_i + \rho_{z_i} \operatorname{sign}(z_i) + B_{i-1}^\top z_{i-1} + A_i \hat{\bar{x}}_i\right] - F_i \hat{f} - S_i \hat{d} \tag{7.57}$$

where $c_i > 0$ is a design constant and ρ_{z_i} is an adaptive parameter to be determined.

Substituting (7.57) into (7.56) gives

$$\dot{V}_{z_{i0}} \leq -c_i \|z_i\|^2 - z_{i-1}^\top B_{i-1} z_i + z_i^\top B_i z_{i+1} + (\rho_i - \rho_{z_i}) \|z_i\| \tag{7.58}$$

where ρ_i is an unknown constant satisfying

$$\rho_i \geq \|A_i e_{\bar{x}_i} + B_i F_i e_f + B_i S_i e_d + B_i e_{x_{i+1}} - \dot{e}_{x_i} - \dot{\alpha}_{i-1}\|,$$

which represents the side effect of the estimation errors on the z_i subsystem.

In order to compensate ρ_i, define $\rho_{z_i} = \hat{\rho}_i + \epsilon_i$, where ϵ_i is a positive design constant and $\hat{\rho}_i$ is used to estimate ρ_i and is generated by

$$\dot{\hat{\rho}}_i = \sigma_i \|z_i\|, \ \hat{\rho}_i(0) = 0 \tag{7.59}$$

where σ_i is a positive design constant.

Define the estimation error of ρ_i as $\tilde{\rho}_i = \rho_i - \hat{\rho}_i$. Consider the Lyapunov function

$$V_{z_i} = V_{z_{i0}} + \frac{1}{2\sigma_i} \tilde{\rho}_i^2.$$

By using (7.58) and (7.59), it is derived that

$$\begin{aligned}\dot{V}_{z_i} &= \dot{V}_{z_{i0}} + \frac{1}{\sigma_i} \tilde{\rho}_i (-\sigma_i \|z_i\|) \\ &\leq -c_i \|z_i\|^2 - z_{i-1}^\top B_{i-1} z_i + z_i^\top B_i z_{i+1}.\end{aligned} \tag{7.60}$$

For the first i steps, consider the composite Lyapunov function

$$V_i = V_{i-1} + V_{z_i}$$

with $V_0 = 0$. Then it follows from (7.60) that

$$\dot{V}_i \leq -\sum_{j=1}^{i} c_j \|z_j\|^2 + z_i^\top B_i z_{i+1}. \tag{7.61}$$

- **Step r:**

The dynamics of z_r is represented by

$$\dot{z}_r = A_r \bar{x}_r + B_r(u + F_r f + S_r d) - \dot{e}_{x_r} - \dot{\alpha}_{r-1}.$$

To analyse its stability, the following Lyapunov function is used:

$$V_{z_{r0}} = \frac{1}{2} z_r^\top z_r.$$

The time derivative of $V_{z_{r0}}$ can be derived as

$$\dot{V}_{z_{r0}} = z_r^\top [A_r \bar{x}_r + B_r(u + F_r f + S_r d)] - z_r^\top (\dot{e}_{x_r} + \dot{\alpha}_{r-1}). \tag{7.62}$$

The FTC controller u is designed as

$$u = -B_r^{-1} \left[c_r z_r + \rho_{z_r} \operatorname{sign}(z_r) + B_{r-1}^\top z_{r-1} + A_r \hat{\bar{x}}_r \right] - F_r \hat{f} - S_r \hat{d} \tag{7.63}$$

where c_r is a design constant and ρ_{z_r} is an adaptive parameter to be designed.
Substituting (7.63) into (7.62) yields

$$\dot{V}_{z_{r0}} \leq -c_r \|z_r\|^2 - z_{r-1}^\top B_{r-1} z_r + (\rho_r - \rho_{z_r}) \|z_r\| \tag{7.64}$$

where ρ_r is an unknown constant such that

$$\rho_r \geq \|A_r e_{\bar{x}_r} + B_r F_r e_f + B_r S_r e_d - \dot{e}_{x_r} - \dot{\alpha}_{r-1}\|.$$

Define $\rho_{z_r} = \hat{\rho}_r + \epsilon_r$, where ϵ_r is a positive design constant and $\hat{\rho}_r$ is the estimate of ρ_r generated by

$$\dot{\hat{\rho}}_r = \sigma_r \|z_r\|, \ \hat{\rho}_r(0) = 0 \tag{7.65}$$

with a positive design constant σ_r.
Define the estimation error of ρ_r as $\tilde{\rho}_r = \rho_r - \hat{\rho}_r$. Consider the Lyapunov function

$$V_{z_r} = V_{z_{r0}} + \frac{1}{2\sigma_r} \tilde{\rho}_r^2.$$

By using (7.64) and (7.65), it is derived that

$$\begin{aligned}\dot{V}_{z_r} &= \dot{V}_{z_{r0}} + \frac{1}{\sigma_r}\tilde{\rho}_r(-\sigma_r\|z_r\|) \\ &\leq -c_r\|z_r\|^2 - z_{r-1}^\top B_{r-1}z_r. \end{aligned} \tag{7.66}$$

Finally, in order to analyse stability of the overall control system, the following Lyapunov function is considered:

$$V_r = V_{r-1} + V_{z_r}.$$

It follows from (7.61) and (7.66) that

$$\dot{V}_r \leq -\sum_{j=1}^{r} c_j\|z_j\|^2.$$

By designing $c_j \geq 0$, $j = 1, 2, \ldots, r$, it is obtained that $\dot{V}_r < 0$. Since V_r is positive definite, it follows from the Barbalat's Lemma (see Sect. 1.7.1) that $V_r(t) \leq V_r(0)$. Hence, $z_j(t)$ and $\tilde{\rho}_j$, $j = 1, 2, \ldots, r$, are bounded, and the system output x_1 tracks the reference x_d with bounded error. Moreover, under the zero initial conditions, i.e. $z_j(0) = 0$ and $\tilde{\rho}_j = 0$, $j = 1, 2, \ldots, r$, one gets $\lim_{t\to\infty} V_r(t) = 0$ and thus $\lim_{t\to\infty} z_j(t) = 0$. This means that the system output x_1 tracks the reference x_d accurately even in the presence of actuator fault and perturbation.

Remark 7.6 Although the adaptive decoupling strategy facilitates FE and FTC designs, the estimation errors inevitably affect the transient performance of the closed-loop system. To improve the transient performance, eigenvalue assignment can be applied to the observer. The FTC system performance can be largely recovered if the observer dynamics are (much) faster than the closed-loop control system dynamics. To reach this, the eigenvalues of matrix N are located into an acceptable LMI region (see Sect. 1.7.1). Specifically, it is achieved by adding a pole placement constraint (7.67) to the existing constraints (7.43) and (7.44) to place the eigenvalues of N into a strip region (a, b), where a and b are negative constants.

$$\begin{bmatrix} \mathrm{He}\left(PT_1\Phi - M\begin{bmatrix} T_2\Phi \\ N_2 \end{bmatrix}\right) - 2bP & 0 \\ \star & -\mathrm{He}\left(PT_1\Phi - M\begin{bmatrix} T_2\Phi \\ N_2 \end{bmatrix}\right) + 2aP \end{bmatrix} \prec 0. \tag{7.67}$$

It is worth noting that after incorporating the above pole placement constraint with (7.43) and (7.44), the resulting LMI formulation is still less complex than the designs in Chaps. 5 and 6. This again implies that the adaptive decoupling strategy enjoys great design freedom.

7.7 Tutorial Example 1

Consider the angular velocity tracking control of a DC motor modelled by

$$\begin{aligned} \dot{x} &= Ax + B(u + f) + D_1 d \\ y &= Cx + D_2 d \end{aligned} \tag{7.68}$$

where $x = [w \ i_a]^\top$ is the state, $u = v_a$ is the control input, y is the output, f is an actuator fault and d is the perturbation. The system matrices are given as

$$A = \begin{bmatrix} -\frac{B_0}{J_i} & \frac{K_m}{J_i} \\ -\frac{K_v}{L_a} & -\frac{R_a}{L_a} \end{bmatrix}, \ B = \begin{bmatrix} 0 \\ \frac{1}{L_a} \end{bmatrix}, \ D_1 = \begin{bmatrix} 0.1 \\ 0.1 \end{bmatrix}, \ C = \begin{bmatrix} 1 & 0 \\ 0 & 1 \end{bmatrix}, \ D_2 = \begin{bmatrix} 1 \\ 0 \end{bmatrix}.$$

The physical parameters of the DC motor are defined as follows. w, i_a, and v_a are the the angular velocity, armature current, and armature voltage, respectively. R_a is the armature resistance. L_a is the inductance. K_v and K_m are the voltage and motor constants, respectively. J_i is the moment of inertia. B_0 is the friction coefficient. Compared with the DC motor model used in Chap. 5, output perturbation is also considered in this simulation.

The reference for w is $x_d = 1$. The system parameters are (Bélanger 1995): $R_a = 1.2, \ L_a = 0.05, \ K_v = 0.6, \ K_m = 0.6, \ J_i = 0.1352, \ B_0 = 0.3$.

Given $\gamma = 10$, $\beta = 0.1$, $a = -20$ and $b = -3$. By solving (7.43) and (7.44) in Theorem 7.2 together with the pole placement constraint (7.67) in Remark 7.6, the observer gains are obtained as

$$P = \begin{bmatrix} 40.049 & 0 & -13.5837 \\ 0 & 26.4654 & 0 \\ -13.5837 & 0 & 40.049 \end{bmatrix},$$

$$N = \begin{bmatrix} -8.3278 & 1.7211 & -5.4846 \\ -3.3484 & -14.0715 & 3.0769 \\ -6.2685 & -1.4495 & -8.0620 \end{bmatrix},$$

$$J = \begin{bmatrix} 0.8666 \\ 10.6204 \\ 0.8684 \end{bmatrix}, \ L = \begin{bmatrix} 0.1043 & 3.0830 \\ 0.0531 & -5.2608 \\ -0.0957 & -4.0886 \end{bmatrix},$$

$$H = \begin{bmatrix} 0 & -0.0433 \\ 0 & 0.4690 \\ 1 & -0.0434 \end{bmatrix}, \ Q = \begin{bmatrix} 22.9588 \\ 281.0727 \end{bmatrix}.$$

The adaptive backstepping FTC controller parameters are chosen as follows: $c_1 = 35, \ \sigma_1 = 0.1, \ \epsilon_1 = 0.1, \ c_2 = 60, \ \sigma_2 = 0.1, \ \epsilon_2 = 0.1, \ \sigma_0 = 5, \ \epsilon = 1$.

Comparative simulations are performed using the following four designs:

- *Nominal design*. It includes a UIO (see Sect. 1.7.3) for state estimation and a state feedback controller, which are designed separately without FE and FTC.
- *Separated FE and FTC design*. The perturbation d is treated as a sensor fault. It includes the ASUIO in Chap. 5 for fault and state estimation and a state feedback FTC controller, which are designed separately with their coupling ignored.
- *Simultaneous FE and FTC design* in Chap. 5. The output perturbation d is treated as a sensor fault. It includes an ASUIO for fault and state estimation and a state feedback FTC controller, which are designed together using a single-step LMI formulation with their coupling considered.
- *Proposed adaptive decoupling FE and FTC design*.

Simulations are carried out with differentiable or non-differentiable actuator faults, using the same observer and controller gains given above and the same zero initial conditions.

7.7.1 Case 1: Differentiable Fault

Suppose the system has differentiable fault f and perturbation d characterized by

$$d(t) = \begin{cases} 0.05\sin(\pi t), & 0\text{ s} \le t \le 10\text{ s} \\ 3\sin(4\pi t) + [0.1\ 0.5]x, & 10\text{ s} < t \le 15\text{ s} \end{cases},$$

$$f(t) = \begin{cases} 0, & 0\text{ s} \le t \le 2\text{ s} \\ 0.04(t-2)^2 + \sin(\pi(t-2)), & 2\text{ s} < t \le 7\text{ s} \\ 1, & 7\text{ s} < t \le 10\text{ s} \\ \cos(3\pi(t-10)) + 1, & 10\text{ s} < t \le 13\text{ s} \\ -1, & 13\text{ s} < t \le 15\text{ s} \end{cases}.$$

The above f and d have different characteristics in different time periods, which are used to test the system performance under different fault and perturbation scenarios. Moreover, a Gaussian noise w with zero mean and variance 1.0×10^{-6} is added to the measured outputs in the time interval $t \in (10, 15]$ s.

The simulation results are depicted in Figs. 7.2, 7.3, 7.4 and 7.5. It is seen from Fig. 7.2 that, among the four designs simulated, only the proposed adaptive decoupling design achieves good tracking performance in the presence of actuator fault. It is shown in Fig. 7.3 that the control efforts of the adaptive decoupling and simultaneous designs are similar but much smaller than those of the other two designs. As observed from Figs. 7.4 and 7.5, the proposed adaptive decoupling design has a better fault and perturbation estimation performance than the separated and simultaneous designs.

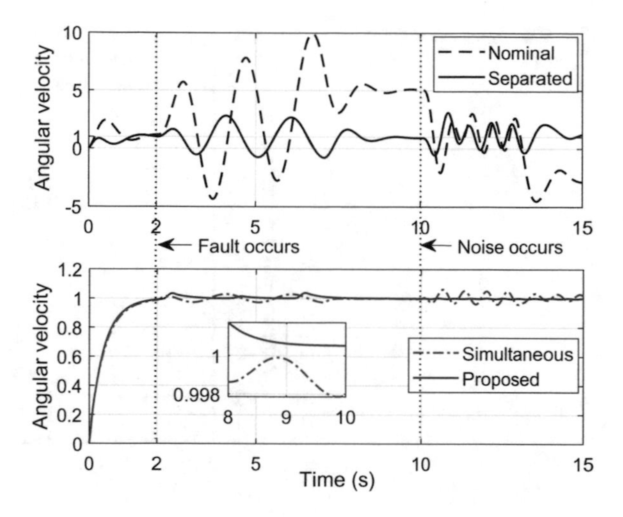

Fig. 7.2 Angular velocity: Case 1

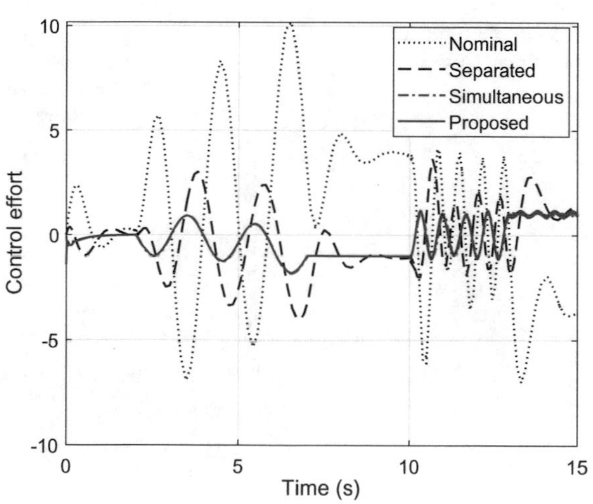

Fig. 7.3 Control effort: Case 1

7.7.2 Case 2: Non-differentiable Fault

This subsection further verifies the efficacy of the proposed adaptive decoupling design for the DC motor system with a non-differentiable fault. The simulations use the non-differentiable fault f and the perturbation d in the forms of

$$f(t) = \sum_{k=0}^{50} 0.5^k \cos(3^k \pi t),\ 0 \text{ s} \leq t \leq 5 \text{ s},$$

$$d(t) = 2 \sin(2\pi t),\ 0 \text{ s} \leq t \leq 5 \text{ s}.$$

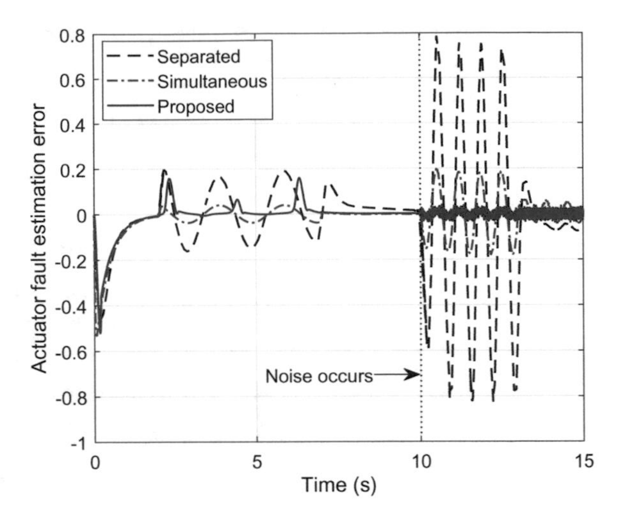

Fig. 7.4 Fault estimation: Case 1

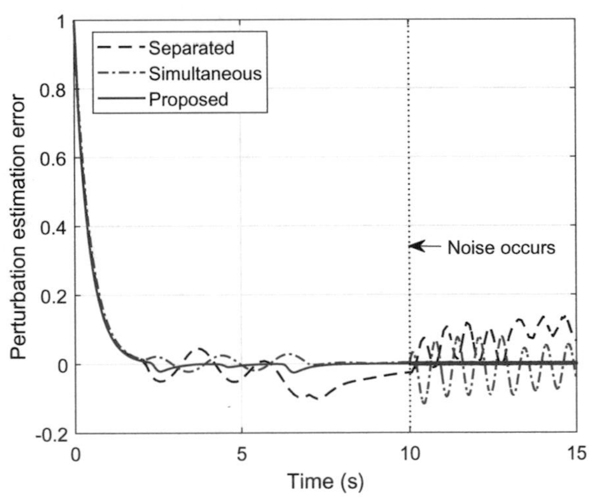

Fig. 7.5 Perturbation estimation: Case 1

The above fault $f(t)$ is a Weierstrass function that is smooth but nowhere differentiable (Hardy 1916).

The comparative simulation results are reported in Figs. 7.6, 7.7, 7.8 and 7.9. It is seen from Fig. 7.6 that neither of the nominal and separated designs achieves angular velocity tracking. Although the angular velocity of the simultaneous design tracks the reference with small error, it has an oscillatory dynamic response. Only the adaptive decoupling design achieves good tracking performance. As shown in Fig. 7.7, the control efforts of the decoupling and simultaneous designs are similar but much smaller than those of the other two designs. It is observed from Figs. 7.8 and 7.9 that the adaptive decoupling design has a much better fault and perturbation estimation performance than the other two designs.

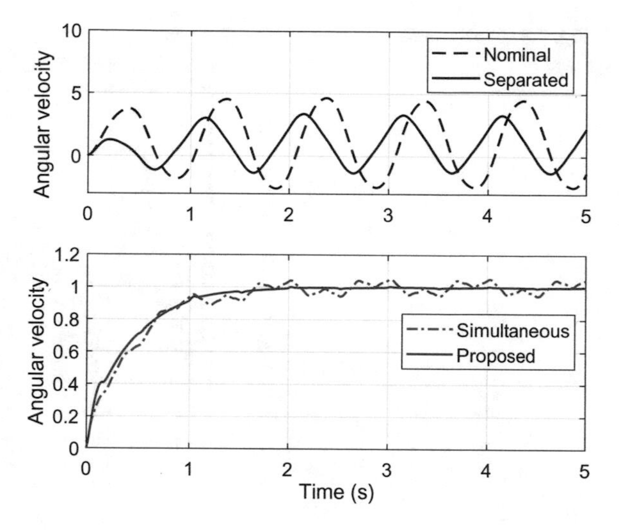

Fig. 7.6 Angular velocity: Case 2

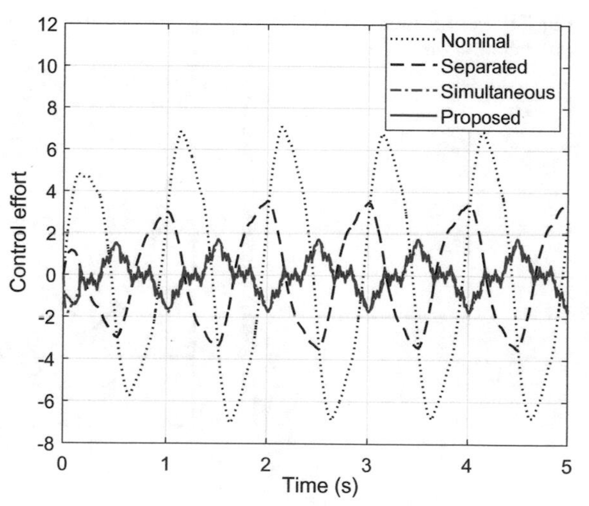

Fig. 7.7 Control effort: Case 2

Summarizing the above two simulation cases for the DC motor (7.68) subject to actuator faults (differentiable or non-differentiable) and perturbations, the superiority ranking of the four control designs from low to high, in terms of robust system performance, is that (1) the nominal design, (2) the separated design, (3) the simultaneous design and (4) the adaptive decoupling design.

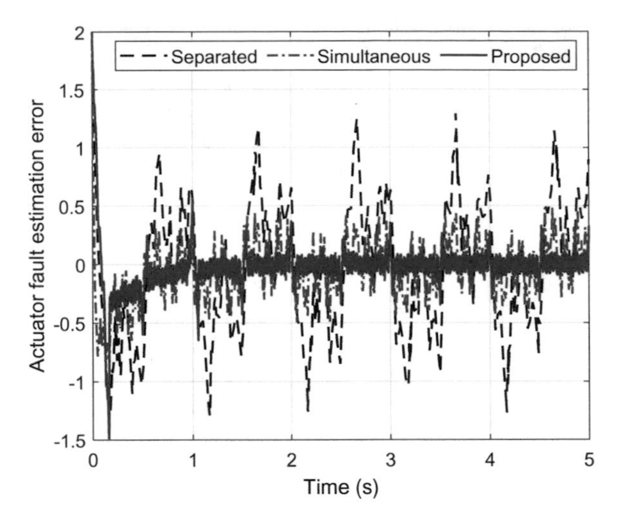

Fig. 7.8 Fault estimation: Case 2

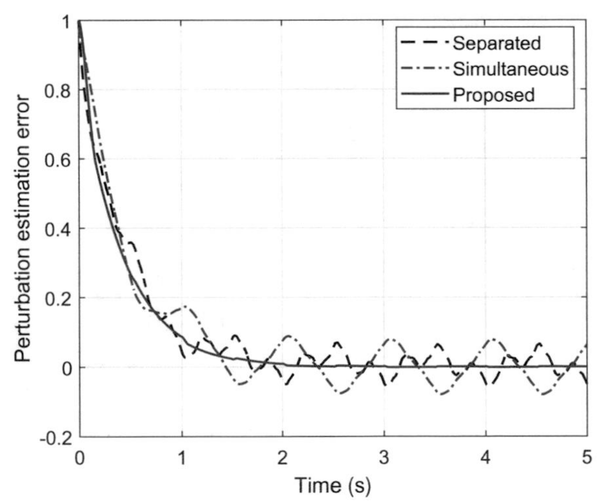

Fig. 7.9 Perturbation estimation: Case 2

7.8 Tutorial Example 2

This section provides a tutorial example to demonstrate the FE observer design described in Sect. 7.5.

Consider the following normalized aircraft lateral dynamics adopted from Eugene et al. (2013):

$$\underbrace{\begin{bmatrix} \dot{\beta} \\ \dot{p} \\ \dot{r} \end{bmatrix}}_{\dot{x}} = \underbrace{\begin{bmatrix} -0.025 & 1.04 & -9.94 \\ 57.47 & 0 & 0 \\ 1.62 & 0 & 0 \end{bmatrix}}_{A} \underbrace{\begin{bmatrix} \beta \\ p \\ r \end{bmatrix}}_{x}$$

$$+ \underbrace{\begin{bmatrix} 0.122 & -0.276 \\ -5.361 & 3.325 \\ 19.55 & -52.94 \end{bmatrix}}_{B} \left(\underbrace{\begin{bmatrix} \delta_a \\ \delta_r \end{bmatrix}}_{u} + \underbrace{\begin{bmatrix} 1 \\ 0 \end{bmatrix}}_{F_a} f + \underbrace{\begin{bmatrix} 0 \\ 1 \end{bmatrix}}_{D_a} d_a \right)$$

$$y = \underbrace{\begin{bmatrix} 1 & 0 & 0 \\ 0 & 1 & 0 \\ 0 & 0 & 1 \end{bmatrix}}_{C} \underbrace{\begin{bmatrix} \beta \\ p \\ r \end{bmatrix}}_{x} + \underbrace{\begin{bmatrix} 0.1 \\ 0.2 \\ 1 \end{bmatrix}}_{D_s} d_s \qquad (7.69)$$

where the state vector x contains the slideslip angle in deg, normalized roll and yaw rates in deg/s; the control u contains the aileron in deg and rudder commands in deg; f is an actuator fault acting on the aileron input; d_a is an attack acting on the rudder input; and d_s is an attack on the measured outputs.

Before proceeding to the FE observer design, satisfaction of required assumptions are verified. The dynamics (7.69) can be rewritten in the form of (7.1) with $F = BF_a$, $D_1 = [0\ BD_a]$ and $D_2 = [D_s\ 0]$, and the system dimensions are $n = 3, m = 2, l = 1$, $q = 2$ and $p = 3$.

It is obvious that the example system satisfies Assumptions 7.1 and 7.2. Since d_a and d_s are independent, it holds that $\text{rank}[D_1^\top\ D_2^\top]^\top = 2$. Hence, Assumption 7.4 is satisfied. It can also be derived that $\text{rank}[D_2\ CF\ CD_1] = 3$ and thus Assumption 7.5 is met.

Since $\text{rank}(D_2) = r = 1$, the singular decomposition of D_2 is given by $D_2 = U_d \Sigma_d V_d^\top$ with

$$U_d = \begin{bmatrix} -0.0976 & -0.1952 & -0.9759 \\ -0.1952 & 0.9653 & -0.1735 \\ -0.9759 & -0.1735 & 0.1323 \end{bmatrix}, \ \Sigma_d = \begin{bmatrix} 1.0247 & 0 & 0 \\ 0 & 0 & 0 \\ 0 & 0 & 0 \end{bmatrix}, \ V_d = \begin{bmatrix} -1 & 0 \\ 0 & 1 \end{bmatrix},$$

where $\Sigma_r = 1.0247$. Rearranging D_1 and C gives

$$[D_{11}\ D_{12}] = \left[\begin{array}{c|c} 0 & -0.276 \\ 0 & 3.325 \\ 0 & -52.94 \end{array}\right], \ \tilde{C} = \begin{bmatrix} -0.0976 & -0.1952 & -0.9759 \\ -0.1952 & 0.9653 & -0.1735 \\ -0.9759 & -0.1735 & 0.1323 \end{bmatrix}.$$

This results in the descriptor system (7.47) with

$$\mathcal{F} = \begin{bmatrix} 0.122 & -0.276 \\ -5.361 & 3.325 \\ 19.550 & -52.94 \end{bmatrix}, \ \mathcal{E} = \begin{bmatrix} 1 & 0 & 0 & 0 \\ 0 & 1 & 0 & 0 \\ 0 & 0 & 1 & 0 \end{bmatrix}, \ \mathcal{A} = \begin{bmatrix} -0.025 & 1.04 & -9.94 & 0 \\ 57.47 & 0 & 0 & 0 \\ 1.62 & 0 & 0 & 0 \end{bmatrix},$$

$$\mathcal{B} = B, \ C = \begin{bmatrix} -0.0976 & -0.1952 & -0.9759 & 1.0247 \\ -0.1952 & 0.9653 & -0.1735 & 0 \\ -0.9759 & -0.1735 & 0.1323 & 0 \end{bmatrix}.$$

The next step is to verify the properties in Corollary 7.1. The property (7.48) is satisfied because

$$\text{rank} \begin{bmatrix} \mathcal{E} \\ C \end{bmatrix} = \left[\begin{array}{ccc|c} 1 & 0 & 0 & 0 \\ 0 & 1 & 0 & 0 \\ 0 & 0 & 1 & 0 \\ \hline -0.0976 & -0.1952 & -0.9759 & 1.0247 \\ -0.1952 & 0.9653 & -0.1735 & 0 \\ -0.9759 & -0.1735 & 0.1323 & 0 \end{array}\right] = 4.$$

The property (7.49) holds since

$$\text{rank} \begin{bmatrix} \mathcal{E} & \mathcal{F} \\ C & 0 \end{bmatrix} = \left[\begin{array}{cccc|cc} 1 & 0 & 0 & 0 & 0.122 & -0.276 \\ 0 & 1 & 0 & 0 & -5.361 & 3.325 \\ 0 & 0 & 1 & 0 & 19.55 & -52.94 \\ -0.0976 & -0.1952 & -0.9759 & 1.0247 & 0 & 0 \\ \hline -0.1952 & 0.9653 & -0.1735 & 0 & 0 & 0 \\ -0.9759 & -0.1735 & 0.1323 & 0 & 0 & 0 \end{array}\right] = 6.$$

Finally, the property (7.50) is satisfied, since for all $s \in \mathbb{C}_+$,

$$\text{rank} \begin{bmatrix} s\mathcal{E} - \mathcal{A} & \mathcal{F} \\ C & 0 \end{bmatrix} = \left[\begin{array}{cccc|cc} s + 0.025 & -1.04 & 9.94 & 0 & 0.122 & -0.276 \\ -57.47 & s & 0 & 0 & -5.361 & 3.325 \\ -1.62 & 0 & s & 0 & 19.55 & -52.94 \\ -0.0976 & -0.1952 & -0.9759 & 1.0247 & 0 & 0 \\ \hline -0.1952 & 0.9653 & -0.1735 & 0 & 0 & 0 \\ -0.9759 & -0.1735 & 0.1323 & 0 & 0 & 0 \end{array}\right] = 6.$$

Therefore, there is an asymptotic observer for the descriptor system (7.6) to estimate accurately χ and $\bar{f}$, and subsequently x, f, d_a and d_s of the lateral system (7.69).

Comparative simulations for the following three observers are carried out to demonstrate the performance of the proposed adaptive sliding mode ASUIO:

- The *attenuation approach full-order observer* in Chap. 5, where f is extended as auxiliary state, and the attacks d_a and d_s are attenuated through H_∞ optimization with the performance gain $\gamma = 9$.
- The *partial decoupling approach observer* in Gao et al. (2016), where f and $\dot{f}$ are extended as state variables, d_a is decoupled and d_s is attenuated with the H_∞ performance gain $\gamma = 0.1$.

- The *proposed observer*, where d_a and d_s are estimated together with x and f. The observer parameters are designed as $\sigma_0 = 2500$, $\epsilon = 0.001$, $\tau = 0.001$ and

$$N = \begin{bmatrix} 12.3986 & -14.9744 & -6.8673 & -4.5017 \\ 32.0494 & -22.9446 & -11.0537 & -5.7266 \\ 140.0155 & -107.84 & -59.861 & -33.4186 \\ 140.1216 & -108.2241 & -58.5872 & -35.1452 \end{bmatrix},$$

$$J = \begin{bmatrix} -0.1745 & 0.9133 \\ -2.7397 & 4.0369 \\ 0.5777 & -0.984 \\ -0.013 & 0.0864 \end{bmatrix}, \; L = \begin{bmatrix} 0 & -16.5 & -130.3 \\ 0 & -24.3 & -196.5 \\ 0 & -128 & -1002.7 \\ 0 & -129.1 & -1010.8 \end{bmatrix},$$

$$W = \begin{bmatrix} -0.0976 & -0.1952 & -0.9759 \\ -0.1952 & 0.9653 & -0.1736 \\ -0.9764 & -0.1735 & 0.1323 \\ 1.0251 & 0.0001 & -0.0001 \end{bmatrix}, \; H = \begin{bmatrix} 0 & 0.0913 & 0.3182 \\ 0 & 1.0523 & 1.8893 \\ 0 & 1.8435 & 10.2453 \\ 0.9759 & 1.9838 & 10.2028 \end{bmatrix}.$$

The observers are simulated under initial state $x(0) = [1 \; -2 \; -1]^\top$, zero observer state and a state feedback H_∞ control law $u(t) = -Kx(t)$ with

$$K = \begin{bmatrix} -6.841 & -0.2919 & 0.1227 \\ 7.9536 & 0.1866 & -0.3261 \end{bmatrix}.$$

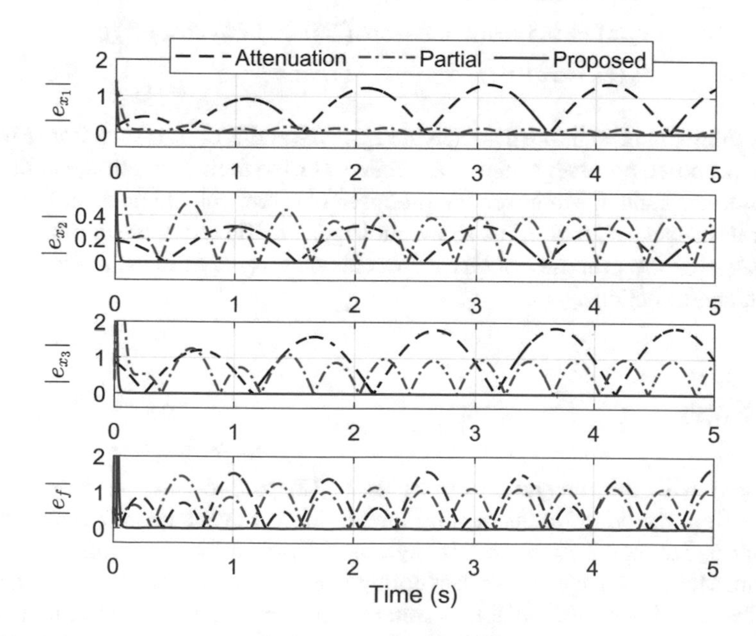

Fig. 7.10 Comparison of state and fault estimation performances

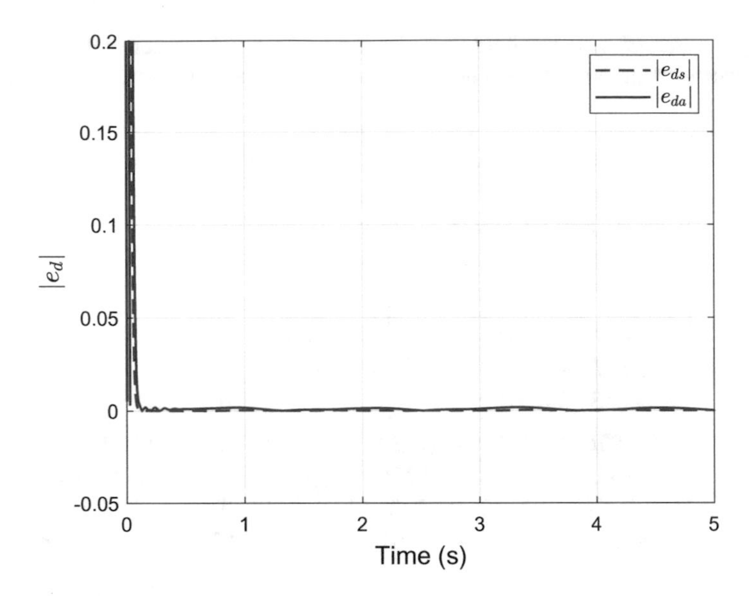

Fig. 7.11 Estimation of d_a and d_s using the proposed observer

The simulated fault and attacks are characterized by

$$
\begin{aligned}
f(t) &= 0.5 + \sin(2.5\pi t),\\
d_a(t) &= 0.5\cos(2.5t) + \cos(2t)\sin(\beta(t))\cos(p(t)),\\
d_s(t) &= \cos(\pi t).
\end{aligned}
$$

The simulation results are depicted in Figs. 7.10 and 7.11. It is seen from Fig. 7.10 that the proposed observer achieves accurate state and fault estimation, but the other two observers cannot. Moreover, the proposed observer can obtain accurate estimation of the attacks d_a and d_s, as shown in Fig. 7.11. These results demonstrate the advantages of the proposed observer in achieving asymptotic estimation of state, fault and perturbations.

7.9 Notes

Compared with the integration strategies in Chaps. 3–6 that are based on H_∞ optimization, the adaptive decoupling integration is advantageous because the FE observer is decoupled from the FTC system. The decoupling affords great design freedom. Moreover, the adaptive decoupling integration strategy can handle actuator faults that are either differentiable or non-differentiable and matched or unmatched.

In the special cases when the perturbation and fault are matched, they can be compensated directly by the control action using their estimates. In such cases, it is unnecessary to reformulate the original system into the triangular form (7.55). FTC can then be achieved by the standard state-feedback FTC controllers in Chaps. 3, 4 and 6, or the sliding mode FTC controllers in Chap. 5.

Section 7.5 considers the FE design when D_2 is not full rank. If, in this case, the original system is still augmented as in (7.6), the resulting descriptor system will be infinitely unobservable and for which no asymptotic observer exists. This is thus the driving force for the system reformulation, where partial perturbations are defined as virtual faults and the rest part as auxiliary state. A similar reformulation has been used in Chan et al. (2017), Ooi et al. (2017) to design FE for infinitely unobservable linear descriptor systems. The systems studied in their works are subject to faults but free from perturbations. The original systems are reformulated into infinitely observable descriptor systems by regarding the partial state as virtual faults. However, the matrix $\mathcal{E}$ in the reformulated descriptor system (7.47) is rectangular, while it is square in Chan et al. (2017), Ooi et al. (2017).

The adaptive decoupling strategy requires the availability of an observable model of the perturbation. Several methods have been described in Chen and Patton (1999) to construct offline the perturbation model using system historical data. However, it can be conservative to use such a model for systems (especially complex systems) with time-varying unmodelled uncertainties.

It is also worth noting that, in the absence of faults, the proposed strategy is reduced to be a disturbance-observer-based control method that has been researched extensively and relates to significant potential industrial applications (Chen et al. 2016).

References

Bélanger PR (1995) Control engineering: a modern approach. Oxford University Press, Inc, Oxford

Chan JCL, Tan CP, Trinh H (2017) Robust fault reconstruction for a class of infinitely unobservable descriptor systems. Int J Syst Sci 48(8):1646–1655

Chen J, Patton RJ (1999) Robust model-based fault diagnosis for dynamic systems. Kluwer Academic Publishers, London

Chen WH, Yang J, Guo L, Li S (2016) Disturbance-observer-based control and related methods-an overview. IEEE Trans Ind Electron 63(2):1083–1095

Corless M, Tu J (1998) State and input estimation for a class of uncertain systems. Automatica 34(6):757–764

Dai L (1989) Singular control systems. Springer, Berlin

Edwards C, Spurgeon SK, Patton RJ (2000) Sliding mode observers for fault detection and isolation. Automatica 36(4):541–553

Eugene L, Kevin W, Howe D (2013) Robust and adaptive control with aerospace applications. Springer, London

Gao Z, Liu X, Chen MZ (2016) Unknown input observer-based robust fault estimation for systems corrupted by partially decoupled disturbances. IEEE Trans Ind Electron 63(4):2537–2547

Hardy GH (1916) Weierstrass's non-differentiable function. Trans Amer Math Soc 17(3):301–325

Isidori A (1995) Nonlinear control systems. Springer, London

Jiang B, Staroswiecki M, Cocquempot V (2006) Fault accommodation for nonlinear dynamic systems. IEEE Trans Autom Control 51(9):1578–1583

Krstic M, Kokotovic PV, Kanellakopoulos I (1995) Nonlinear and adaptive control design. Wiley, New York

Ooi JHT, Tan CP, Nurzaman SG, Ng KY (2017) A sliding mode observer for infinitely unobservable descriptor systems. IEEE Trans Autom Control 62(7):3580–3587

Polyakov A (2012) Nonlinear feedback design for fixed-time stabilization of linear control systems. IEEE Trans Autom Control 57(8):2106–2110

Utkin VI (1992) Sliding modes in control and optimization. Springer Science & Business Media, Berlin

Yu J, Liu Z (2009) Fault reconstruction based on sliding mode observer for linear descriptor systems. In: Proceedings of Asian control conference, IEEE, pp 1132–1137

Zames G (1966) On the input-output stability of time-varying nonlinear feedback systems part one: Conditions derived using concepts of loop gain, conicity, and positivity. IEEE Trans Autom Control 11(2):228–238

Part III
Extension and Application

Chapter 8
Fault-Tolerant Wind Turbine Pitch Control

8.1 Introduction

This chapter serves as a tutorial for implementing FE-based FTC function on the wind turbine pitch control system. A new FE observer is also developed for reconstructing the real shapes of component faults, rather than their fictitious replacement as in Chap. 5.

Wind turbines have contributed to a large proportion of the world's power generation. Meanwhile, there is a strong demand for enhanced reliability of the wind turbine control system to guarantee power generation and reduce cost in operation and maintenance. Within a wind turbine system, hydraulic pitch control subsystems play a critical role because pitch actuation is important for (1) limiting power capture under high wind situations, i.e. in operation Region 3 with effective wind speeds of $12.5 \sim 25$ m/s and (2) mitigating operational load, stalling and aerodynamic braking (Burton et al. 2011). At lower wind speeds, i.e. in operation Region 2 with effective wind speeds of $3 \sim 12.5$ m/s, the pitch actuation system is inactive and the turbine conversion efficiency is regulated by rotor speed control.

In real operations, pitch systems may have actuator faults caused by a pressure drop in the hydraulic supply system or high air content in the oil. These faults can lead to slow pitch action and make it impossible for the pitch control to maintain the rotor at rated speed, and thus cause fluctuations in the generator speed and power and degrade wind turbine system stability (Odgaard et al. 2013; Ribrant and Bertling 2007). This study considers a low-pressure pitch actuator fault, which is faster and more severe than the high air content fault. In the presence of low-pressure actuator faults, the existing nominal (baseline) control alone cannot maintain a desired pitch action, which gives rise to the need for FTC for pitch systems. FTC can be used to achieve baseline control system performance in fault-free cases and compensate fault effects to maintain robust system performance when faults occur. Moreover, the FTC strategy can also be used to provide fault information for effective use in subsequent maintenance schedules by incorporating a fault diagnosis module.

© The Author(s), under exclusive license to Springer Nature Switzerland AG 2021

J. Lan and R. J. Patton, *Robust Integration of Model-Based Fault Estimation and Fault-Tolerant Control*, Advances in Industrial Control,
https://doi.org/10.1007/978-3-030-58760-4_8

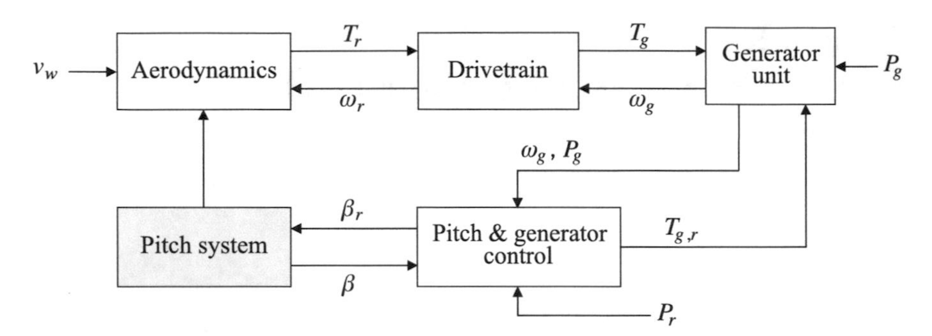

Fig. 8.1 The 4.8 MW benchmark wind turbine control system

8.2 Problem Description

This chapter studies the representative 4.8 MW benchmark wind turbine, whose control architecture is outlined in Fig. 8.1. The wind turbine closed-loop system comprises five subsystems (Odgaard et al. 2013): aerodynamics, pitch system, drivetrain, generator unit and pitch & generator control. The system variables are defined as follows: v_w is the wind speed acting on the turbine blades, β is the pitch angle, T_r is the rotor torque, ω_r is the rotation speed of the rotor, T_g is the generator torque, ω_g is the rotation speed of the generator, P_g is the generator power, $T_{g,r}$ is the torque reference to the generator, P_r is the power reference to the wind turbine and β_r is the angle reference to the pitch actuator system. It is assumed throughout this chapter that the yaw control has a normal operation with no yaw misalignment.

The control subsystem includes both pitch and generator speed controls, used to generate the angle and torque references β_r and $T_{g,r}$, respectively. At wind speeds below the rated value (<12.5 m/s), the pitch control signal β_r is set zero and the generator control is used to adjust the rotor speed to maximize power capture. In Region 3, pitch control is required to keep the rotor at rated speed. Therefore, this work considers the pitch control design in Region 3 and it is assumed that there are no stuck actuator faults so that the proposed control can also work well in Region 2.

In the benchmark model, the three pitch systems each has an individual pitch actuator and the three individual actuators are assumed to have the same dynamic structure. All the pitch systems have the same control input (i.e. collective pitch control). For the sake of simplicity, a single pitch system is considered in the design procedure. In principle, the pitch system is a piston servo modelled by the following second-order system (Odgaard et al. 2013):

$$\begin{aligned} \dot{x}_1 &= x_2 \\ \dot{x}_2 &= -\omega_n^2 x_1 - 2\xi\omega_n x_2 + \omega_n^2 u \\ y &= x_1 \end{aligned} \tag{8.1}$$

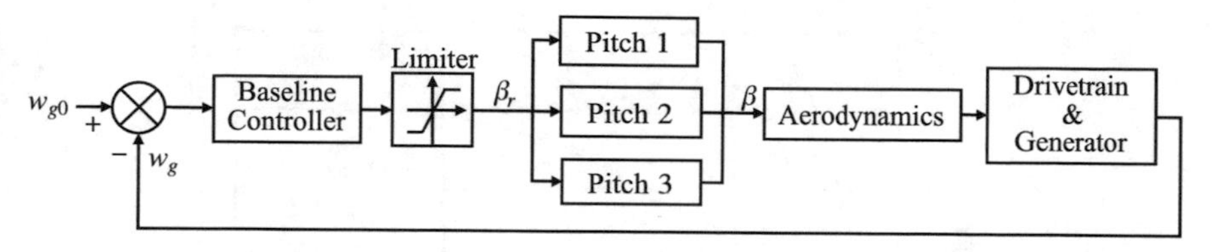

Fig. 8.2 Nominal pitch control system

where $[x_1\ x_2]^\top = [\beta\ \dot{\beta}]^\top$ are the system state variables, $u = \beta_r$ is the pitch command input and y is the output angle. β and $\dot{\beta}$ are the pitch angle and angular velocity, respectively. ω_n and ξ are the natural frequency and damping factor, respectively.

The collective pitch control scheme used here has a single command input β_r acting on the three pitch systems to actuate a common desired angle β, as outlined in Fig. 8.2. In the scheme, w_{g0} and w_g are the nominal (rated) and real speeds of the generator, respectively. Considering the physical constraints of each actuator system, a magnitude and rate limiter is implemented in the benchmark.

Each pitch system may have faults resulting from dynamical changes (variations of the values of ω_n and ξ), due to a drop in hydraulic pressure (Esbensen and Sloth 2009). Under this low pressure fault, the parameters ω_n^2 and $\xi\omega_n$ in the pitch system (8.1) can be modelled as convex combinations of their values at the nominal and low-pressure situations and given by

$$\begin{aligned} \omega_n^2 &= \omega_{n_0}^2 + (\omega_{n_f}^2 - \omega_{n_0}^2) f \\ \xi\omega_n &= \xi_0\omega_{n_0} + (\xi_f\omega_{n_f} - \xi_0\omega_{n_0}) f \end{aligned} \tag{8.2}$$

where ω_{n_0} and ξ_0 are the nominal values of ω_n and ξ, while ω_{n_f} and ξ_f are their values at low pressure. The unknown function $f \in [0, 1]$ is a fault indicator. $f = 0$ corresponds to the normal pressure with $\omega_{n_0} = 11.11$ rad/s and $\xi_0 = 0.6$ rad/s, and $f = 1$ corresponds to the low pressure with $\omega_{n_f} = 3.42$ rad/s and $\xi_f = 0.9$ rad/s. It is rational to assume that $\|\dot{f}\| \le \bar{f}_0 < \infty$ for some unknown constant $\bar{f}_0$.

Unit step responses of a pitch system under different values of f are shown in Fig. 8.3. It can be seen that the pressure drop slows down the pitch actuator dynamics. Therefore, when suffering from a low-pressure fault, the pitch actuator dynamics become slow and the pitching performance is degraded. To recover the pitch action, an ideal solution is to compensate for the fault by implicitly changing the dynamic characteristics of the actuator.

A projection-based FTC design is proposed in Jain et al. (2013) for pitch systems without explicit use of the fault information (detection or estimation). An FTC pitch system is developed in Sloth et al. (2011) based on FDI. However, it is difficult or impossible for this FDI strategy to get the pitch actuator fault magnitude. Observer-based FE approaches have been applied to replace the FDI approach to reconstruct the pitch actuator fault based on the fault model in (8.2). In Chen et al. (2013), fault shapes are attained using an adaptive observer whose gains are solved via LMIs

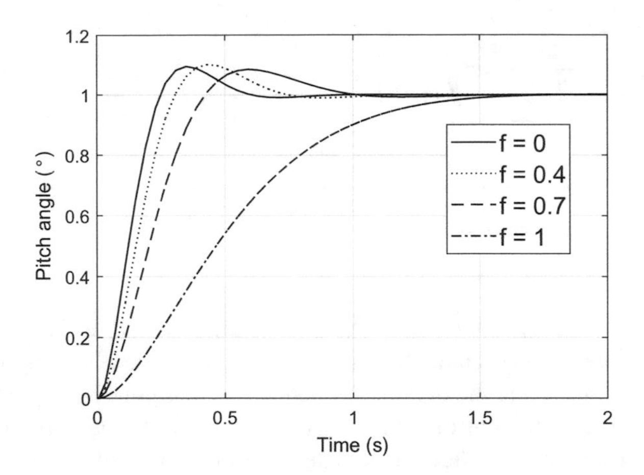

Fig. 8.3 Step response of a pitch system under different fault conditions

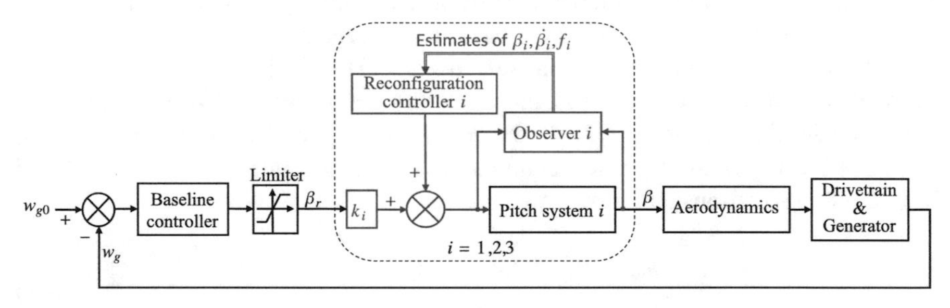

Fig. 8.4 Fault-tolerant pitch system control scheme

dependent on the fault indicator. A reconfigurable FTC design is proposed in Shi and Patton (2015) based on a linear parameter varying system modelling of the wind turbine system, using an extended state observer. Although their FTC design is proved to be effective, a fictitious actuator fault rather than the real fault indicator function is reconstructed, which is seen to be a disadvantage.

In this chapter, an FE-based FTC pitch system is designed, following the structure outlined in Fig. 8.4, to maintain the desired pitching action despite the existence of actuator faults. The FTC pitch system involves (1) a baseline controller, (2) observers for estimating the states and faults and (3) reconfigurable controllers for fault compensation. Besides, the gains k_i, $i = 1, 2, 3$, are also to be designed. In the presence of actuator faults, the three pitch systems have the same baseline controller as the fault-free case in Fig. 8.2. It should also be noted that in the fault-free case this FTC system reverts to the baseline system in Fig. 8.2.

8.3 Adaptive Step-by-Step SMO FE Design

By using the convex model (8.2) of the pitch actuator faults, the original parametric faults are converted into a form of additive faults. Hence, the pitch system (8.1) with actuator faults and uncertainty can be represented by

$$\begin{aligned} \dot{x}_1 &= x_2 \\ \dot{x}_2 &= G_0(x) + B_0 u + F(x) f + d \\ y &= x_1 \end{aligned} \tag{8.3}$$

where

$$x = [x_1 \; x_2]^\top, \; G_0(x) = -\omega_{n_0}^2 x_1 - 2\xi_0 \omega_{n_0} x_2, \; B_0 = \omega_{n_0}^2,$$
$$F(x) = (\omega_{n_0}^2 - \omega_{n_f}^2)(x_1 - u) + 2(\xi_0 \omega_{n_0} - \xi_f \omega_{n_f}) x_2,$$

and d is some unknown uncertainty.

As shown in (8.3), the fault distribution function $F(x)$ is nonlinear, making it difficult or impossible to achieve FE using the existing observers. If the term $F(x)f$ as a whole is regarded as a fictitious actuator fault, then the faulty wind pitch system is in a similar form of the systems studied in Chaps. 3–6. Therefore, the strategies developed in those chapters can be directly applied to achieve robust integration of FE and FTC for wind turbine pitch systems. This chapter considers reconstructing directly the shape of the real fault signal, rather its fictitious replacement. Moreover, inspired by the adaptive decoupling strategy described in Chap. 7, it is desirable to also estimate and compensate the uncertainty d to improve the FE and FTC robustness.

A step-by-step SMO to estimate the state, fault and uncertainty of the system (8.3) is designed in the form of

$$\begin{aligned} \dot{\hat{x}}_1 &= \hat{x}_2 + v_1 \\ \dot{\hat{x}}_2 &= G_0(\hat{x}) + B_0 u + F(\hat{x}) \hat{f} + \hat{d} + v_2 \\ \dot{\hat{f}} &= \eta_f \text{sign}(e_f) \\ \dot{\hat{d}} &= \eta_d \text{sign}(e_d) \end{aligned} \tag{8.4}$$

where $\hat{x} = [\hat{x}_1 \; \hat{x}_2]^\top$ are the state estimates, $\hat{f}$ is the estimate of f and $\hat{d}$ is the estimate of d. Define the estimation errors as $e_{x_1} = x_1 - \hat{x}_1$, $e_{x_2} = x_2 - \hat{x}_2$, $e_f = f - \hat{f}$ and $e_d = d - \hat{d}$, respectively. The switching functions v_1 and v_2 are defined as

$$v_1 = \eta_{v_1} \text{sign}(e_{x_1}), \; v_2 = \eta_{v_2} \text{sign}(\tilde{x}_2 - \hat{x}_2) \tag{8.5}$$

where $\tilde{x}_2 = \hat{x}_2 + v_1$, and the scalars η_f, η_d, η_{v_1} and η_{v_2} are to be designed.

The existence of the asymptotically stable observer (8.4) is proved below based on the Lyapunov stability theory.

Theorem 8.1 *For the pitch system (8.3), there exists an adaptive step-by-step SMO (8.4) to achieve asymptotic estimation of state x, fault f and uncertainty d.*

Proof **(1) Step 1**

Define $\tilde{G_0} = G_0(x) - G_0(\hat{x})$ and $\tilde{F} = F(x) - F(\hat{x})$, then $F(x)f - F(\hat{x})\hat{f} = \tilde{F}f + F(\hat{x})e_f$. By using (8.3) and (8.4), the estimation error system is obtained as

$$\dot{e}_{x_1} = e_{x_2} - v_1 \tag{8.6}$$

$$\dot{e}_{x_2} = \tilde{G_0} + \tilde{F}f + F(\hat{x})e_f + e_d - v_2 \tag{8.7}$$

$$\dot{e}_f = \dot{f} - \eta_f \text{sign}(e_f) \tag{8.8}$$

$$\dot{e}_d = \dot{d} - \eta_d \text{sign}(e_d). \tag{8.9}$$

A Lyapunov function for the subsystem (8.6) is defined as

$$V_1 = \frac{1}{2}e_{x_1}^2.$$

By using (8.5) and (8.6), it can be derived that

$$\begin{aligned} \dot{V}_1 &= e_{x_1}(e_{x_2} - v_1) \\ &\leq (\|e_{x_2}\| - \eta_{v_1})\|e_{x_1}\|. \end{aligned} \tag{8.10}$$

All the signals in (8.7) are bounded in finite time, so is e_{x_2}. Hence, there exists a positive scalar satisfying $\rho_{v_1} \geq \|e_{x_2}\|$. In order to compensate the effect of the unknown estimation error e_{x_2}, design $\eta_{v_1} = \rho_{v_1} + \varepsilon_{v_1}$, where ε_{v_1} is a positive constant. Substituting η_{v_1} into (8.10) yields

$$\dot{V}_1 \leq -\varepsilon_{v_1}\|e_{x_1}\|. \tag{8.11}$$

It follows from (8.11) that $\dot{V}_1 \leq -\alpha_1\sqrt{V_1}$ with $\alpha_1 = \sqrt{2}\varepsilon_{v_1}$. Hence, (8.11) satisfies the standard reachability condition (Edwards and Spurgeon 1998) and the sliding surface $s_1 = e_{x_1} = 0$ is reached in finite time $t_r \leq 2\sqrt{V(0)}/\alpha_1$.

(2) Step 2

According to (8.6) and the proof in Step 1, the dynamics of e_{x_1} reach the sliding surface $s_1 = 0$ in finite time t_r and remain on it thereafter. After t_r, by using the equivalent control input concept, the subsystem (8.6) is reduced to be

$$0 = e_{x_2} - v_{eq,1},$$

where $v_{eq,1}$ is the equivalent control signal. Thus, after a finite time t_r, it holds that $\tilde{x}_2 = \hat{x}_2 + e_{x_2}$ and $v_2 = \eta_{v_2}\text{sign}(e_{x_2})$.

A Lyapunov function for the subsystems (8.7)–(8.9) is defined as

$$\dot{V}_{20} = \frac{1}{2}e_{x_2}^2 + \frac{1}{2}e_f^2 + \frac{1}{2}e_d^2. \tag{8.12}$$

It can be derived that

$$\begin{aligned}\dot{V}_{20} &= e_{x_2}[\tilde{G_0} + \tilde{F}f + F(\hat{x})e_f - v_2] + e_f(\dot{f} - \eta_f \text{sign}(e_f)) + e_f(\dot{d} - \eta_d \text{sign}(e_d)) \\ &\le [\|\tilde{G_0}\| + \|\tilde{F}f\| + \|F(\hat{x})e_f\| - \eta_{v_2}]\|e_{x_2}\| + (\rho_f - \eta_f)\|e_f\| + (\rho_d - \eta_d)\|e_d\| \\ &\le (\rho_{v_2} - \eta_{v_2})\|e_{x_2}\| + (\rho_f - \eta_f)\|e_f\| + (\rho_d - \eta_d)\|e_d\| \end{aligned} \tag{8.13}$$

where $\rho_{v_2} \ge \|\tilde{G_0}\| + \|\tilde{F}f\| + \|F(\hat{x})e_f\|$, $\rho_f \ge \|\dot{f}\|$ and $\rho_d \ge \|\dot{d}\|$.

Using $\hat{\rho}_{v_2}$, $\hat{\rho}_f$ and $\hat{\rho}_d$ to estimate ρ_{v_2}, ρ_f and ρ_d, respectively, with the update laws

$$\dot{\hat{\rho}}_{v_2} = \sigma_{v_2}\|e_{x_2}\|, \; \dot{\hat{\rho}}_f = \sigma_f\|e_f\|, \; \dot{\hat{\rho}}_d = \sigma_d\|e_d\| \tag{8.14}$$

where σ_{v_2}, σ_f and σ_d are positive design constants.

Design $\eta_{v_2} = \hat{\rho}_{v_2} + \varepsilon_{v_2}$, $\eta_f = \hat{\rho}_f + \varepsilon_f$ and $\eta_d = \hat{\rho}_d + \varepsilon_d$ with positive design constants ε_{v_2}, ε_f and ε_d. Define the estimation errors of ρ_{v_2}, ρ_f and ρ_d as $\tilde{\rho}_{v_2} = \rho_{v_2} - \hat{\rho}_{v_2}$, $\tilde{\rho}_f = \rho_f - \hat{\rho}_f$ and $\tilde{\rho}_d = \rho_d - \hat{\rho}_d$, respectively. A Lyapunov function for the composite estimation error system, including $e_{x_2}, e_f, e_d, \tilde{\rho}_{v_2}, \tilde{\rho}_f$ and $\tilde{\rho}_d$, is defined as

$$V_2 = V_{20} + \frac{1}{2\sigma_{v_2}}\tilde{\rho}_{v_2}^2 + \frac{1}{2\sigma_f}\tilde{\rho}_f^2 + \frac{1}{2\sigma_d}\tilde{\rho}_d^2.$$

By using (8.13) and (8.14), it is derived that

$$\begin{aligned}\dot{V}_2 &= \dot{V}_{20} + \frac{1}{\sigma_{v_2}}(-\tilde{\rho}_{v_2}\dot{\hat{\rho}}_{v_2}) + \frac{1}{\sigma_f}(-\tilde{\rho}_f\dot{\hat{\rho}}_f) + \frac{1}{\sigma_d}(-\tilde{\rho}_d\dot{\hat{\rho}}_d) \\ &\le (\rho_{v_2} - \tilde{\rho}_{v_2} - \eta_{v_2})\|e_{x_2}\| + (\rho_f - \tilde{\rho}_f - \eta_f)\|e_f\| + (\rho_d - \tilde{\rho}_d - \eta_d)\|e_d\| \\ &\le -\varepsilon_{v_2}\|e_{x_2}\| - \varepsilon_f\|e_f\| - \varepsilon_d\|e_d\| \\ &\le 0. \end{aligned} \tag{8.15}$$

It follows from Barbalat's Lemma (see Sect. 1.7.1) that $\lim_{t\to\infty} V_2(t) = 0$. Therefore, $V_2(t) \le V_2(0)$ and e_{x_2}, e_f, $\tilde{\rho}_{v_2}$, $\tilde{\rho}_f$ and $\tilde{\rho}_d$ are bounded. Furthermore, $|e_{x_2}(t)| \le \sqrt{2V_2(0)}$, $|e_f(t)| \le \sqrt{2V_2(0)}$ and $|e_d(t)| \le \sqrt{2V_2(0)}$. Under zero initial conditions, $e_{x_2}(0) = e_f(0) = \tilde{\rho}_{v_2}(0) = \tilde{\rho}_f(0) = \tilde{\rho}_d(0) = 0$, and thus $V_2(0) = 0$. Hence, it holds that $\lim_{t\to\infty} e_{x_2}(t) = 0$, $\lim_{t\to\infty} e_f(t) = 0$ and $\lim_{t\to\infty} e_d(t) = 0$.

It is concluded here that for the pitch system (8.3), by ensuring satisfaction of the zero initial conditions, then there exist parameters η_{v_1}, η_{v_2}, η_f and η_d such that the observer (8.4) can estimate the system state, fault and uncertainty accurately. □

Remark 8.1 Although e_f and e_d are unavailable in practice, it is shown below that they can be obtained using the equivalent control input concept. Design a sliding surface as $s_2 = e_{x_2} = 0$. Since the switching functions v_1 and v_2 ensure the reachability of s_2, i.e. $s_2 = \dot{s}_2 = 0$, then the equivalent control input signal of v_2 is

$$v_{eq,2} = F_d e_{fd},$$

with $F_d = [F(\hat{x})\ 1]$ and $e_{fd} = [e_f\ e_d]^\top$. Hence, it can be derived that

$$e_{fd} = F_d^\dagger v_{eq,2},\ F_d^\dagger = F_d^\top (F_d F_d^\top)^{-1} = \frac{F_d^\top}{F(\hat{x})^2 + 1}.$$

Therefore, e_f and e_d are calculated by

$$\begin{aligned} e_f &= [1\ 0] e_{fd} = \frac{F(\hat{x})}{F(\hat{x})^2 + 1} v_{eq,2}, \\ e_d &= [0\ 1] e_{fd} = \frac{1}{F(\hat{x})^2 + 1} v_{eq,2}. \end{aligned} \tag{8.16}$$

Furthermore, $v_{eq,2}$ can be attained by passing v_2 through a low-pass filter, i.e.

$$v_{eq,2} \cong \frac{1}{1 + \tau_1 s} v_2,$$

where τ_1 is a time constant.

Since the fault indicator function is known to be $f \in [0, 1]$, a magnitude limiter is used to ensure $\hat{f} \in [0, 1]$.

Remark 8.2 For the pitch system (8.3), the observer design parameters are σ_{v_1}, ε_{v_1}, σ_{v_2}, ε_{v_2}, σ_f, ε_f, σ_d, ε_d and τ_1. All these are positive constants determined offline, while ε_{v_1}, ε_{v_2}, ε_f, ε_d and τ_1 are of small values and should be tuned by trial and error. However, since the three pitch systems have the same dynamics, the same set of parameters can be applied to them, despite that they may have different faults and uncertainties. This can ease the real implementation of the proposed FE design.

Remark 8.3 Alternatively, by defining the fictitious actuator fault $f_a = F(x)f$, then f_a can be estimated using the observers in Chaps. 3–6. Once the estimate of state $\hat{x}$ and fault $\hat{f}_a$ are obtained, the real fault signal f can be reconstructed by $\hat{f} = \hat{f}_a / F(\hat{x})$. In order to avoid singularity when $F(\hat{x}) = 0$, $\hat{f}$ can be approximated by the following function in real-time implementation:

$$\hat{f} = \frac{F(\hat{x})}{F(\hat{x})^2 + \epsilon} \hat{f}_a,$$

where ϵ is a small enough positive scalar. However, it should be noted that by using the observers developed in Chaps. 3–5, the uncertainty d is attenuated through H_∞

optimization. This, together with the approximation above, normally results in a fault estimation that is less accurate than the one obtained in this section.

8.4 FTC Design

The pitch actuator system (8.3) can be rearranged as

$$\begin{aligned}\dot{x}_1 &= x_2\\ \dot{x}_2 &= G_0(x) + \omega_n^2 u + F_1(x) f + d\end{aligned} \tag{8.17}$$

where $F_1(x) = (\omega_{n0}^2 - \omega_{n_f}^2)x_1 + 2(\xi_0\omega_{n_0} - \xi_f\omega_{n_f})x_2$.

An FTC controller for the pitch system (8.3), comprising a baseline controller and a reconfigurable controller (see Fig. 8.4), is designed as

$$u = k_1\beta_r + u_{fd} \tag{8.18}$$

where β_r is the baseline controller to achieve pitch angle control under fault-free case, and u_{fd} is the reconfigurable controller to compensate the fault and uncertainty. The design parameter k_1 is used to modify the baseline controller. The designs of β_r, k_1 and u_{fd} are described below.

8.4.1 Baseline Controller

Proportional-Integral (PI) baseline controllers have been used effectively in wind turbine pitch control in academic research (Boukhezzar et al. 2007; Gao and Gao 2016; Hand 1999) and industrial implementation (Burton et al. 2011) due to its facilitation and robustness. Therefore, in this study, the baseline controller β_r for the three pitch systems is chosen as a PI controller

$$\beta_r(t) = K_P e_{w_g}(t) + K_I \int_0^t e_{w_g}(\tau)\mathrm{d}\tau \tag{8.19}$$

where $e_{w_g} = w_g - w_{g0}$ is the tracking error of the generator speed. K_P and K_I are the proportional and integral gains, respectively.

The modification parameter k_1 is designed as

$$k_1 = \frac{\omega_{n0}^2}{\hat{\omega}_n^2} \tag{8.20}$$

where $\hat{\omega}_n^2 = \omega_{n_0}^2 + (\omega_{n_f}^2 - \omega_{n_0}^2)\hat{f}$.

Clearly, it can be seen that in the absence of fault and uncertainty, $k_1 = 1$ and the FTC controller reverts to a baseline controller.

8.4.2 Reconfiguration Controller

The reconfiguration controller u_{fd} is activated automatically once the fault f or uncertainty d occurs and it is designed as

$$u_{fd} = -\frac{F_1(\hat{x})\hat{f} + \hat{d}}{\hat{\omega}_n^2} \tag{8.21}$$

where $F_1(\hat{x}) = (\omega_{n_0}^2 - \omega_{n_f}^2)\hat{x}_1 + 2(\xi_0\omega_{n_0} - \xi_f\omega_{n_f})\hat{x}_2$.

8.4.3 FTC Performance Analysis

The FTC system performance based on (8.18) is analysed in Theorem 8.2.

Theorem 8.2 *With the asymptotic estimation of state and fault obtained from the observer (8.4), the controller (8.18) can compensate the fault f and the uncertainty d to recover the nominal pitch dynamics.*

Proof Substituting (8.18) with (8.19)–(8.21) into (8.17) yields

$$\begin{aligned} \dot{x}_1 &= x_2 \\ \dot{x}_2 &= G_0(x) + B_0\beta_r + \delta(e, t) \end{aligned} \tag{8.22}$$

where

$$\begin{aligned} \delta(e, t) = &\ (F_1(x) - F_1(\hat{x}))\hat{f} + e_d + F_1(x)e_f \\ &+ (\omega_{n_f}^2 - \omega_{n_0}^2)(\omega_{n_0}^2\beta_r - F_1(\hat{x})\hat{f})e_f/\hat{\omega}_n^2. \end{aligned}$$

Notice that $\delta(e, t)$ is a function of the estimation errors e_{x_1}, e_{x_2}, e_f and e_d. It has been proved in Sect. 8.3 that $\lim_{t\to\infty} e_{x_1}(t) = 0$, and $e_{x_2}(t)$, $e_f(t)$ and $e_d(t)$ are bounded. Hence, $\delta(e, t)$ is bounded. Moreover, under zero initial conditions, i.e. $e_{x_1}(0) = e_{x_2}(0) = e_f(0) = e_d(0) = 0$, then $\lim_{t\to\infty} \delta(e, t) = 0$. In such case, the closed-loop system (8.22) becomes

$$\begin{aligned} \dot{x}_1 &= x_2 \\ \dot{x}_2 &= G_0(x) + B_0\beta_r \end{aligned} \tag{8.23}$$

which is exactly the fault-free pitch system (8.1).

It is concluded here that the controller (8.18) with (8.19)–(8.21) can compensate the fault and uncertainty in the pitch system (8.17) automatically, by using the estimates of state, fault and uncertainty from the observer (8.4). The compensation guarantees that the faulty pitch system performs as a healthy one. □

8.5 Simulation Results

This section demonstrates efficacy of the proposed FE-based FTC design by applying it to a 4.8 MW wind turbine benchmark (Odgaard et al. 2013). The rated generator speed is 162 rad/s. Limits of the pitch angle and pitch rate are $[-2°, 90°]$ and $[-9°/\text{s}, 9°/\text{s}]$, respectively. The measurement noise is a zero mean white Gaussian noise with the variance of 1.0e-10.

The three pitch systems use the same PI baseline controller (8.19) with $K_P = 1$ and $K_I = 4$. Since all the three pitch actuator systems have the same dynamics, the same set of parameters can be applied to them. This facilitates the real implementation. To illustrate this, observers for all the three pitch actuator systems are implemented using the same parameters: $\rho_{v_1} = \sigma_{v_2} = \sigma_f = \sigma_d = 0.1$, $\varepsilon_{v_1} = \varepsilon_{v_2} = \varepsilon_f = \tau_1 = 0.001$.

The other configurations of the wind turbine system are the same as those in the benchmark. However, the three pitch actuator systems are simulated with the following uncertainties: $d_1 = 0.1x_1 \sin(0.01x_2)$, $d_2 = 0.05 \cos(0.01x_2)$ and $d_3 = 0.05 \sin(0.1\pi t)$, respectively.

Simulations are carried out for two cases: (1) only the pitch system three has an actuator fault f_3 and (2) the three pitch systems have faults f_1, f_2 and f_3, respectively. All the simulations use the same wind speed given in Fig. 8.5, which covers almost the whole area of Region 3 (12.5 m/s – 25 m/s) and part of Region 2 (3 m/s – 12.5 m/s). In Region 2, the wind turbine aims at maximizing power capture and the pitch angles are set as zero. Thus, the pitch angle references are always set as zero in Region 2.

In order to demonstrate the efficacy of the proposed FE-based FTC design, the results of the following systems are compared:

- *Fault-free system.* It is the nominal pitch system implemented with the baseline controller but without any fault or uncertainty.
- *Faulty without FTC system.* It is the pitch system subject to fault and uncertainty but implemented with the baseline controller alone.
- *Faulty with FTC system.* It is the pitch system subject to fault and uncertainty and implemented with the proposed FE-based FTC design.

8.5.1 *Case 1: Single Actuator Fault*

This subsection considers simulation of the wind turbine system with a single fault f_3 on pitch system 3.

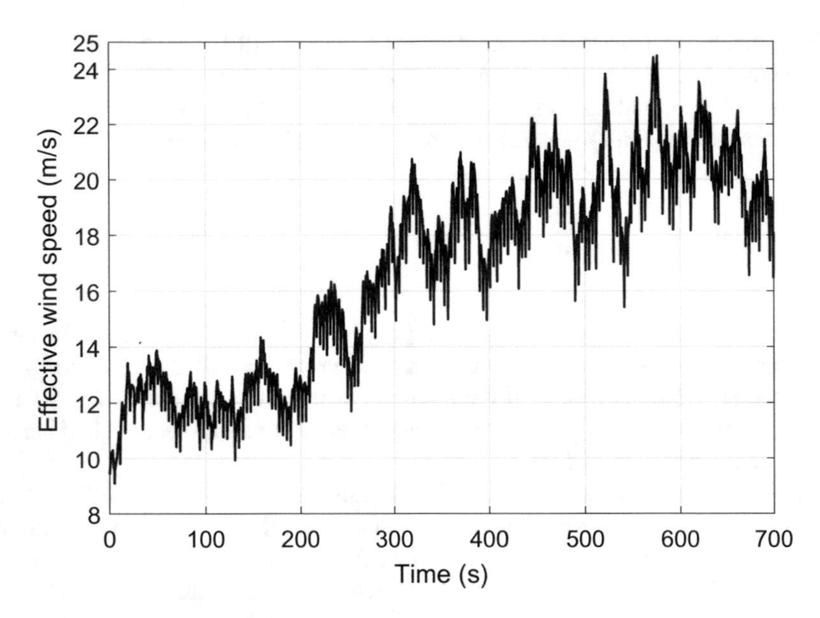

Fig. 8.5 Effective wind speed

It is shown in Fig. 8.6 that the fault f_3 on pitch system 3 is estimated with acceptable accuracy. In the time interval $t \in [0, 250]$ s, the effective wind speed changes between Regions 2 and 3. Note that in Region 2 the pitch angle reference is always set to be zero and the pitch control is inactive. Hence, the fault f_3 is not well estimated within the time period of $t \in [0, 250]$ s.

It is shown in Figs. 8.7, 8.8, 8.9, 8.10, 8.11, 8.12, 8.13 and 8.14 that in the presence of f_3, the pitch actuator 3 has slow pitch rate. This causes undesirable pitch angle responses of the pitch system 3 and subsequently the pitch systems 1 and 2. The undesired pitching results in larger fluctuations in the generator speed and power. However, it can be seen from the results that there is no difference between the performances of fault-free and FTC cases. This means that the proposed FTC design can compensate well the fault f_3 and recover pitch actuator 3 to its nominal situation.

8.5.2 Case 2: Multiple Actuator Faults

This subsection further considers simulation of the wind turbine system with faults in all the three pitch actuators.

It is seen from Figs. 8.15, 8.16 and 8.17 that the three faults are estimated with good accuracy except in the time period $t \in [0, 250]$ s as explained in Case 1. As shown in Figs. 8.18, 8.19, 8.20, 8.21, 8.22, 8.23, 8.24 and 8.25, the faults slow down the dynamics of three pitch systems, and thus the generator cannot be kept

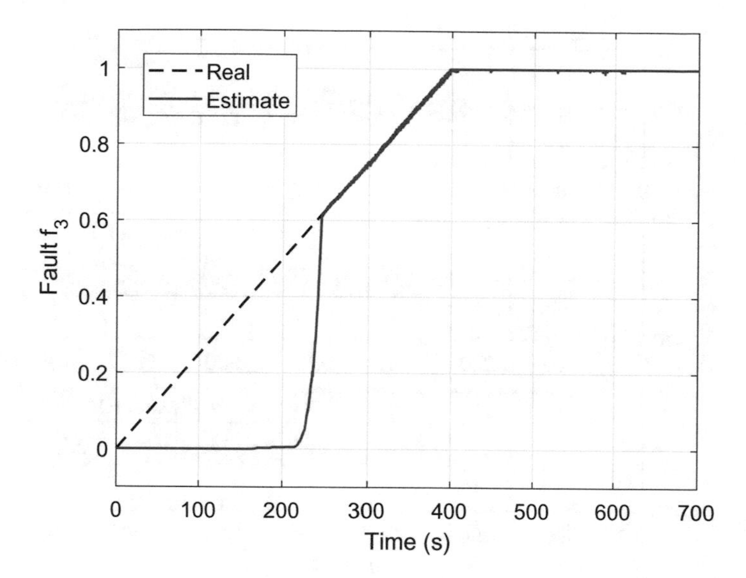

Fig. 8.6 Fault estimation: pitch 3, Case 1

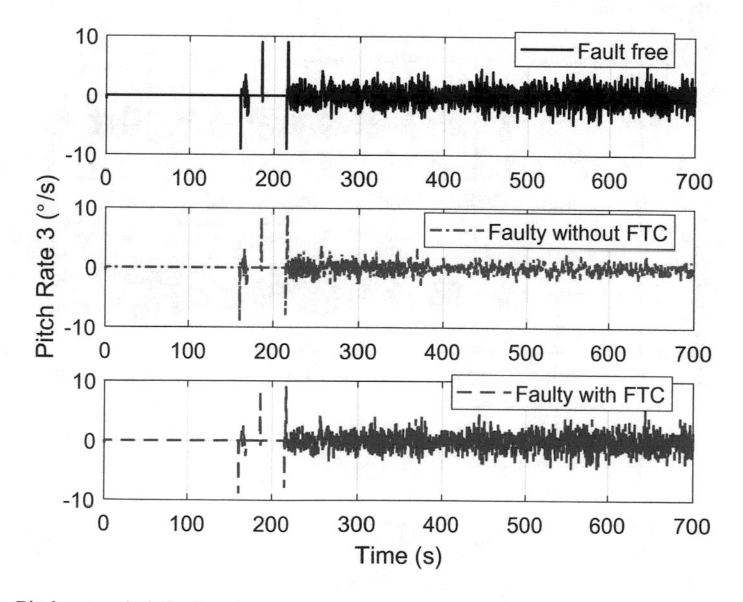

Fig. 8.7 Pitch rate: pitch 3, Case 1

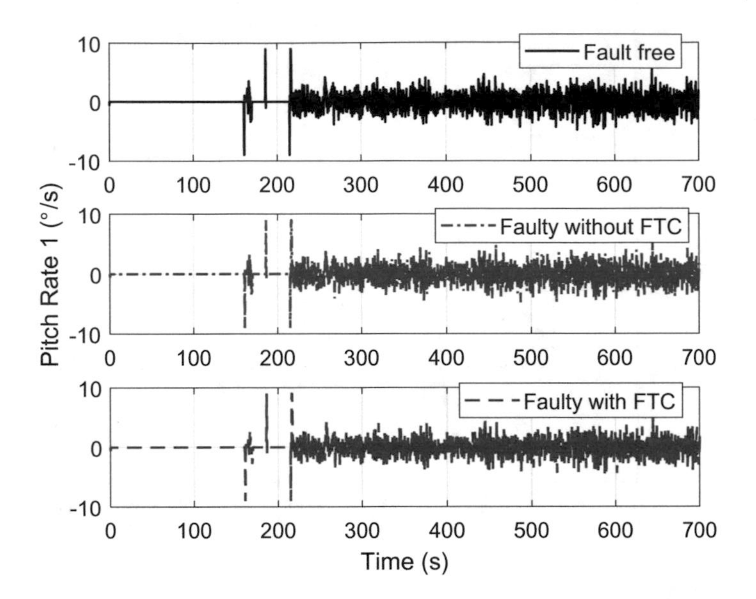

Fig. 8.8 Pitch rate: pitch 1, Case 1

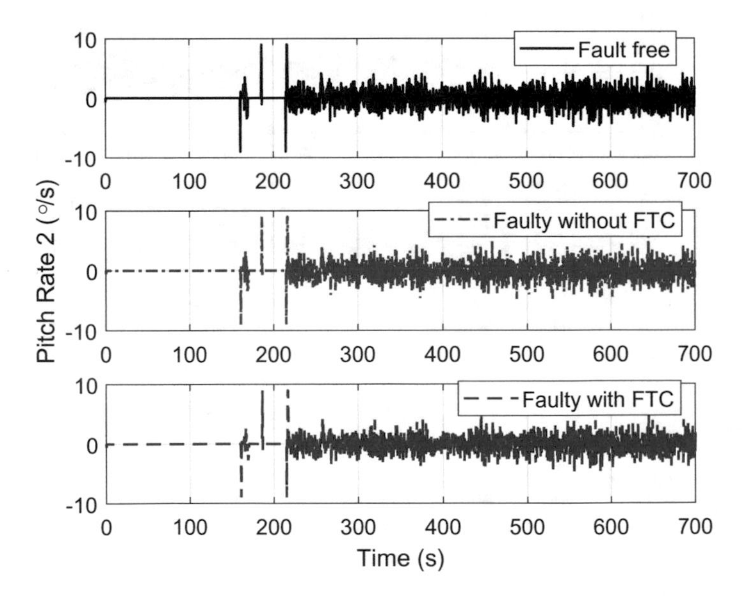

Fig. 8.9 Pitch rate: pitch 2, Case 1

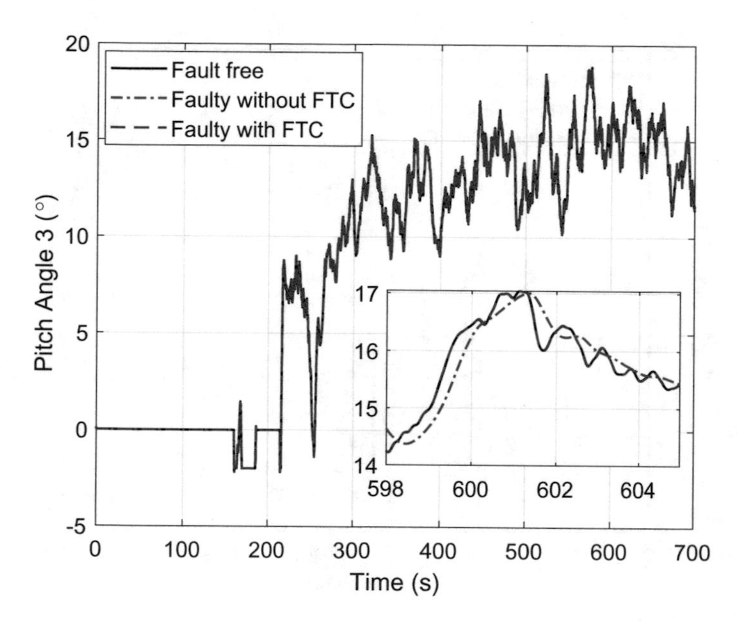

Fig. 8.10 Pitch angle: pitch 3, Case 1

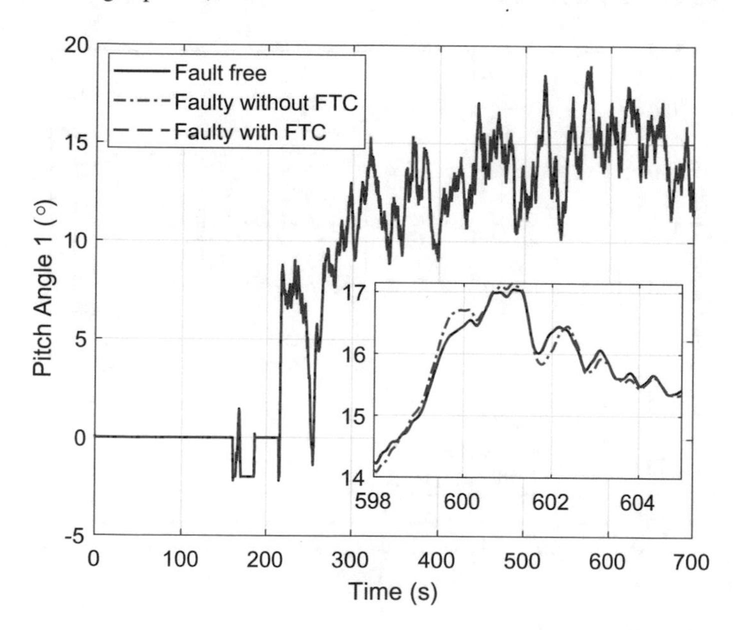

Fig. 8.11 Pitch angle: pitch 1, Case 1

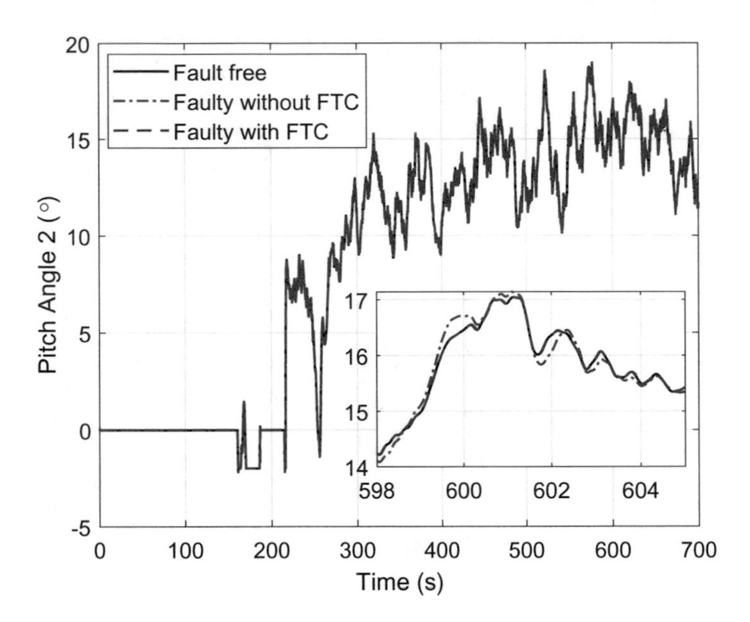

Fig. 8.12 Pitch angle: pitch 2, Case 1

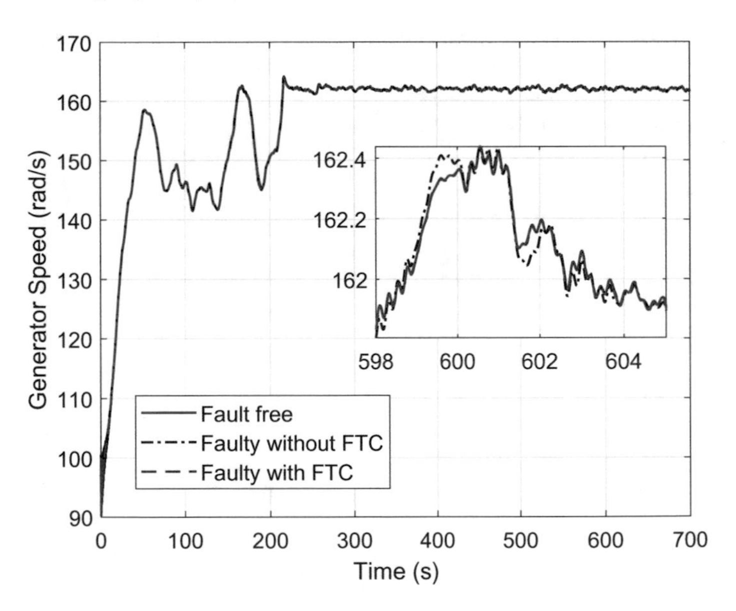

Fig. 8.13 Generator speed: Case 1

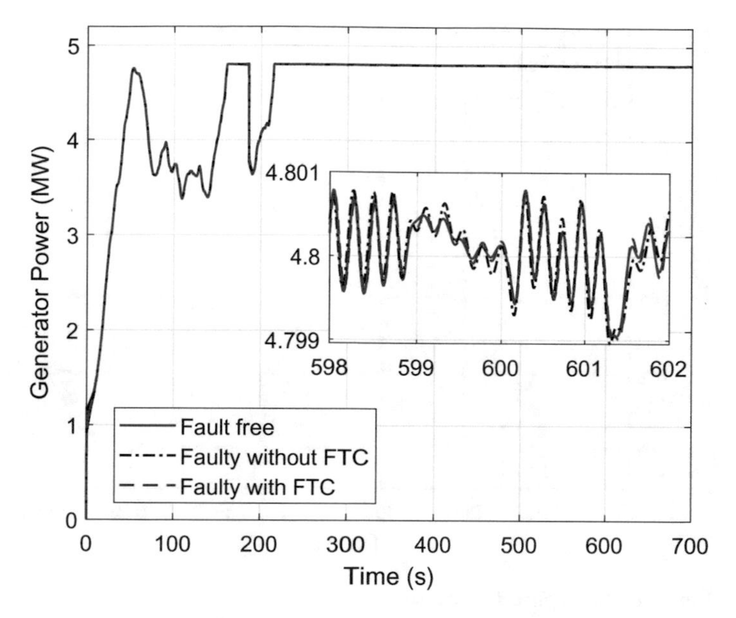

Fig. 8.14 Generator power: Case 1

at the rated speed. The fluctuating generator speed subsequently causes variations in the generated power and leads to pitch angle oscillation and pitch rate saturation. Moreover, the oscillation will unbalance the three pitch systems and increase their blade loads with potential damage to the wind turbine, especially for large turbines (Bossanyi 2003). However, the proposed FTC design compensates the faults well and recovers the pitch actuator dynamics, which ensures desired pitching action and avoids pitch rate saturation.

8.6 Notes

This chapter presents an FE-based FTC for wind turbine pitch control system subject to low-pressure pitch actuator fault. An adaptive step-by-step SMO observer is developed to obtain a simultaneous estimation of the pitch system state, fault and uncertainty. The FTC controller is a combination of the industrial preferred PI baseline controller and a reconfiguration controller to compensate for the fault and uncertainty. The FTC design recovers the nominal pitching behaviour despite the existence of the fault and uncertainty. Moreover, since all the three pitch systems within a wind turbine have the same form of dynamics, their FE-based FTC (parameters) can be the same. This makes the design effective and easily implemented in real operation, which has been demonstrated in the simulations. By using a similar step-by-step observer, fault-tolerant individual pitch control has been further

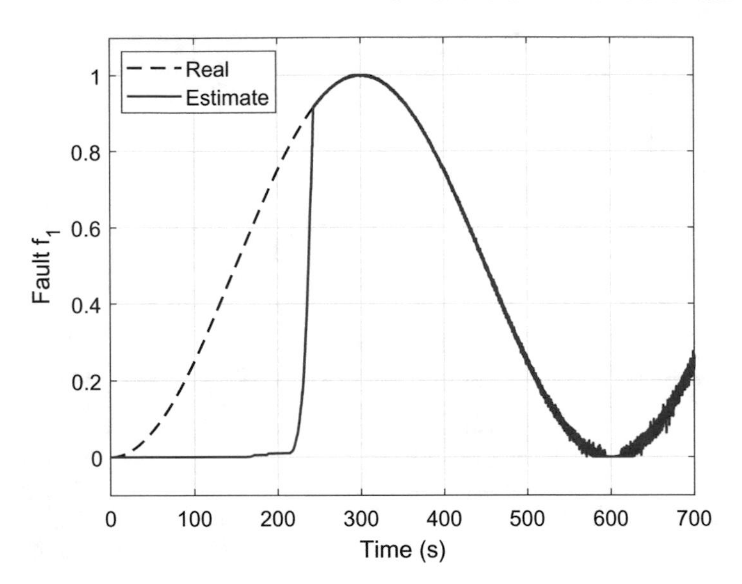

Fig. 8.15 Fault estimation: pitch 1, Case 2

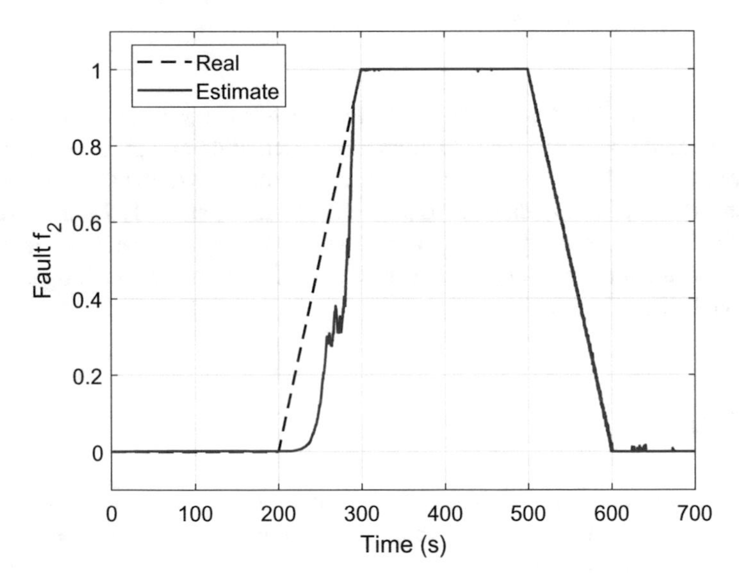

Fig. 8.16 Fault estimation: pitch 2, Case 2

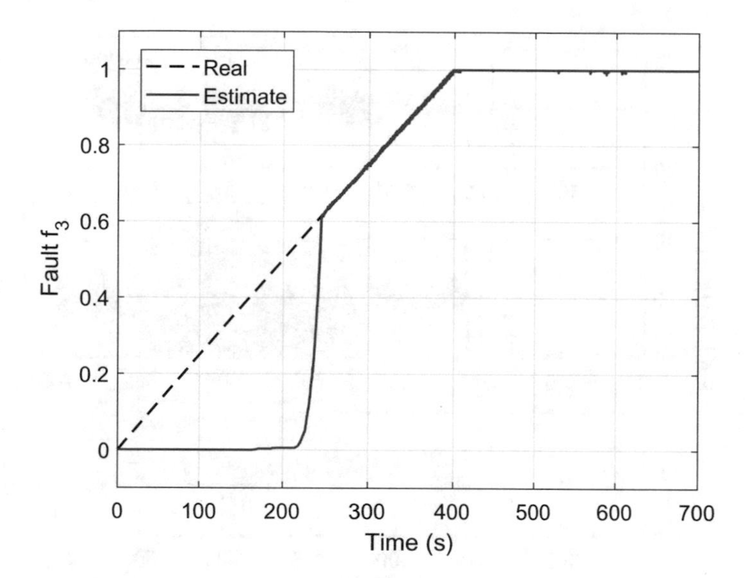

Fig. 8.17 Fault estimation: pitch 3, Case 2

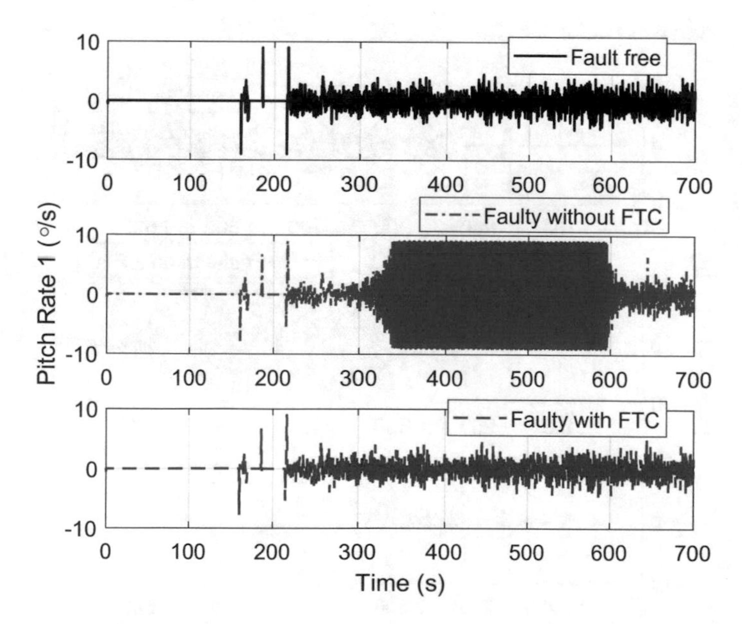

Fig. 8.18 Pitch rate: pitch 1, Case 2

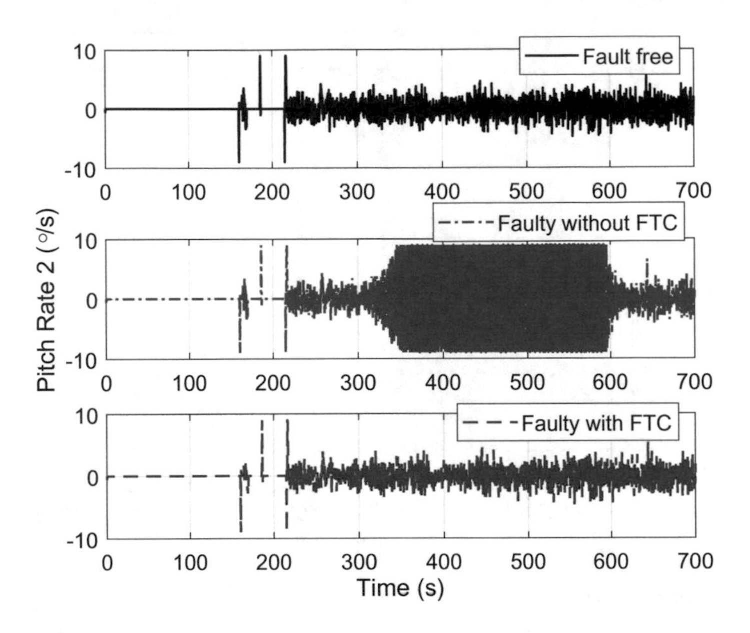

Fig. 8.19 Pitch rate: pitch 2, Case 2

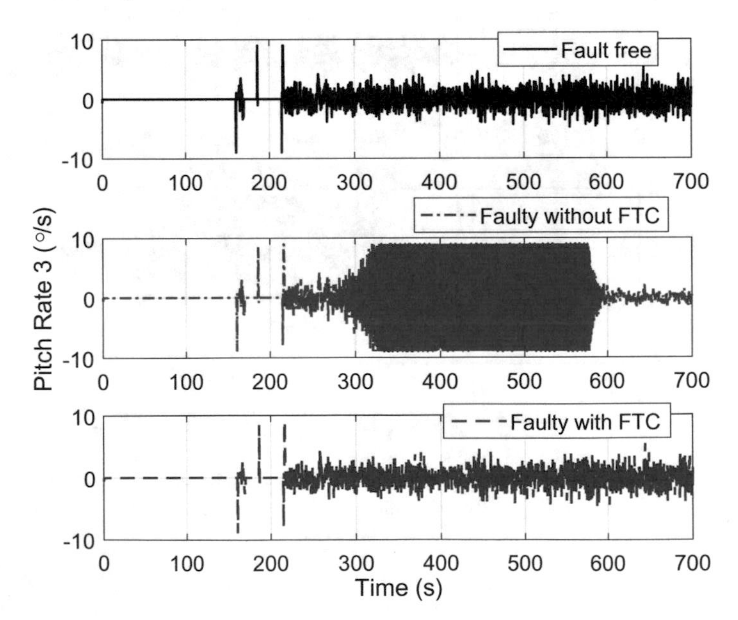

Fig. 8.20 Pitch rate: pitch 3, Case 2

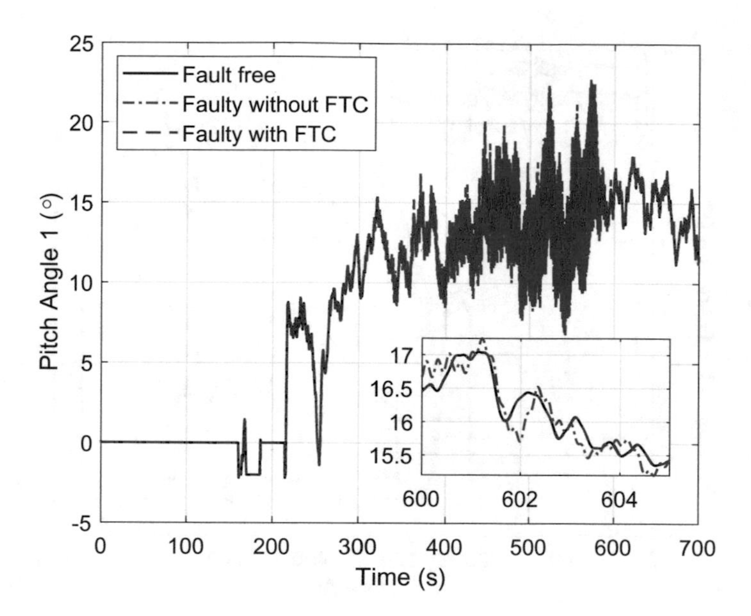

Fig. 8.21 Pitch angle: pitch 1, Case 2

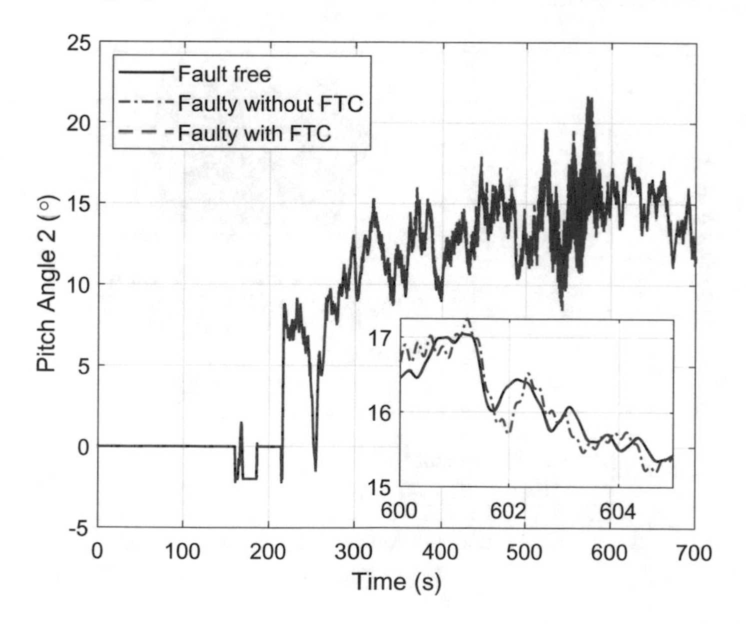

Fig. 8.22 Pitch angle: pitch 2, Case 2

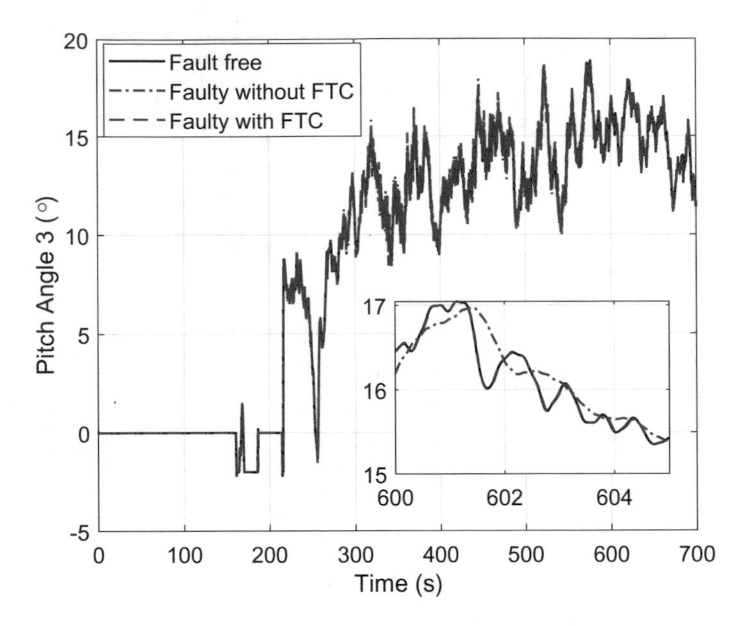

Fig. 8.23 Pitch angle: pitch 3, Case 2

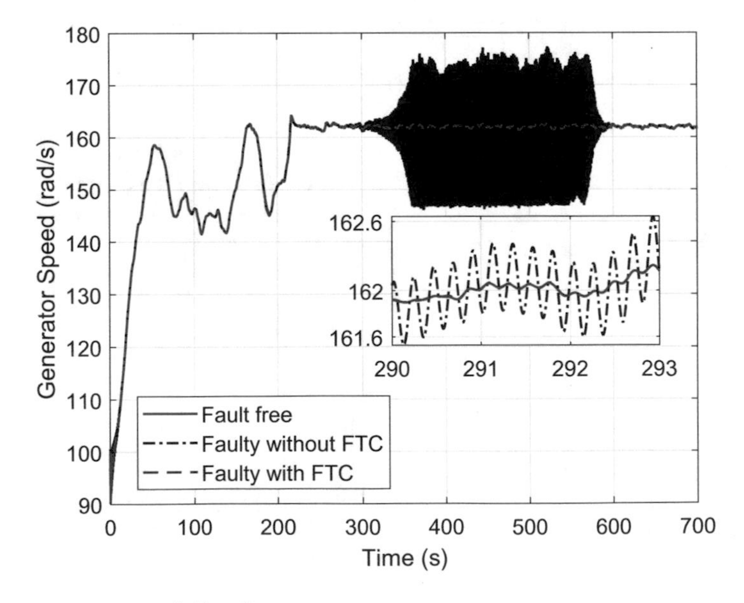

Fig. 8.24 Generator speed: Case 2

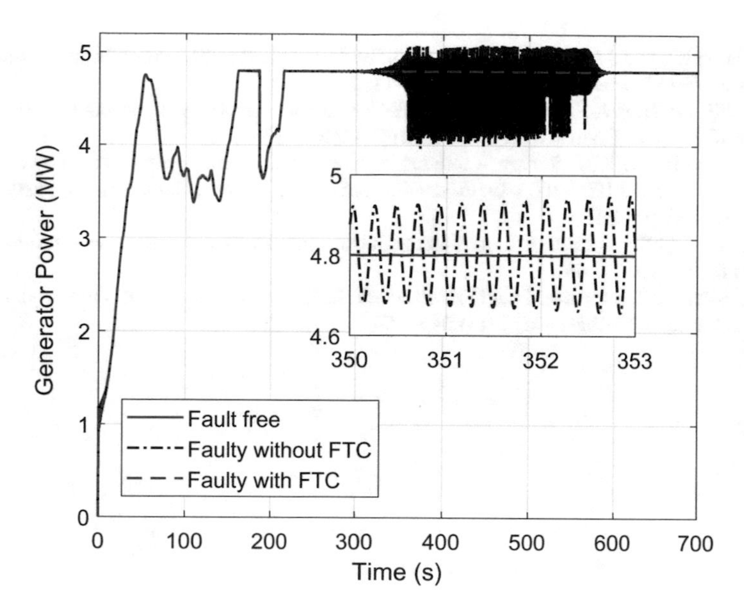

Fig. 8.25 Generator power: Case 2

developed in Liu et al. (2018) to compensate the pitch actuator faults and also mitigate the turbine blade fatigue. This helps to maintain good pitch performance and prolong the wind turbine life.

References

Bossanyi E (2003) Individual blade pitch control for load reduction. Wind Energy 6(2):119–128

Boukhezzar B, Lupu L, Siguerdidjane H, Hand M (2007) Multivariable control strategy for variable speed, variable pitch wind turbines. Renew Energy 32(8):1273–1287

Burton T, Jenkins N, Sharpe D, Bossanyi E (2011) Wind energy handbook. Wiley, New York

Chen L, Shi F, Patton R (2013) Active FTC for hydraulic pitch system for an off-shore wind turbine. In: Proceedings of International Conference on Control and Fault-Tolerant Systems, IEEE, pp 510–515

Edwards C, Spurgeon S (1998) Sliding mode control: theory and applications. CRC Press, Boca Raton

Esbensen T, Sloth C (2009) Fault diagnosis and fault-tolerant control of wind turbines. Master's thesis, Faculty of Engineering, Section for Automation and Control, Aalborg University-Danmark

Gao R, Gao Z (2016) Pitch control for wind turbine systems using optimization, estimation and compensation. Renew Energy 91:501–515

Hand MM (1999) Variable-speed wind turbine controller systematic design methodology: a comparison of non-linear and linear model-based designs. Technical report, National Renewable Energy Lab., Golden, CO (US)

Jain T, Yamé JJ, Sauter D (2013) A novel approach to real-time fault accommodation in NREL's 5-MW wind turbine systems. IEEE Trans Sustain Energy 4(4):1082–1090

Liu Y, Patton RJ, Lan J (2018) Fault-tolerant individual pitch control using adaptive sliding mode observer. IFAC-PapersOnLine 51(24):1127–1132
Odgaard PF, Stoustrup J, Kinnaert M (2013) Fault-tolerant control of wind turbines: A benchmark model. IEEE Trans Control Syst Technol 21(4):1168–1182
Ribrant J, Bertling L (2007) Survey of failures in wind power systems with focus on Swedish wind power plants during 1997-2005. In: Proceedings of the IEEE Power Engineering Society General Meeting, IEEE, pp 1–8
Shi F, Patton R (2015) An active fault tolerant control approach to an offshore wind turbine model. Renew Energy 75:788–798
Sloth C, Esbensen T, Stoustrup J (2011) Robust and fault-tolerant linear parameter-varying control of wind turbines. Mechatronics 21(4):645–659

Chapter 9
Integration of FE and FTC for Nonlinear Systems with 3-DOF Helicopter Application

9.1 Introduction

This chapter further extends the simultaneous strategy in Chap. 5 to address the robust integration of FE and FTC for a class of Lipschitz nonlinear systems. A nonlinear ASUIO FE observer is used to estimate the system state and faults and an adaptive sliding mode FTC controller using the fault and state estimates is designed to compensate for the fault and robustly stabilize the system. The robust integration problem is solved via a new single-step LMI formulation without equality constraint, which is otherwise required in Chap. 5 and imposes conservativeness. The extended strategy is further applied to a nonlinear unmanned 3-DOF (degree-of-freedom) helicopter system affecting by both actuator faults and saturation.

Aircraft may suffer from certain system faults (e.g. actuator, sensor and component faults) that prevent them from achieving manoeuvre tasks and cause problems in stability and safety. In order to ensure a reliable and safe flight, it is necessary to operate the aircraft with fault-tolerant capability. Moreover, actuator saturation has the effect of paralysing the action of the control system, and hence in order to achieve the full control performance, it is necessary to take the saturation into account as a form of malfunction of the system.

For manned aircraft systems, it is usual to use hardware redundancy such as duplicate copies of actuators and sensors. However, for unmanned aircraft, the possibility of using hardware redundancy is rather limited due to size and weight restrictions. For such aircraft, analytic forms of redundancy become essential. FTC provides a way for recovering the acceptable aircraft performance and stability in the presence of certain faults. One way to achieve FTC is to have an FDI unit to detect and isolate the presence of faults with an additional system for managing the switching of different feedback controllers to maintain acceptable aircraft system performance, see for example, Ducard (2009), Edwards et al. (2010) and Zolghadri et al. (2014). Instead of using FDI, the work in this chapter focuses on FE.

© The Author(s), under exclusive license to Springer Nature Switzerland AG 2021

J. Lan and R. J. Patton, *Robust Integration of Model-Based Fault Estimation and Fault-Tolerant Control*, Advances in Industrial Control,
https://doi.org/10.1007/978-3-030-58760-4_9

9.2 Problem Description

Consider a class of nonlinear systems described by

$$\begin{aligned} \dot{x} &= (A + \Delta A)x + Bu + Ff + g(x) + Dd \\ y &= Cx \end{aligned} \tag{9.1}$$

where $x(t) \in \mathbb{R}^n$, $u(t) \in \mathbb{R}^m$, $y(t) \in \mathbb{R}^p$, $f(t) \in \mathbb{R}^q$ and $d(t) \in \mathbb{R}^l$ are the state, control input, measured output, actuator fault and external disturbance, respectively. The constant matrices A, B, F, D and C are known. ΔA is the unknown uncertainty.

Assumption 9.1 The pair (A, B) is controllable, and the triple (A, F, C) has no invariant zero in the closed right-half complex plane. The actuator fault is matched (see Sect. 1.7), i.e. $\text{rank}[B\ F] = \text{rank}(B) = m$.

Assumption 9.2 The uncertainty matrix ΔA is norm-bounded (energy-bounded) with the form $\Delta A = \mathcal{M}\mathcal{F}(t)\mathcal{N}$, where $\mathcal{M}$ and $\mathcal{N}$ are known matrices with appropriate dimensions, and $\mathcal{F}(t)$ is an unknown matrix satisfying $\mathcal{F}^\top(t)\mathcal{F}(t) \preceq I$.

Assumption 9.3 There exist some unknown positive constants f_0 and d_0 such that $\|f\| \leq f_0$ and $\|d\| \leq d_0$. Moreover, f has bounded differentials.

Assumption 9.4 The nonlinear function $g(x)$ satisfies the Lipschitz constraint

$$\|g(x_t) - g(x)\| \leq L_f\|x_t - x\|,\ \forall\, x,\ x_t \in \mathbb{R}^n,$$

where L_f is the Lipschitz constant independent of x and x_t.

This chapter aims to stabilize the system (9.1) through an FTC strategy, involving designs of (1) an observer to estimate the system state and actuator fault and (2) an FTC controller based on the estimates to compensate the fault to ensure system stability.

9.3 FE Observer Design

This section presents a nonlinear ASUIO for estimating the system state x and actuator fault f. Define f as auxiliary state, then the system (9.1) is augmented as

$$\begin{aligned} \dot{\bar{x}} &= (\bar{A} + \Delta\bar{A})\bar{x} + \bar{g}(A_0\bar{x}) + \bar{B}u + \bar{D}\bar{d} \\ y &= \bar{C}\bar{x} \end{aligned} \tag{9.2}$$

where

$$\bar{x} = \begin{bmatrix} x \\ f \end{bmatrix}, \bar{d} = \begin{bmatrix} d \\ \dot{f} \end{bmatrix}, \bar{A} = \begin{bmatrix} A & B \\ 0 & 0 \end{bmatrix}, \Delta\bar{A} = \begin{bmatrix} \Delta A & 0 \\ 0 & 0 \end{bmatrix}, \bar{B} = \begin{bmatrix} B \\ 0 \end{bmatrix},$$

$$\bar{D} = \begin{bmatrix} D & 0 \\ 0 & I_q \end{bmatrix}, \bar{g}(A_0\bar{x}) = \begin{bmatrix} g(A_0\bar{x}) \\ 0 \end{bmatrix}, A_0 = [I_n \ 0], \ \bar{C} = [C \ 0].$$

It can be verified that the augmented system (9.2) is observable under Assumption 9.1. Hence, a nonlinear ASUIO to estimate the augmented state $\bar{x}$ is designed as

$$\begin{aligned} \dot{z} &= Mz + Gu + N\bar{g}(A_0\hat{\bar{x}}) + Ly \\ \hat{\bar{x}} &= z + Hy \end{aligned} \tag{9.3}$$

where $z \in \mathbb{R}^{n+q}$ is the observer system state and $\hat{\bar{x}} \in \mathbb{R}^{n+q}$ is the estimate of $\bar{x}$. The constant matrices M, G, N, L and H are to be designed. The estimates of x and f are obtained as $\hat{x} = [I_n \ 0_{n\times q}]\hat{\bar{x}}$ and $[0_{q\times n} \ I_q]\hat{\bar{x}}$, respectively.

Define the estimation error as $e = \bar{x} - \hat{\bar{x}}$. It follows from (9.2) and (9.3) that

$$\begin{aligned} \dot{e} &= (\Xi\bar{A} - L_1\bar{C})e + (\Xi\bar{A} - L_1\bar{C} - M)z \\ &+(\Xi\bar{B} - G)u + [(\Xi\bar{A} - L_1\bar{C})H - L_2]y \\ &+\Xi\bar{g}(A_0\bar{x}) - N\bar{g}(A_0\hat{\bar{x}}) + \Xi\Delta\bar{A}\bar{x} + \Xi\bar{D}\bar{d} \end{aligned} \tag{9.4}$$

where $\Xi = I_{n+q} - H\bar{C}$ and $L = L_1 + L_2$.

Define the matrices M, G, N and L_2 as

$$M = \Xi\bar{A} - L_1\bar{C}, \ G = \Xi\bar{B}, \ N = \Xi, \ L_2 = (\Xi\bar{A} - L_1\bar{C})H. \tag{9.5}$$

Substituting (9.5) into (9.4) yields

$$\dot{e} = (\Xi\bar{A} - L_1\bar{C})e + \Xi\Delta\bar{g} + \Xi\Delta\bar{A}\bar{x} + \Xi\bar{D}\bar{d} \tag{9.6}$$

where $\Delta\bar{g} = \bar{g}(A_0\bar{x}) - \bar{g}(A_0\hat{\bar{x}})$.

Now the design of the observer (9.3) becomes a problem of designing the matrices L_1 and H such that (9.6) is robustly stable. Once L_1 and H are designed, the other matrices M, G, N and L_2 can be directly obtained by using (9.5).

9.4 FTC Controller Design

This section describes the design of a sliding mode FTC controller, based on estimated state and fault, to compensate the actuator faults and stabilize the system (9.1). The proposed switching function is in the form of

$$s = N_1\hat{x} \tag{9.7}$$

where $s \in \mathbb{R}^m$ and $N_1 = B^\dagger - Y_1(I_n - BB^\dagger)$ with a design matrix $Y_1 \in \mathbb{R}^{m\times n}$.

The first step in designing the sliding mode controller is to establish the reachability of $\hat{x}$ to the sliding surface $s = 0$. This is achieved by designing the controller u to satisfy the reachability condition $s^\top \dot{s} \leq 0$. To this end, the dynamics of s are derived as

$$\dot{s} = N_1(A + \Delta A)x + u + N_1Ff + N_1g(x) + N_1Dd - N_1\dot{e}_x \tag{9.8}$$

where $e_x = x - \hat{x}$ is the estimation error of x.

The FTC controller is designed as

$$u = u_l + u_n \tag{9.9}$$

where u_l is the linear feedback component given by $u_l = -K\hat{\bar{x}}$, with a design matrix $K = [K_x\ K_f]$. $K_x \in \mathbb{R}^{m\times n}$ is to be determined, while K_f is chosen as $K_f = B^\dagger F$. The nonlinear component u_n is designed as $u_n = -\rho\text{sign}(s)$, with a scalar function ρ to be determined.

Consider the Lyapunov function

$$V_s = \frac{1}{2}s^\top s.$$

The time derivative of V_s along (9.8) is

$$\begin{aligned}\dot{V}_s &= s^\top [N_1(A + \Delta A)x + u + N_1Ff + N_1g(x) + N_1Dd - N_1\dot{e}_x] \\ &\leq s^\top \left[\Delta_e - \rho\text{sign}(s)\right] \\ &\leq (\eta - \rho)\|s\|\end{aligned} \tag{9.10}$$

where

$$\Delta_e = \|[N_1(A + \Delta A) - K_x]x + K_xe_x + e_f + N_1g(x) + N_1D\tilde{d} - N_1\dot{e}_x\|,$$

$e_f = f - \hat{f}$, and η is an unknown constant satisfying $\eta \geq \Delta_e$.

Define $\rho = \hat{\eta} + \epsilon$, where ϵ is a positive design constant. The scalar $\hat{\eta}$ is the estimate of η defined as

$$\dot{\hat{\eta}} = \sigma\|s\|,\ \hat{\eta}(0) \geq 0 \tag{9.11}$$

with a positive design constant σ.

Define the estimation error of η as $\tilde{\eta} = \eta - \hat{\eta}$. Consider a Lyapunov function

$$V = V_s + \frac{1}{2\sigma}\tilde{\eta}^2.$$

It follows from (9.10) and (9.11) that

$$\begin{aligned} \dot{V} &= \dot{V}_s - \frac{1}{\sigma}\tilde{\eta}\dot{\hat{\eta}} \\ &\leq (\eta - \rho - \tilde{\eta})\|s\| \\ &\leq 0. \end{aligned} \tag{9.12}$$

Since V is positive definite, it follows from (9.12) and the Barbalat's Lemma (see Sect. 1.7.1) that $V_s(t) \leq V_s(0)$. Therefore, $s(t)$ and $\tilde{\eta}(t)$ are bounded. This means that the designed controller (9.9) can maintain the sliding motion around $s = 0$. Moreover, in the case of zero initial condition (i.e. $V_s(0) = 0$), $\lim_{t\to\infty} s(t) = 0$ and thus the ideal sliding motion can be maintained.

Consider next the system stability analysis corresponding to the sliding motion. From (9.8), the equivalent control input can be defined as

$$u_{eq} = -N_1\left[(A + \Delta A)x + g(x) + Dd\right] + u_l. \tag{9.13}$$

Substituting (9.13) into (9.1) gives the equivalent closed-loop system

$$\dot{x} = (\Theta A - BK_x)x + \Theta\Delta Ax + BKe + \Theta g(x) + \Theta Dd \tag{9.14}$$

where $\Theta = I_n - BN_1$.

Therefore, the system (9.1) is maintained on the sliding mode with the equivalent control (9.13) by designing K_x such that (9.14) is stable. The closed-loop system (9.14) contains the uncertainty ΔAx, disturbance d and nonlinearity $g(x)$, whose effects must be minimized to achieve a suitable degree of robustness. This is achieved by using H_∞ optimization in the next section.

9.5 Integration of FE and FTC

To obtain the FE observer and FTC controller parameters, a way widely used in the literature is the separated FE and FTC design strategy, where the FE observer and FTC controller are designed separately. This approach follows the Separation Principle (see Sect. 1.7.2) by neglecting the effects of system uncertainty and nonlinearity on the FE performance and effect of estimation errors on the FTC system. This section first presents the traditional separated synthesis strategy with an analysis of its restriction and drawbacks. It then describes a simultaneous robust integration strategy based on the strategy proposed in Chap. 5.

9.5.1 Traditional Separated Strategy

This subsection presents a separated strategy for FE-based FTC design, which is similar to the separated strategy described in Sect. 2.4.2. By neglecting the system uncertainty and nonlinearity, the error system (9.6) is reduced to be

$$\begin{aligned} \dot{e} &= (\Xi \bar{A} - L_1 \bar{C})e + \Xi \bar{D}\bar{d} \\ z_{s1} &= C_{s1} e \end{aligned} \tag{9.15}$$

where $z_{s1} \in \mathbb{R}^{n+q}$ is the measured output with a given coefficient matrix C_{s1}.

The following theorem is given to design the matrices H and L_1 to make the error system (9.15) robustly stable.

Theorem 9.1 *The error system (9.15) is stable with H_∞ performance $\|G_{z_{s1}\bar{d}}\|_\infty < \gamma_{s1}$, if the following optimization problem is feasible:*

$$\text{Find } \gamma_{s_1} \tag{9.16}$$

$$\text{s.t.} \begin{bmatrix} \text{He}\left(Q_s \bar{A} - M_{s1}\bar{C}\bar{A} - M_{s2}\bar{C}\right) & (Q_s - M_{s1}\bar{C})\bar{D} & C_{s1}^\top \\ \star & -\gamma_{s1} I & 0 \\ \star & \star & -\gamma_{s1} I \end{bmatrix} \prec 0,$$

$$Q_s = Q_s^\top \succ 0, \ \gamma_{s1} > 0.$$

Then the gains are obtained as $H = Q_s^{-1} M_{s1}$ and $L_1 = Q_s^{-1} M_{s2}$.

Proof By using the Bounded Real Lemma (see Sect. 1.7.1) and defining $M_{s1} = Q_s H$ and $M_{s2} = Q_s L_1$, the proof is trivial and thus is omitted here. □

Similarly, in the separated strategy the FTC system is assumed to be unaffected by the estimation error, thus the closed-loop control system (9.14) becomes

$$\begin{aligned} \dot{x} &= (\Theta A - BK_x)x + \Theta \Delta A x + \Theta g(x) + \Theta D d \\ z_{s2} &= C_{s2} x \end{aligned} \tag{9.17}$$

where $z_{s2} \in \mathbb{R}^n$ is the performance output with a given coefficient matrix C_{s2}.

The following theorem is given to design K_x to ensure that (9.17) is robustly stable.

Theorem 9.2 *The closed-loop control system (9.17) is stable with H_∞ performance $\|G_{z_{s2}d}\|_\infty < \gamma_{s2}$, if the following optimization problem is feasible:*

$$
\begin{aligned}
&\text{Find } \gamma_{s2} \\
\text{s.t.} \quad &\begin{bmatrix} \hat{\Pi}_{1,1} & D & P_s C_{s2}^\top & P_s & P_s \mathcal{N}^\top \\ \star & -\gamma_{s2} I & 0 & 0 & 0 \\ \star & \star & -\gamma_{s2} I & 0 & 0 \\ \star & \star & \star & -(\varepsilon_s L_f^2)^{-1} I & 0 \\ \star & \star & \star & \star & -I \end{bmatrix} \prec 0, \\
&P_s = P_s^\top \succ 0, \ \gamma_{s2} > 0, \ \varepsilon_s > 0,
\end{aligned}
\tag{9.18}
$$

where $\hat{\Pi}_{1,1} = \text{He}(\Theta A P_s - B M_{s3}) + \varepsilon_s^{-1}\Theta\Theta^\top + (\Theta\mathcal{M})(\Theta\mathcal{M})^\top$. *Then the control gain is obtained as* $K_x = M_{s3} P_s^{-1}$.

Proof Define the Lyapunov function $V_s = x^\top Z_s x$ with a s.p.d. matrix $Z_s \in \mathbb{R}^{n\times n}$. Since the uncertainty satisfies Assumption 9.2, it holds that

$$
\text{He}(x^\top Z_s \Theta \Delta A x) \le x^\top [Z_s(\Theta\mathcal{M})(\Theta\mathcal{M})^\top Z_s + \mathcal{N}^\top \mathcal{N}] x.
$$

Since $g(0) = 0$, then $\|g(x)\| \le L_f \|x\|, \forall x \in \mathbb{R}^n$. It holds that, for some positive scalar ε_s,

$$
\text{He}(x^\top Z_s \Theta g(x)) \le \varepsilon_s^{-1} x^\top Z_s \Theta\Theta^\top Z_s x + \varepsilon_s L_f^2 \|x\|^2.
$$

By using Bounded Real Lemma, the closed-loop system (9.17) is stable with H_∞ performance $\|G_{z_{s_2}d}\|_\infty < \gamma_{s2}$, if

$$
\begin{bmatrix} \Pi_{1,1} & Z_s D & C_{s_2}^\top \\ \star & -\gamma_{s2} I & 0 \\ \star & \star & -\gamma_{s2} I \end{bmatrix} \prec 0 \tag{9.19}
$$

where $\Pi_{1,1} = \text{He}\,[Z_s(\Theta A - B K_x)] + \varepsilon_s^{-1} Z_s \Theta\Theta^\top Z_s + Z_s(\Theta\mathcal{M})(\Theta\mathcal{M})^\top Z_s + \mathcal{N}^\top\mathcal{N} + \varepsilon_s L_f^2 I_n$.

Define $P_s = Z_s^{-1}$ and $M_{s3} = K_x P_s$. Pre- and post-multiplying both sides of (9.19) with $\text{diag}(P_s, I, I)$, respectively, and using the Schur Complement (see Sect. 1.7.1), then (9.19) becomes

$$
\begin{bmatrix} \hat{\Pi}_{1,1} & D & P_s C_{s_2}^\top & P_s & P_s \mathcal{N}^\top \\ \star & -\gamma_{s2} I & 0 & 0 & 0 \\ \star & \star & -\gamma_{s2} I & 0 & 0 \\ \star & \star & \star & -(\varepsilon_s L_f^2)^{-1} I & 0 \\ \star & \star & \star & \star & -I \end{bmatrix} \prec 0,
$$

where $\hat{\Pi}_{1,1} = \text{He}(\Theta A P_s - B M_{s3}) + \varepsilon_s^{-1}\Theta\Theta^\top + (\Theta\mathcal{M})(\Theta\mathcal{M})^\top$. Now the controller design can be formulated as the optimization problem (9.18). □

The separated strategy outlined in Theorems 9.1 and 9.2 allows great design freedom for the FE and FTC design for nonlinear systems, where the observer and controller can be optimized independently. However, it can be seen from the error system (9.6) and the control system (9.14) that the system uncertainty, nonlinearity and disturbance affect the estimation, and in turn, the estimation error has effect on the closed-loop control system. This means that there are *bidirectional robustness interactions* between the FE and FTC functions, which breaks down the Separation Principle on which the separated strategy is based. Therefore, it is necessary to introduce a robust integration strategy to achieve robust closed-loop performance, by taking into account of the *bidirectional robustness interactions*.

9.5.2 Simultaneous Robust Integration Strategy

The composite closed-loop system consisting of (9.6) and (9.14) is given as

$$\begin{aligned} \dot{x} &= (\Theta A - BK_x)x + \Theta \Delta A x + \Theta g(x) + D_1 \bar{d} \\ \dot{e} &= (\Xi \bar{A} - L_1 \bar{C})e + \Xi \Delta \bar{g} + \Xi \Delta \bar{A} \bar{x} + \Xi \bar{D} \bar{d} \\ z_c &= \mathrm{diag}(C_x x, C_e e) \end{aligned} \tag{9.20}$$

where $z_c \in \mathbb{R}^{2n+q}$ is the measured output used to verify the closed-loop system performance with matrices C_x and C_e, and $D_1 = [\Theta D \; 0]$.

Theorem 9.3 provides a simultaneous strategy for robust integration of the FE observer and FTC controller gains using a single-step LMI formulation.

Theorem 9.3 *The closed-loop system (9.20) is stable with H_∞ performance $\|G_{z_c\bar{d}}\|_\infty < \gamma$, if the following optimization problem is feasible:*

$$\min \gamma \tag{9.21}$$

$$\text{s.t.} \begin{bmatrix} J_{1,1} & B\hat{K}_f & D_1 & ZC_x^\top & 0 & J_{1,2} \\ \star & J_{2,2} & (Q - M_2\bar{C})\bar{D} & 0 & C_e^\top & J_{2,3} \\ \star & \star & -\gamma I & 0 & 0 & 0 \\ \star & \star & \star & -\gamma I & 0 & 0 \\ \star & \star & \star & \star & -\gamma I & 0 \\ \star & \star & \star & \star & \star & -\Lambda \end{bmatrix} \prec 0 \tag{9.22}$$

$$Z = Z^\top \succ 0, \; Q = Q^\top \succ 0, \; \gamma > 0 \tag{9.23}$$

with

$$J_{1,1} = \text{He}(\Theta AZ - BM_1) + \varepsilon_2^{-1}\Theta\Theta^\top + (\Theta\mathcal{M})(\Theta\mathcal{M})^\top, \; \hat{K}_f = [0 \; K_f],$$
$$J_{1,2} = [BM_1 \; 0 \; Z \; Z\mathcal{N}^\top \; 0 \; 0], \; J_{2,2} = \text{He}(Q\bar{A} - M_2\bar{C}\bar{A} - M_3\bar{C}) + \varepsilon_1 L_f^2 A_0^\top A_0,$$
$$J_{2,3} = [0 \; A_0^\top \; 0 \; 0 \; (Q - M_2\bar{C}) \; (Q - M_2\bar{C})\bar{\mathcal{M}}],$$
$$\Lambda = \text{diag}(\varepsilon_3 Z, \varepsilon_3^{-1} Z, (\varepsilon_2 L_f^2)^{-1} I, 0.5I, \varepsilon_1 I, I),$$

where ε_i, $i = 1, 2, 3$, *are given positive scalars. Then the design gains are obtained as follows:* $K_x = M_1 Z^{-1}$, $H = Q^{-1}M_2$, $L_1 = Q^{-1}M_3$.

Proof The proof is similar to that in Chap. 5 and it is only sketched below. Consider the Lyapunov function $V_e = e^\top Qe$ with a s.p.d. matrix Q. It holds that

$$\text{He}(e^\top Q\Xi\Delta\bar{A}\bar{x}) \le e^\top(Q\Xi\bar{\mathcal{M}})(Q\Xi\bar{\mathcal{M}})^\top e + x^\top\mathcal{N}^\top\mathcal{N}x,$$

where $\bar{\mathcal{M}} = [\mathcal{M}^\top \; 0]^\top$.

Under Assumption 9.4, for any scalar $\varepsilon_1 > 0$, the following inequality holds

$$\text{He}(e^\top Q\Xi\Delta\bar{g}) \le \varepsilon_1^{-1} e^\top Q\Xi\Xi^\top Qe + \varepsilon_1 L_f^2 \|A_0 e\|^2.$$

Consider the Lyapunov function $V_x = x^\top Px$ with a s.p.d. matrix P. It is true that

$$\text{He}(x^\top P\Theta\Delta Ax) \le x^\top[P(\Theta\mathcal{M})(\Theta\mathcal{M})^\top P + \mathcal{N}^\top\mathcal{N}]x.$$

It also holds that, for some positive scalar ε_2,

$$2x^\top P\Theta g(x) \le \varepsilon_2^{-1} x^\top P\Theta\Theta^\top Px + \varepsilon_2 L_f^2 \|x\|^2.$$

Define the overall Lyapunov function as $V = V_e + V_x$. By using the Bounded Real Lemma, the closed-loop system (9.14) is stable with $\|G_{z_c\bar{d}}\|_\infty < \gamma$ if

$$\begin{bmatrix} J_{1,1} & PBK & PD_1 & C_x^\top & 0 \\ \star & J_{2,2} & Q\Xi\bar{D} & 0 & C_e^\top \\ \star & \star & -\gamma I & 0 & 0 \\ \star & \star & \star & -\gamma I & 0 \\ \star & \star & \star & \star & -\gamma I \end{bmatrix} \prec 0 \tag{9.24}$$

where

$$J_{1,1} = \text{He}[P(\Theta A - BK_x)] + \varepsilon_2^{-1} P\Theta\Theta^\top P + \varepsilon_2 L_f^2 I + P(\Theta\mathcal{M})(\Theta\mathcal{M})^\top P + 2\mathcal{N}^\top\mathcal{N},$$
$$J_{2,2} = \text{He}[Q(\Xi\bar{A} - L_1\bar{C})] + \varepsilon_1^{-1} Q\Xi\Xi^\top Q + \varepsilon_1 L_f^2 A_0^\top A_0 + (Q\Xi\bar{\mathcal{M}})(Q\Xi\bar{\mathcal{M}})^\top.$$

Define $Z = P^{-1}$. Pre- and post-multiplying both sides of (9.24) with $\text{diag}(Z, I, I, I, I)$ and its transpose, respectively, the it gives

$$\begin{bmatrix} J_{1,1} & BK & D_1 & ZC_x^\top & 0 \\ \star & J_{2,2} & Q\Xi\bar{D} & 0 & C_e^\top \\ \star & \star & -\gamma I & 0 & 0 \\ \star & \star & \star & -\gamma I & 0 \\ \star & \star & \star & \star & -\gamma I \end{bmatrix} \prec 0 \tag{9.25}$$

where

$$J_{1,1} = \mathrm{He}[(\Theta A - BK_x)Z] + \varepsilon_2^{-1}\Theta\Theta^\top + \varepsilon_2 L_f^2 ZZ + (\Theta\mathcal{M})(\Theta\mathcal{M})^\top + 2Z\mathcal{N}^\top\mathcal{N}Z,$$
$$J_{2,2} = \mathrm{He}[Q(\Xi\bar{A} - L_1\bar{C})] + \varepsilon_1^{-1}Q\Xi\Xi^\top Q + \varepsilon_1 L_f^2 A_0^\top A_0 + (Q\Xi\bar{\mathcal{M}})(Q\Xi\bar{\mathcal{M}})^\top.$$

Define $V_1 = [(BK_x)^\top\, 0\,0\,0\,0]^\top$ and $V_2^\top = [0\, A_0\, 0\,0\,0]$. By using Young inequality (see Sect. 1.7.1), there exists some positive scalar ε_3 such that

$$\mathrm{He}(V_1V_2^\top) \le \varepsilon_3^{-1}(V_1Z)Z^{-1}(V_1Z)^\top + \varepsilon_3 V_2 Z^{-1} V_2^\top. \tag{9.26}$$

Applying (9.26) to (9.25) and using Schur Complement and $BK = B[K_x\ K_f]$ yields

$$\begin{bmatrix} J_{1,1} & B\hat{K}_f & D_1 & ZC_x^\top & 0 & J_{1,2} \\ \star & J_{2,2} & Q\Xi\bar{D} & 0 & C_e^\top & J_{2,3} \\ \star & \star & -\gamma I & 0 & 0 & 0 \\ \star & \star & \star & -\gamma I & 0 & 0 \\ \star & \star & \star & \star & -\gamma I & 0 \\ \star & \star & \star & \star & \star & -\Lambda \end{bmatrix} \prec 0 \tag{9.27}$$

$$J_{1,1} = \mathrm{He}[(\Theta A - BK_x)Z] + \varepsilon_2^{-1}\Theta\Theta^\top + (\Theta\mathcal{M})(\Theta\mathcal{M})^\top,\ \hat{K}_f = [0\ K_f],$$
$$J_{1,2} = [BK_xZ\ 0\ Z\ Z\mathcal{N}^\top\ 0\ 0],\ J_{2,2} = \mathrm{He}[Q(\Xi\bar{A} - L_1\bar{C})] + \varepsilon_1 L_f^2 A_0^\top A_0,$$
$$J_{2,3} = [0\ A_0^\top\ 0\ 0\ Q\Xi\ \ Q\Xi\bar{\mathcal{M}}],\ \Lambda = \mathrm{diag}(\varepsilon_3 Z, \varepsilon_3^{-1}Z, (\varepsilon_2 L_f^2)^{-1}I, 0.5I, \varepsilon_1 I, I).$$

Define $M_1 = K_xZ$, $M_2 = QH$ and $M_3 = QL_1$, then (9.27) is converted into (9.22). □

9.6 Application to 3-DOF Helicopter System

9.6.1 Background

Unmanned aerial vehicles (UAVs) have numerous applications in military and civilian domains, due to their small size and features of long air hovering, vertical take-off and landing capability, low-speed/-altitude and flexible flight. The control designs for unmanned helicopters have been researched extensively, see for example, Chen et al. (2010), Alexis et al. (2012), Li et al. (2015) and Izaguirre-Espinosa et al. (2016).

Considering reliability and safety, FE and FTC designs for helicopter control systems have also attracted much attention, see Qi et al. (2014) and the references therein.

The implementation of FTC for most UAVs becomes very challenging due to the lack of actuator or sensor (hardware) redundancy in these systems. An exception to this for UAVs is the actuator redundancy that exists in hexrotor and octorotor systems. However, in this study, all forms of hardware redundancy are excluded as a deliberate exercise to test the potential of FE-based FTC.

This section considers the Quanser 3-DOF helicopter model with twin rotors (Apkarian 2006). This model has been used by many researchers as a benchmark representative of the rigid body dynamics of a full-size tandem rotor transport and rescue helicopter. Many studies focus on the use of it to verify control designs. This system can also be representative of a rigid body UAV system. There is no hardware redundancy and the FTC must be based fully on the analytical or functional redundancy concept, e.g. using FE with combined fault and state estimation.

Many FE-based FTC designs for the Quanser 3-DOF helicopter model have also been published. Chen et al. (2016a) propose an SMO for actuator fault estimation for a Lipschitz nonlinear helicopter model without uncertainty and external disturbance. In their work, the faults are estimated with bounded errors and FTC is not considered. Afonso and Galvão (2010) present a robust model predictive FTC design considering a linear 3-DOF helicopter system with uncertainty, disturbance and an actuator fault. However, model predictive control involves online optimization and their work does not include FE. Zheng et al. (2014) and Chen et al. (2016b) describe a number of adaptive FTC schemes for uncertain nonlinear 3-DOF helicopter systems, however, they also exclude FE in their studies. An FE-based FTC output tracking strategy is developed in de Loza et al. (2015) for a linearized 3-DOF helicopter with perturbations, oscillatory and drift actuator faults.

Consideration of actuator saturation is important in flight system design and, for full-size aircraft, it is always taken into account. It is thus necessary to include a study of actuator saturation effect on the FTC performance for a UAV. Actuator saturation problems for 3-DOF helicopters have been considered in Kiefer et al. (2010) using an inversion-based control approach, and in Zheng et al. (2015) using an anti-windup compensator. However, neither of these studies pays attention to actuator faults. Qi et al. (2016) consider the self-healing control for a single-rotor UAV with actuator fault and constraints using an anti-windup compensator (not FTC).

This section considers an uncertain Lipschitz nonlinear 3-DOF helicopter with both actuator faults and saturation. A composite fault function is defined to lump the effects of actuator faults and saturation. The obtained composite fault function is non-differentiable and thus further approximated by a differentiable function with a sufficiently small error. The differentiable approximation is then treated as auxiliary system state and estimated by a nonlinear ASUIO. Unlike the adaptive SMO (Chen et al. 2016a) or high-order SMO (de Loza et al. 2015) FE methods, the proposed nonlinear ASUIO can achieve fault estimation with no need for system output derivatives.

An adaptive sliding mode FTC controller is proposed to compensate the effects of actuator faults and saturation and stabilize the elevation and pitch motions of the

Table 9.1 Definition of physical parameters

Parameter	Physical meaning
ε, p	Elevation and pitch angles
F_f, F_b	Control voltages of the front and back motors
J_ε, J_p	Moments of inertia of elevation and pitch axes
L_a, L_h	Distances between travel axis and helicopter body, pitch axis and each motor
m_h, K_f, g	Mass of the helicopter, propeller force-thrust constant, gravity constant
w_ε, w_p	Unknown bounded external disturbances

3-DOF helicopter. SMC is known as a robust control method, since once sliding motion is reached the system is insensitive to any matched perturbation. Moreover, the adaptive method is incorporated with the SMC to avoid the requirement of a priori knowledge of the perturbation bounds. Compared with the model predictive FTC (Afonso and Galvão 2010), adaptive FTCs (Zheng et al. 2014; Chen et al. 2016b) and backstepping sliding mode FTC (de Loza et al. 2015), the proposed FTC is easier to design and implement without requiring online optimization and system decomposition.

In the absence of actuator faults, the proposed integrated FTC design reverts to a new anti-windup control method for compensating the input saturation effect to recover the non-saturated system performance.

9.6.2 *Problem Formulation*

This work considers the Quanser 3-DOF helicopter with the dynamic model

$$\begin{aligned} J_\varepsilon \ddot{\varepsilon} &= K_f L_a \cos(p)\left(F_f + F_b\right) - m_h g L_a \sin(\varepsilon) + w_\varepsilon \\ J_p \ddot{p} &= K_f L_h \left(F_f - F_b\right) + w_p \end{aligned} \tag{9.28}$$

where the physical parameters are defined in Table 9.1.

Define the state vector as $x = [x_1\ x_2\ x_3\ x_4]^\top = [\varepsilon\ p\ \dot{\varepsilon}\ \dot{p}]^\top$, the input vector as $u = [u_1\ u_2]^\top = [F_f\ F_b]^\top$ and the output vector as $y = [\varepsilon\ p\ \dot{\varepsilon}\ \dot{p}]^\top$. Suppose the front and back motors suffer from saturation and unknown bounded actuator faults f_{a1} and f_{a2}, respectively. The actuator faults may be oscillatory faults (Goupil 2010) or drift faults (de Loza et al. 2015) acting on the flight or helicopter control systems. Without loss of generality, f_{a1} and f_{a2} are assumed to have first-order time derivatives $\dot{f}_{a1}$ and $\dot{f}_{a2}$, respectively. Moreover, f_{a1}, f_{a2}, $\dot{f}_{a1}$ and $\dot{f}_{a2}$ are bounded. Hence, the control inputs applied to the helicopter have the structure in Fig. 9.1.

The control input can be mathematically represented as

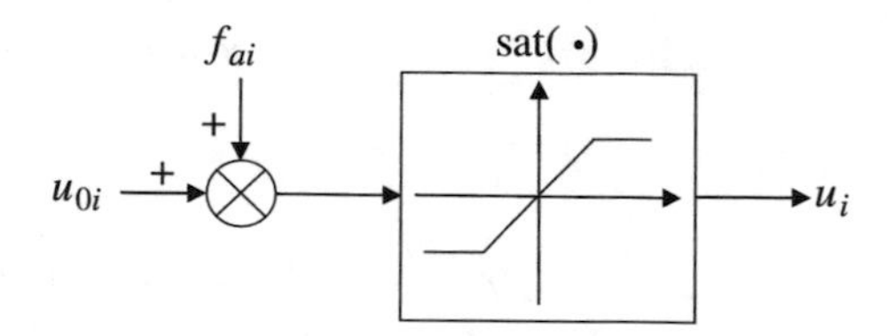

Fig. 9.1 The actuator model with both fault and saturation

$$u_i = \text{sat}(u_{0i} + f_{ai}),\ i = 1, 2,$$

where u_{0i} is the designed control input and sat(·) is a saturation function defined by

$$\text{sat}(v) = \begin{cases} \text{sign}(v)\bar{u}, & |v| \geq \bar{u} \\ v, & |v| < \bar{u} \end{cases}$$

with v the input to the actuator and $\bar{u}$ the maximum control magnitude allowed by the actuator. The control input u of the system (9.28) can then be rearranged into

$$u = u_0 + f_0 \tag{9.29}$$

where $u_0 = [u_{01}\ u_{02}]^\top$ is the designed control input vector and $f_0 = [f_{01}\ f_{02}]^\top$ is the composite actuator fault vector with $f_{0i} = \text{sat}(u_{0i} + f_{ai}) - u_{0i}$, $i = 1, 2$.

It can be seen from (9.29) that f_0 is a function of the designed control input u_0 and the saturation function sat(v). In this chapter, u_0 will be designed as a state-feedback controller that is differentiable. However, the saturation function sat(v) is known to be non-differentiable. Therefore, f_0 is non-differentiable and cannot be treated as an auxiliary system state. To overcome this, a differentiable approximation of f_0 needs to be attained before designing the nonlinear ASUIO.

If the saturation function sat(v) can be approximated by a differentiable function $\overline{\text{sat}}(v)$, then f_0 is modelled as a new function consisting of the differentiable u_0 and $\overline{\text{sat}}(v)$. Hence, f_0 is differentiable. The saturation function sat(v) is approximated by a differentiable function $\overline{\text{sat}}(v)$ with the form of Freidovich and Khalil (2008):

$$\overline{\text{sat}}(v) = \begin{cases} v, & 0 \leq |v| \leq \bar{u} \\ v - [v - \bar{u}\text{sign}(v)]^2 \frac{\text{sign}(v)}{2\epsilon_0}, & \bar{u} \leq |v| \leq \bar{u} + \epsilon_0 \\ (\bar{u} + \frac{\epsilon_0}{2})\text{sign}(v), & |v| \geq \bar{u} + \epsilon_0 \end{cases} \tag{9.30}$$

where ϵ_0 is a positive constant.

It can be shown that the function $\overline{\text{sat}}(v)$ satisfies continuity across $|v| = \bar{u}$ as well as $|v| = \bar{u} + \epsilon_0$. Furthermore, the left and right derivatives of $\overline{\text{sat}}(v)$ with respect to v at the above boundaries are equal. It follows that $\overline{\text{sat}}(v)$ is differentiable. Moreover, it is bounded uniformly in ϵ_0 on any bounded interval of ϵ_0 and $|\overline{\text{sat}}(v) - \text{sat}(v)| \leq \epsilon_0/2$ and $0 \leq d\overline{\text{sat}}(v)/dv \leq 1$ for all $v \in \mathbb{R}$. Hence, the approximation error of sat(v) is small if ϵ_0 is sufficiently small.

By using (9.30), the control input (9.29) can be further modelled as

$$u = u_0 + f + \Delta u \tag{9.31}$$

with

$$f = [f_1\ f_2]^\top,\ f_i = \overline{\text{sat}}(u_{0i} + f_{ai}) - u_{0i},\ i = 1, 2,$$
$$\Delta u = [\Delta u_1\ \Delta u_2]^\top,\ \Delta u_i = \text{sat}(u_{0i} + f_{ai}) - \overline{\text{sat}}(u_{0i} + f_{ai}),\ i = 1, 2.$$

By using the aforementioned definitions, a state-space model of (9.28) is given as

$$\begin{aligned} \dot{x} &= Ax + B(u_0 + f_0) + g(x) + Dd \\ y &= Cx \end{aligned} \tag{9.32}$$

with

$$A = \begin{bmatrix} 0 & 0 & 1 & 0 \\ 0 & 0 & 0 & 1 \\ 0 & 0 & 0 & 0 \\ 0 & 0 & 0 & 0 \end{bmatrix}, B = \begin{bmatrix} 0 & 0 \\ 0 & 0 \\ b_1 & b_1 \\ b_2 & -b_2 \end{bmatrix}, g(x) = \begin{bmatrix} 0 \\ 0 \\ g_1(x) \\ 0 \end{bmatrix}, D = \begin{bmatrix} 0 & 0 \\ 0 & 0 \\ 1 & 0 \\ 0 & 1 \end{bmatrix}, d = \begin{bmatrix} d_1 \\ d_2 \end{bmatrix}, C = I_4,$$
$$b_1 = K_f L_a / J_\varepsilon,\ b_2 = K_f L_h / J_p,\ g_1(x) = -m_h g L_a \sin(x_1)/J_\varepsilon,$$
$$d_1 = w_\varepsilon / J_\varepsilon + b_1(\cos(x_2) - 1)\,(u_{01} + f_{01} + u_{02} + f_{02}) + b_1 \Delta u_1 + b_1 \Delta u_2,$$
$$d_2 = w_p / J_p + b_2 \Delta u_1 - b_2 \Delta u_2,$$

where d is a bounded lumped uncertainty including external disturbances (w_ε and w_p), the system uncertainty $b_1(\cos(x_2) - 1)(u_{01} + f_{01} + u_{02} + f_{02})$ and the approximation errors Δu_i of $\text{sat}(u_{0i} + f_{ai})$, $i = 1, 2$.

The presence of actuator faults and saturation can affect the stability of the helicopter system and prevent it from performing prescribed tasks. This section aims to stabilize the elevation and pitch motions of the system (9.32) through an FTC strategy, involving the designs of (1) an observer to estimate the system state and the composite actuator fault and (2) an FTC controller based on the estimation to compensate the faults and saturation effect to ensure system stability.

Remark 9.1 In real operations, the helicopter actuators may suffer from both stuck and partial loss of effectiveness faults. In such case, the control inputs applied to the helicopter are represented by

$$u_i = \text{sat}(\theta_i u_{0i} + u_{si}),\ i = 1, 2 \tag{9.33}$$

where $0 < \theta_i \le 1$ is the partial loss of actuator effectiveness fault and u_{si} is the stuck fault. Assume that both θ_i and u_{si} are unknown bounded and differentiable, then the actuator model (9.33) can be rearranged as

$$u = u_0 + f_0 \tag{9.34}$$

where $u_0 = [u_{01}\ u_{02}]^\top$ is the designed control input vector and $f_0 = [f_{01}\ f_{02}]^\top$ is the actuator fault vector with $f_{0i} = \text{sat}(\theta_i u_{0i} + u_{si}) - u_{0i}, i = 1, 2$. Since (9.34) and (9.29) are in the same form, the FE-based FTC design in this section can be directly applied to the estimation and compensation of the total effect of saturation, stuck fault and partial loss of effectiveness fault.

9.6.3 FE-Based FTC Design

The helicopter model (9.32) is in the same form of the system (9.1) with $\Delta Ax = 0$, $u = u_0, F = B, f = f_0, n = 4, m = 2, p = 4, q = 2, l = 2$. Moreover, it is verified that the system (9.32) is observable and controllable with the nonlinear function $g(x)$ satisfying Assumption 9.1 with $L_f = m_h g L_a / J_\varepsilon$. The fault and disturbance also satisfy Assumption 9.3. Therefore, the FE-based FTC design in Sects. 9.3–9.5 can be directly applied to the 3-DOF helicopter system. The observer and controller gains are obtained by solving the following optimization problem:

$$\min \gamma \tag{9.35}$$

$$\text{s.t.} \begin{bmatrix} J_{1,1} & B\hat{K}_f & D_1 & ZC_x^\top & 0 & J_{1,2} \\ \star & J_{2,2} & (Q - M_2\bar{C})\bar{D} & 0 & C_e^\top & J_{2,3} \\ \star & \star & -\gamma I & 0 & 0 & 0 \\ \star & \star & \star & -\gamma I & 0 & 0 \\ \star & \star & \star & \star & -\gamma I & 0 \\ \star & \star & \star & \star & \star & -\Lambda \end{bmatrix} \prec 0,$$

$$Z = Z^\top \succ 0,\ Q = Q^\top \succ 0,\ \gamma > 0,$$

with

$$\begin{aligned} J_{1,1} &= \text{He}(\Theta AZ - BM_1) + \varepsilon_2^{-1}\Theta\Theta^\top,\ \hat{K}_f = [0\ K_f], \\ J_{1,2} &= [BM_1\ 0\ Z\ 0],\ J_{2,2} = \text{He}(Q\bar{A} - M_2\bar{C}\bar{A} - M_3\bar{C}) + \varepsilon_1 L_f^2 A_0^\top A_0, \\ J_{2,3} &= [0\ A_0^\top\ 0\ (Q - M_2\bar{C})],\ \Lambda = \text{diag}(\varepsilon_3 Z, \varepsilon_3^{-1} Z, (\varepsilon_2 L_f^2)^{-1} I, \varepsilon_1 I), \end{aligned}$$

where $\varepsilon_i, i = 1, 2, 3$, are given positive scalars. Then the gains are obtained as follows: $K_x = M_1 Z^{-1}, H = Q^{-1}M_2, L_1 = Q^{-1}M_3$.

To facilitate implementation of the proposed FE-based FTC design, the sign function sign(s) used in the controller is approximated by a smooth function $\overline{\text{sign}}(s, \theta_0) = \frac{s}{\|s\|+\theta_0}$, with a sufficiently small positive constant θ_0. It is a differentiable approximation of sign(s) ensuring that the control function u_0 is also differentiable. Define the approximation error as $\Delta_{\text{sign}} = \text{sign}(s) - \overline{\text{sign}}(s, \theta_0)$, then it can be verified that $\|\Delta_{\text{sign}}\| \leq \frac{1}{\|s\|/\theta_0+1} \leq 1$ and for $\|s\| \neq 0$, $\|\Delta_{\text{sign}}\|$ is small by selecting a sufficiently small θ_0.

9.7 Simulation Results

This section provides comparative simulations for the elevation and pitch motions of the Quanser 3-DOF helicopter system (9.1) with single or multiple actuator faults, using (1) the nominal strategy (without FE or FTC and the state observer and controller are designed separately), (2) the separated strategy for FE-based FTC and (3) the proposed simultaneous strategy.

The 3-DOF helicopter system parameters are set as $J_\varepsilon = 0.91$ kg$\cdot$ m^2, $J_p = 0.0364$ kg$\cdot$ m^2, $K_f = 0.5$ N/V, $m_h = 1.01$ kg, $L_a = 0.66$ m, $L_h = 0.177$ m, $g = 9.81$ m/s^2. Due to the mechanical limits, the elevation angle is constrained within ± 31.75 deg and the pitch angle is within ± 32.0 deg. The voltage limits of the front and back motors are ± 12 V. The external disturbances acting on the helicopter are supposed to be $w_\varepsilon = 0.01\sin(10t)$ and $w_p = 0.01\sin(5t)$. The output measurement noise is a Gaussian noise with zero mean and variance 1.0×10^{-6}.

Solving the optimization problem (9.35) with $Y_1 = 0.1_{2\times 4}$, $C_x = I_4$, $C_e = I_6$, $\varepsilon_1 = 0.1$, $\varepsilon_2 = 10$ and $\varepsilon_3 = 15$, then the design gains are obtained as

$$K_x = \begin{bmatrix} 25.4895 & 3.0675 & 20.7441 & 3.0393 \\ 25.5063 & -3.3641 & 20.7190 & -3.3667 \end{bmatrix}, \quad N_1 = \begin{bmatrix} -0.1 & -0.1 & 1.3788 & 0.2056 \\ -0.1 & -0.1 & 1.3788 & -0.2056 \end{bmatrix},$$

$$M = \begin{bmatrix} -0.4977 & 0 & 0 & 0 & 0 & 0 \\ 0 & -0.4976 & 0 & 0 & 0 & 0 \\ 0 & 0 & -0.4979 & 0 & 0 & 0 \\ 0 & 0 & 0.0002 & -0.497 & 0 & 0 \\ 0 & 0 & 0.0003 & 0.0002 & -8.8460 & 8.1499 \\ 0 & 0 & 0.0001 & -0.0011 & 7.9079 & -8.5965 \end{bmatrix},$$

$$N = \begin{bmatrix} 0 & 0 & 0 & 0 & 0 & 0 \\ 0 & 0 & 0 & 0 & 0 & 0 \\ 0 & 0 & 0 & 0 & 0 & 0 \\ 0 & 0 & 0 & 0 & 0 & 0 \\ 0 & 0 & -0.9597 & -3.4952 & 1 & 0 \\ 0 & 0 & -0.9597 & 3.4952 & 0 & 1 \end{bmatrix}, \quad G = \begin{bmatrix} 0 & 0 \\ 0 & 0 \\ 0 & 0 \\ 0 & 0 \\ -8.8460 & 8.1499 \\ 7.9079 & -8.5965 \end{bmatrix},$$

$$L = \begin{bmatrix} 0 & 0 & 0 & 0 \\ 0 & 0 & 0 & 0 \\ 0 & 0 & 0 & 0 \\ 0 & 0 & 0 & 0 \\ 0 & 0 & -0.7524 & -58.5805 \\ 0 & 0 & -0.5718 & 56.8174 \end{bmatrix}, \quad H = \begin{bmatrix} 1 & 0 & 0 & 0 \\ 0 & 1 & 0 & 0 \\ 0 & 0 & 1 & 0 \\ 0 & 0 & 0 & 1 \\ 0 & 0 & 0.9597 & 3.4952 \\ 0 & 0 & 0.9597 & -3.4952 \end{bmatrix}.$$

The other control parameters are chosen as $\epsilon = 0.1$, $\sigma = 0.01$ and $\theta_0 = 0.001$.

For the separated strategy, the observer and controller gains are obtained by solving the optimization problems in Theorems 9.1 and 9.2 with $\gamma_{s1} = 1$, $\gamma_{s2} = 0.06$ and $\varepsilon_s = 0.01$. All simulations are performed with $\varepsilon(0) = 30$ deg and $p(0) = 18$ deg,

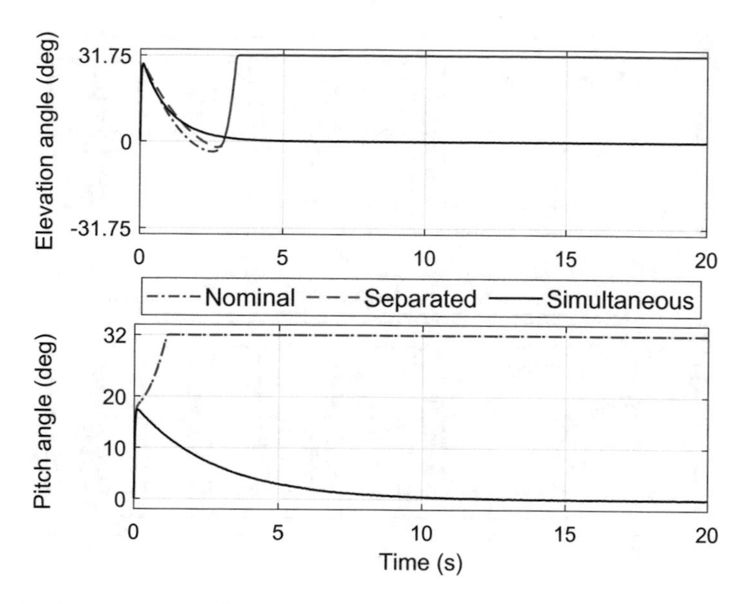

Fig. 9.2 Angle response: Case 1

and zero initial values of other parameters. A first-order low-pass filter $1/(2\pi f_0 s + 1)$ with a frequency $f_0 = 7$ Hz is used to filter each of the measured outputs.

9.7.1 Case 1: Fault-Free

Without faults, the separated and simultaneous strategies revert to nominal observer-based state-feedback robust controls. It is seen from Fig. 9.2 that only the proposed simultaneous strategy can stabilize the elevation and pitch angles. The angles are saturated at their maximum values under the nominal and separated strategies. It is also seen from Fig. 9.3 that the actuators are not saturated under the proposed simultaneous strategy, but saturated and oscillated between the maximum and minimum values under the other two strategies.

9.7.2 Case 2: Single Actuator Fault

In this case, the back actuator of the helicopter is healthy, while the front actuator has a fault characterized by

$$f_{a1}(t) = \begin{cases} 0.1t + 0.08t^2, & 0\ \text{s} \leq t \leq 10\ \text{s} \\ 2\cos(0.5\pi(t-10)) + 7, & 10\ \text{s} < t \leq 20\ \text{s} \end{cases}.$$

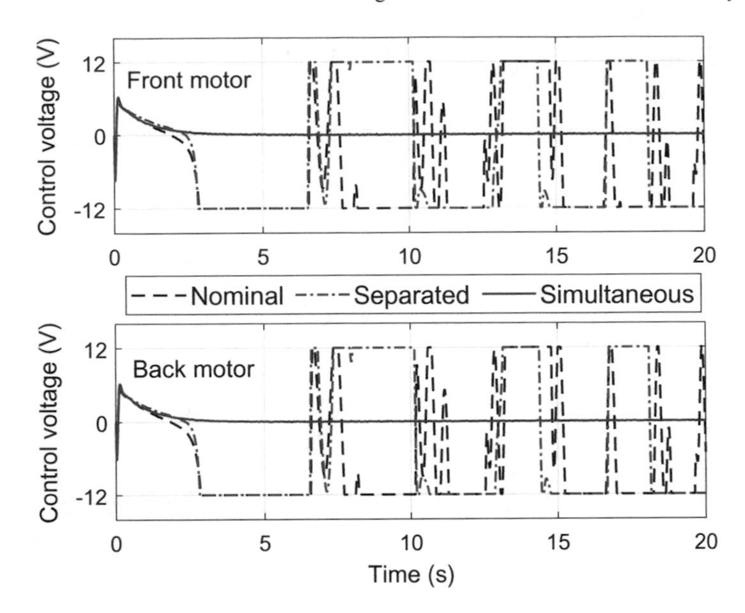

Fig. 9.3 Control effort: Case 1

It is shown in Fig. 9.4 that the proposed simultaneous strategy achieves better FE performance than the separated strategy. The angle responses and control efforts depicted in Figs. 9.5 and 9.6 show that only the simultaneous strategy can stabilize both the elevation and pitch angles in the presence of a single actuator fault. The nominal and separated strategies suffer from saturations in the angles and control inputs.

9.7.3 Case 3: Multiple Actuator Faults

In this case, the front and back actuators suffer from the following oscillatory faults f_{a1} and f_{a2}, respectively:

$$f_{a1}(t) = \begin{cases} 0.1t + 0.08t^2, & 0\text{ s} \leq t \leq 10\text{ s} \\ 2\cos(0.5\pi(t-10)) + 7, & 10\text{ s} < t \leq 20\text{ s} \end{cases},$$
$$f_{a2}(t) = \sin(0.5t) + 0.5\sin(t),\ 0\text{ s} \leq t \leq 20\text{ s}.$$

Similar to Case 2, the results depicted in Figs. 9.7, 9.8 and 9.9 show that, compared with the other two strategies, the proposed simultaneous strategy achieves better fault estimation and output stabilization without input saturations. However, both the nominal and separated strategies have saturations in the angles and control inputs.

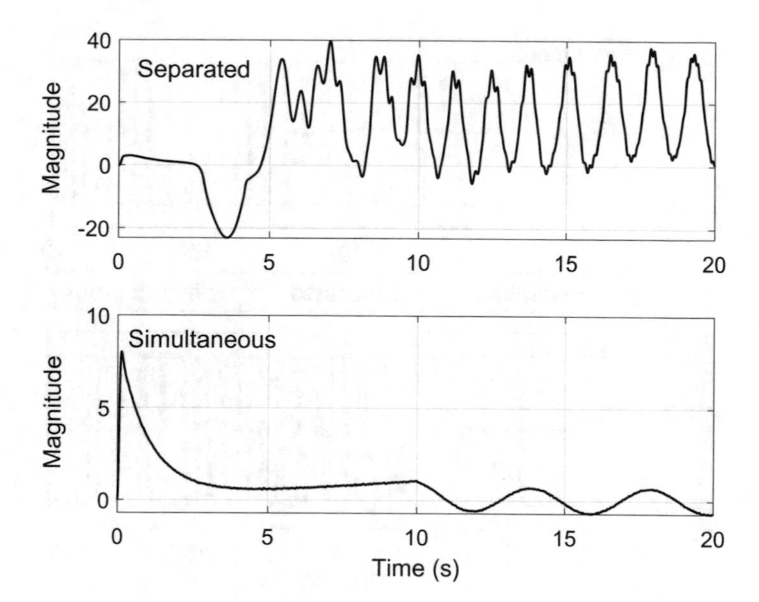

Fig. 9.4 Fault estimation error: Case 2

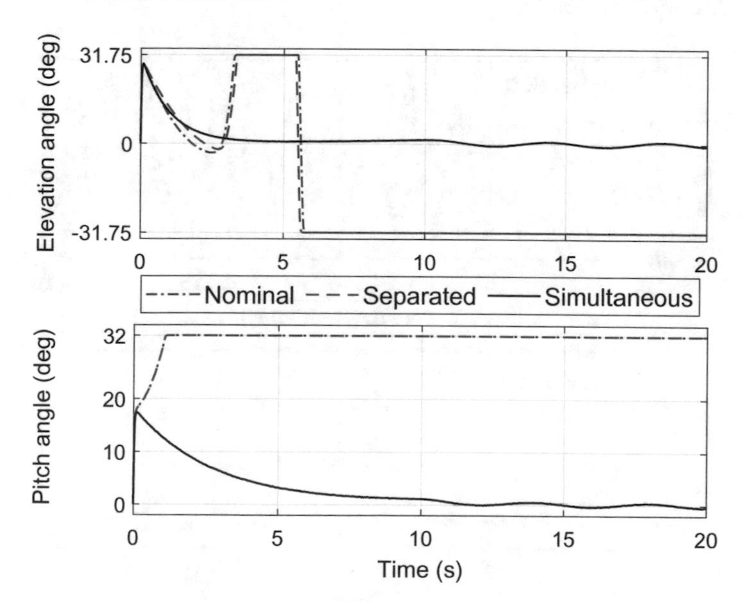

Fig. 9.5 Angle response: Case 2

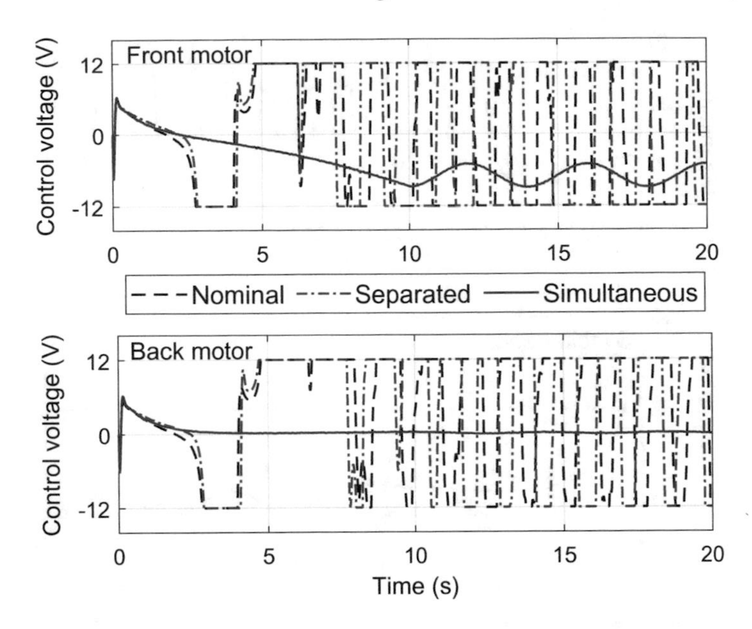

Fig. 9.6 Control effort: Case 2

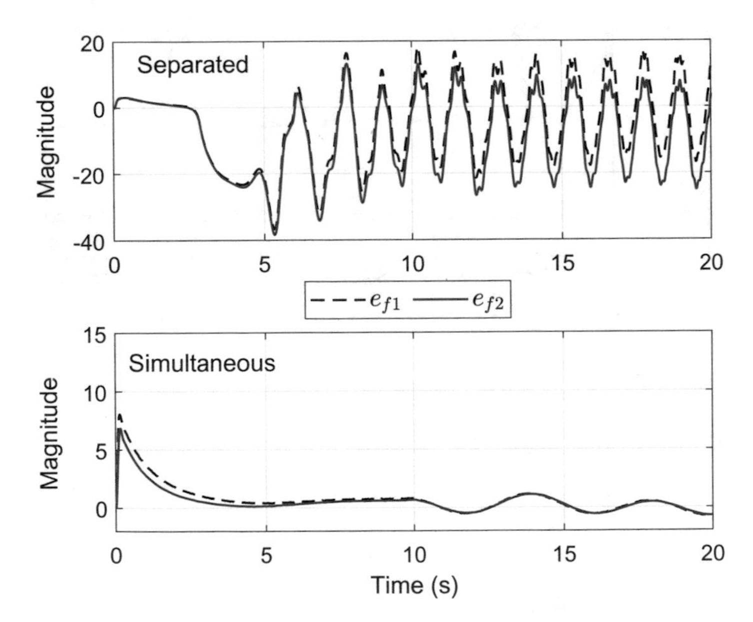

Fig. 9.7 Fault estimation error: Case 3

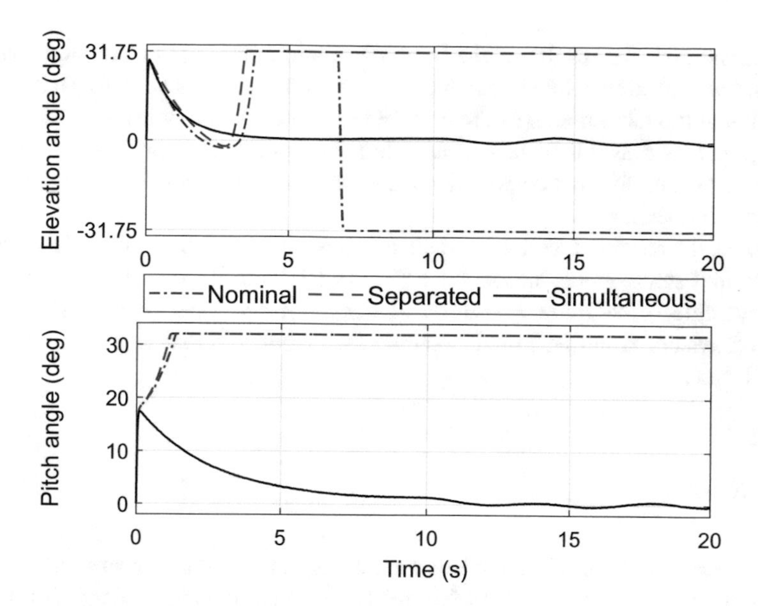

Fig. 9.8 Angle response: Case 3

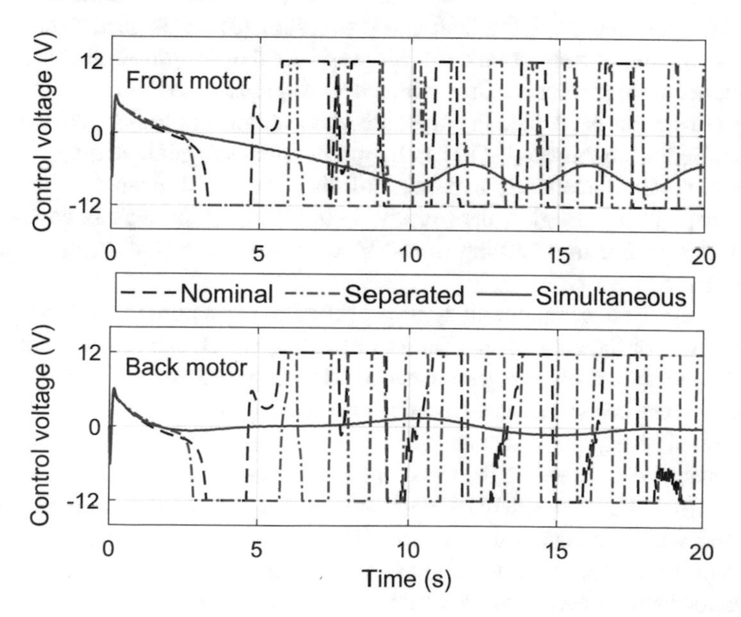

Fig. 9.9 Control effort: Case 3

Summarizing the results of the three simulation cases: (1) Compared with the nominal and separated strategies, the proposed simultaneous strategy stabilizes the elevation and pitch motions of the 3-DOF helicopter system with the best transient performance and avoids actuator saturation, no matter actuator faults exist or not. (2) The proposed simultaneous strategy achieves more accurate fault estimation than the separated strategy.

The results represent well the expected behaviour of the three strategies, since (1) the nominal strategy does not include FE and FTC functions and (2) the separated strategy neglects the *bidirectional robustness interactions* between the FE observer and FTC system, resulting from inaccurate estimation and system performance with low robustness.

9.8 Notes

This chapter extends the simultaneous robust integration strategy proposed in Chap. 5 to address the FE-based FTC problem for Lipschitz nonlinear systems. A nonlinear ASUIO is developed to estimate the system state and faults. An adaptive sliding mode FTC controller using the fault and state estimates is designed to compensate the fault and saturation effects and robustly stabilize the elevation and pitch motions. A simultaneous integration strategy with a new single-step LMI formulation without equality constraints is adopted to obtain the observer and controller gains. The strategy is applied to a nonlinear 3-DOF helicopter system subject to actuator faults and saturation, system uncertainty and external disturbance. Compared with the nominal and separated strategies, the proposed simultaneous strategy is demonstrated to be more effective in achieving robust FTC performance (stabilization and fault compensation).

In the presence of actuator faults, the proposed design estimates and compensates the total effect of the actuator faults and saturation, which can robustly recover the nominal non-saturated system performance. Hence, where there is no fault, the proposed design can be used as an alternative anti-windup control framework to recover non-saturated system performance. Adaptive anti-windup controls have been developed in many works based on the use of an auxiliary system and a switched control, see for example He et al. (2016). Other mainstream anti-windup methods incorporate an anti-windup compensator as part of the normal control function (Tarbouriech and Turner 2009). However, the above anti-windup designs are implemented with the actuator output measurement, which is unavailable and undesirable, especially if the actuator has fast unmodelled dynamics. Compared with the existing literature, the proposed strategy for anti-windup control is convenient in the sense that (1) the observer achieves simultaneous estimation of the system state and saturation effect without measuring the actuator output and (2) all the observer and controller gains are obtained by solving the optimization problem (9.35) in a single step.

References

Afonso RJM, Galvão RKH (2010) Predictive control of a helicopter model with tolerance to actuator faults. In: Proceedings of the conference on control and fault-tolerant systems, IEEE, pp 744–751

Alexis K, Nikolakopoulos G, Tzes A (2012) Model predictive quadrotor control: attitude, altitude and position experimental studies. IET Control Theory & Appl 6(12):1812–1827

Apkarian J (2006) 3-DOF helicopter reference manual. Quanser Consulting Inc, Canada

Chen M, Ge S, Ren B (2010) Robust attitude control of helicopters with actuator dynamics using neural networks. IET Control Theory & Appl 4(12):2837–2854

Chen F, Zhang K, Jiang B, Wen C (2016a) Adaptive sliding mode observer-based robust fault reconstruction for a Helicopter with actuator fault. Asian J Control 18(4):1558–1565

Chen M, Shi P, Lim CC (2016b) Adaptive neural fault-tolerant control of a 3-DOF model helicopter system. IEEE Trans Syst Man Cybern: Syst 46(2):260–270

Ducard GJ (2009) Fault-tolerant flight control and guidance systems: practical methods for small unmanned aerial vehicles. Springer Science & Business Media, Berlin

Edwards C, Lombaerts T, Smaili H (2010) Fault tolerant flight control: a benchmark challenge. Springer Science & Business Media, Berlin

Freidovich LB, Khalil HK (2008) Performance recovery of feedback-linearization-based designs. IEEE Trans Autom Control 53(10):2324–2334

Goupil P (2010) Oscillatory failure case detection in the A380 electrical flight control system by analytical redundancy. Control Eng Pract 18(9):1110–1119

He W, Dong Y, Sun C (2016) Adaptive neural impedance control of a robotic manipulator with input saturation. IEEE Trans Syst Man Cybern: Syst 46(3):334–344

Izaguirre-Espinosa C, Muñoz-Vázquez AJ, Sánchez-Orta A, Parra-Vega V, Castillo P (2016) Attitude control of quadrotors based on fractional sliding modes: theory and experiments. IET Control Theory & Appl 10(7):825–832

Kiefer T, Graichen K, Kugi A (2010) Trajectory tracking of a 3-DOF laboratory helicopter under input and state constraints. IEEE Trans Control Syst Technol 18(4):944–952

Li Z, Yu J, Xing X, Gao H (2015) Robust output-feedback attitude control of a three-degree-of-freedom helicopter via sliding-mode observation technique. IET Control Theory & Appl 9(11):1637–1643

de Loza AF, Cieslak J, Henry D, Zolghadri A, Fridman LM (2015) Output tracking of systems subjected to perturbations and a class of actuator faults based on HOSM observation and identification. Automatica 59:200–205

Qi X, Qi J, Theilliol D, Zhang Y, Han J, Song D, Hua C (2014) A review on fault diagnosis and fault tolerant control methods for single-rotor aerial vehicles. J Intell & Robot Syst 73(1–4):535–555

Qi X, Qi J, Theilliol D, Song D, Zhang Y, Han J (2016) Self-healing control design under actuator fault occurrence on single-rotor unmanned helicopters. J Intell & Robot Syst 84(1–4):21–35

Tarbouriech S, Turner M (2009) Anti-windup design: an overview of some recent advances and open problems. IET Control Theory & Appl 3(1):1–19

Zheng W, Fuyang C, Bin J (2014) An improved nonlinear model and adaptive fault-tolerant control for a twin rotor helicopter. In: Proceedings of the 33rd Chinese control conference, IEEE, pp 3208–3212

Zheng Z, Sun L, Zou Y (2015) Attitude tracking control of a 3-DOF helicopter with actuator saturation and model uncertainties. In: Proceedings of the 34th Chinese control conference, IEEE, pp 5641–5646

Zolghadri A, Henry D, Cieslak J, Efimov D, Goupil P (2014) Fault diagnosis and fault-tolerant control and guidance for aerospace vehicles. Springer, Berlin

Chapter 10
Integration of FE and FTC for Large-Scale Interconnected Systems

10.1 Introduction

Chapters 3–9 deal with FE-based FTC designs for small-scale linear or nonlinear systems. However, the complexity of industrial, process, banking and IT systems increases rapidly as modern technology makes more and more use of interconnected, embedded, networked and distributed architectures (Bakule and Rossel 2008; Ikeda 1989; Šiljak and Zečević 2005). It is stated in Šiljak (1991) that the complexity of real systems might not be well organized, while for control to be effective a good structural system organization is required. This can be achieved using a decentralized system structure in which local interconnected subsystems are well defined. Decentralized control is economical and can be reliable. However, the disturbance from interactions should be handled by combined use of state estimation and control. Some researchers use decentralized observer-based control to achieve stability and robustness control goals, e.g. Bakule and Rodellar (1996), Benigni et al. (2010), Kalsi et al. (2010), Pagilla and Zhu (2004), Shafai et al. (2011), Tlili and Braiek (2009). A further challenge arises when there exist actuator, sensor or process faults. The design problem is further complicated by the combined presence of faults and system uncertainties.

Existing works on FTC for large-scale interconnected systems are categorized below.

- *PFTC (without fault detection/estimation)*. Jin and Yang (2009) propose an adaptive model matching control for interconnected systems with actuator faults. Panagi and Polycarpou (2011) address the decentralized FTC problems for interconnected nonlinear systems subject to connection faults (faults on interaction functions). Amani et al. (2014) describe a large-scale cooperative FTC system design considering actuator faults. Naghavi et al. (2014) propose a decentralized fault tolerant predictive control for discrete-time interconnected nonlinear systems with connection faults. Huang and Patton (2015) develop an output feedback sliding mode

© The Author(s), under exclusive license to Springer Nature Switzerland AG 2021

J. Lan and R. J. Patton, *Robust Integration of Model-Based Fault Estimation and Fault-Tolerant Control*, Advances in Industrial Control,
https://doi.org/10.1007/978-3-030-58760-4_10

FTC design for interconnected systems with actuator faults. Yang et al. (2015) develop a fault recovery and FTC strategy for interconnected nonlinear systems with both actuator and sensor faults. Adaptive decentralized FTCs are proposed in Chen et al. (2016), Hashemi et al. (2016) for large-scale interconnected nonlinear systems with actuator faults.

- *AFTC (with FDI)*. In Sauter et al. (2006), a decentralized FDI/FTC system is designed for networked systems considering actuator faults. Patton et al. (2007) propose an FTC design with a distributed hierarchical structure for network control systems. Patton and Klinkhieo (2009) provide a two-level sliding mode FTC scheme for distributed and interconnected systems. Khalili et al. (2015) deal with the decentralized fault accommodation problem for multi-agents systems using FDI.

However, few works have considered decentralized FTC via FE, instead of FDI. FE directly estimates the fault signal, avoiding complex procedures of threshold setting and fault isolation required in FDI, which can significantly facilitate the FTC design. This chapter aims to develop a decentralized FTC for large-scale linear systems subject to uncertain nonlinear interactions and actuator or sensor faults, using a decentralized ASUIO for simultaneous state and fault estimation. There exist *bidirectional robustness interactions* between the FE observer and FTC system for each subsystem. This is because (1) the uncertain nonlinear interactions affect the state/fault estimation performance and (2) the estimation errors affect the FTC system performance. This chapter extends the simultaneous robust integration strategy in Chap. 5 for small-scale systems to the considered large-scale interconnected systems. An integrated decentralized FE and FTC design is proposed based on H_∞ optimization with a single-step LMI formulation.

10.2 Problem Description

Consider a large-scale system consisting of n subsystems, where the ith ($i = 1, 2, \ldots, n$) subsystem is represented by

$$\begin{aligned} \dot{x}_i &= A_i x_i + B_i u_i + h_i(x, t) \\ y_i &= C_i x_i \end{aligned} \tag{10.1}$$

where $x_i \in \mathbb{R}^{n_i}$, $u_i \in \mathbb{R}^{m_i}$ and $y_i \in \mathbb{R}^{p_i}$ are the state, control input and system output, respectively. The constant matrices A_i, B_i and C_i are known and of compatible dimensions. $h_i(x, t) \in \mathbb{R}^{n_i}$ is the uncertain nonlinear interaction with $x = [x_1^\top, \ldots, x_n^\top]^\top$. The system (10.1) satisfies the following assumptions:

Assumption 10.1 The pairs (A_i, B_i) and (A_i, C_i) are controllable and observable, respectively.

Assumption 10.2 The interaction function $h_i(x, t)$ satisfies the quadratic constraint

$$h_i^\top(x,t)h_i(x,t) \le \alpha_i x^\top H_{0i}^\top H_{0i} x,$$

where H_{0i} is a known constant matrix and α_i is some positive scalar defined as the uncertain interaction bound.

Remark 10.1 By defining $h(x,t) = [h_1^\top(x,t), \ldots, h_n^\top(x,t)]^\top$ as the interaction term of the overall large-scale system, it can be derived that

$$h^\top(x,t)h(x,t) \le x^\top H_0^\top H_0 x,$$

where $H_0 = \left[\sqrt{\alpha_1} H_{01}^\top, \ldots, \sqrt{\alpha_n} H_{0n}^\top\right]^\top$.

In the real operation of the large-scale system (10.1), there may be actuator faults and/or sensor faults which will degrade the control performance or even destabilize the system. Hence, this chapter aims to address the following problem:

Problem 10.1 For the large-scale interconnected system (10.2) with uncertain non-linear interactions and faults, design an integrated FE-based FTC strategy including (1) a decentralized ASUIO to estimate the system state and faults and (2) a decentralized FTC controller to guarantee robust stability of the overall closed-loop system.

10.3 Decentralized FE and FTC with Actuator Fault

This section considers the FE-based FTC design for the system (10.2) suffering from actuator faults. In this case, the ith ($i = 1, 2, \ldots, n$) subsystem is represented by

$$\begin{aligned} \dot{x}_i &= A_i x_i + B_i u_i + F_i f_i + h_i(x,t) \\ y_i &= C_i x_i \end{aligned} \tag{10.2}$$

where $f_i \in \mathbb{R}^{q_i}$ denotes the actuator fault with the known constant distribution matrix $F_i \in \mathbb{R}^{n_i \times q_i}$. The other variables and matrices are the same as in (10.1). The subsystem (10.2) satisfies Assumption 10.2 and the following assumptions:

Assumption 10.3 The pairs (A_i, B_i) is controllable, and the triple (A_i, F_i, C_i) has no invariant zero in the closed right-half complex plane.

Assumption 10.4 The actuator fault f_i and its first-order derivative are norm-bounded, and the fault is matched, i.e. $\text{rank}(B_i, F_i) = \text{rank}(B_i)$.

The proposed decentralized ASUIO-based FTC for large-scale systems follows the structure outlined in Fig. 10.1, where Si, Oi, and Ci are the ith subsystem, observer, and controller, respectively. In this structure, each subsystem has its own observer and controller. All the observer and controller gains will be designed together but implemented separately at each subsystem.

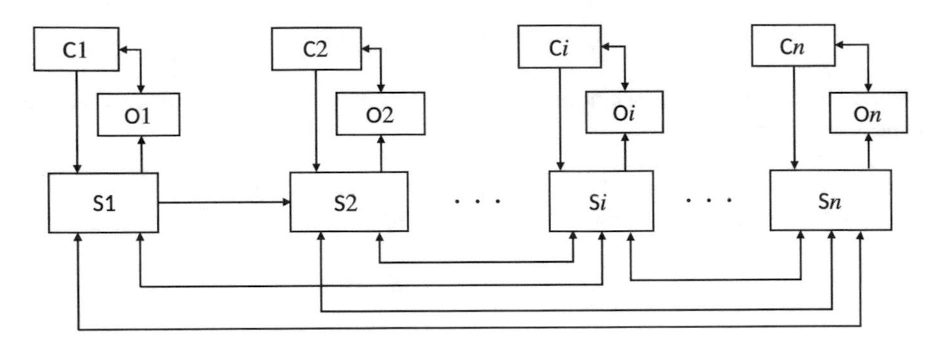

Fig. 10.1 Decentralized integrated FE-based FTC design for large-scale systems

10.3.1 Decentralized FE Observer Design

Define f_i as auxiliary state, then the ith subsystem (10.2) is augmented as

$$\begin{aligned} \dot{\bar{x}}_i &= \bar{A}_i \bar{x}_i + \bar{h}_i(x,t) + \bar{B}_i u_i + \bar{D}_i \bar{d}_i \\ y_i &= \bar{C}_i \bar{x}_i \end{aligned} \tag{10.3}$$

where

$$\bar{x}_i = \begin{bmatrix} x_i \\ f_i \end{bmatrix}, \ \bar{A}_i = \begin{bmatrix} A_i & F_i \\ 0 & 0 \end{bmatrix}, \ \bar{h}_i(x,t) = \begin{bmatrix} h_i(x,t) \\ 0 \end{bmatrix}, \ \bar{B}_i = \begin{bmatrix} B_i \\ 0 \end{bmatrix},$$
$$\bar{D}_i = \begin{bmatrix} 0 \\ I_{q_i} \end{bmatrix}, \ \bar{d}_i = \dot{f}_i, \ \bar{C}_i = [C_i \ 0].$$

The augmented state vector $\bar{x}_i$ is estimated by the ASUIO in the form of

$$\begin{aligned} \dot{z}_i &= M_i z_i + G_i u_i + L_i y_i \\ \hat{\bar{x}}_i &= z_i + H_i y_i \end{aligned} \tag{10.4}$$

where $z_i \in \mathbb{R}^{n_i+q_i}$ is the observer state and $\hat{\bar{x}}_i \in \mathbb{R}^{n_i+q_i}$ is the estimate of $\bar{x}_i$. The matrices M_i, G_i, N_i, L_i and H_i are of appropriate dimensions and to be designed.

Define the estimation error as $e_i = \bar{x}_i - \hat{\bar{x}}_i$. By using (10.3) and (10.4), the error system is derived as

$$\begin{aligned} \dot{e}_i = &(\Xi_i \bar{A}_i - L_{i1}\bar{C}_i)e_i + (\Xi \bar{A}_i - L_{i1}\bar{C}_i - M_i)z_i + (\Xi_i \bar{B}_i - G_i)u_i \\ &+ \left[(\Xi_i \bar{A}_i - L_{i1}\bar{C}_i)H_i - L_{i2}\right] y_i + \Xi_i \bar{h}_i(x,t) + \Xi_i \bar{D}_i \bar{d}_i \end{aligned} \tag{10.5}$$

where $\Xi_i = I_{n_i+q_i} - H_i \bar{C}_i$ and $L_i = L_{i1} + L_{i2}$.

Define the following relations:

$$\Xi_i \bar{A}_i - L_{i1}\bar{C}_i - M_i = 0 \tag{10.6}$$

$$\Xi_i \bar{B}_i - G_i = 0 \tag{10.7}$$

$$(\Xi_i \bar{A}_i - L_{i1}\bar{C}_i)H_i - L_{i2} = 0. \tag{10.8}$$

By using (10.6)–(10.8), the error system (10.5) is rearranged into

$$\dot{e}_i = (\Xi_i \bar{A}_i - L_{i1}\bar{C}_i)e_i + \Xi_i \bar{h}_i(x,t) + \Xi_i \bar{D}_i \bar{d}_i. \tag{10.9}$$

The matrices M_i, G_i and L_{i2} are calculated from (10.6)–(10.8) once L_{i1} and H_i are obtained. Hence, the main task in the following is to design the matrices L_{i1} and H_i.

10.3.2 Decentralized FTC Controller Design

An FTC controller for the ith subsystem (10.2) is designed as

$$u_i = -K_i \hat{\bar{x}}_i \tag{10.10}$$

where $K_i = [K_{xi} \;\; K_{fi}]$, $K_{xi} \in \mathbb{R}^{m_i \times n_i}$ is the nominal controller gain, and $K_{fi} \in \mathbb{R}^{m_i \times q_i}$ is the actuator fault compensation gain designed as $K_{fi} = B_i^{\dagger} F_i$.

Substituting (10.10) into (10.2) gives the FTC closed-loop system

$$\dot{x}_i = (A_i - B_i K_{xi})x_i + B_i K_i e_i + h_i(x,t). \tag{10.11}$$

10.3.3 Robust Integration of FE and FTC

The FE-based FTC closed-loop system composed of (10.9) and (10.11) is given by

$$\begin{aligned} \dot{x}_i &= (A_i - B_i K_{xi})x_i + B_i K_i e_i + h_i(x,t) \\ \dot{e}_i &= (\Xi_i \bar{A}_i - L_{i1}\bar{C}_i)e_i + \Xi_i \bar{D}_i \bar{d}_i. \end{aligned} \tag{10.12}$$

Therefore, the composite closed-loop system of the large-scale system is

$$\begin{aligned} \dot{x} &= \tilde{A}x + \tilde{F}e + h(x,t) \\ \dot{e} &= \tilde{A}_e e + \Xi \bar{h}(x,t) + \Xi \bar{D}\bar{d} \\ z &= \mathrm{diag}(C_x x, C_e e) \end{aligned} \tag{10.13}$$

where $z \in \mathbb{R}^{2\bar{n}+\bar{q}}$ is the performance output, $C_x \in \mathbb{R}^{\bar{n}\times\bar{n}}$, $C_e \in \mathbb{R}^{(\bar{n}+\bar{q})\times(\bar{n}+\bar{q})}$, $\bar{n} = \sum_{i=1}^{n} n_i$, $\bar{q} = \sum_{i=1}^{n} q_i$, and

$$e = \begin{bmatrix} e_1 \\ \vdots \\ e_n \end{bmatrix}, \ \bar{d} = \begin{bmatrix} \bar{d}_1 \\ \vdots \\ \bar{d}_n \end{bmatrix}, \ \bar{h}(x,t) = \begin{bmatrix} \bar{h}_1 \\ \vdots \\ \bar{h}_n \end{bmatrix},$$

$$\tilde{A} = \mathrm{diag}(A_1 - B_1 K_{x1}, \ldots, A_n - B_n K_{xn}),$$
$$\tilde{F} = \mathrm{diag}(B_1 K_1, \ldots, B_n K_n),$$
$$\Xi = \mathrm{diag}(\Xi_1, \ldots, \Xi_n), \ \bar{D} = \mathrm{diag}(\bar{D}_1, \ldots, \bar{D}_n),$$
$$\tilde{A}_e = \mathrm{diag}(\Xi_1 \bar{A}_1 - L_{11}\bar{C}_1, \ldots, \Xi_n \bar{A}_n - L_{n1}\bar{C}_n).$$

Now the integrated design problem can be stated as follows: design the controller gains K_{xi}, $i = 1, \ldots, n$, and observer gains H_i and L_{i1}, $i = 1, \ldots, n$, to ensure robust stability of the overall closed-loop system (10.13). This design problem is solved using Theorem 10.1 with a single-step LMI formulation.

Theorem 10.1 *Under Assumptions 10.2, 10.3 and 10.4, given positive scalars β_1, β_2, ε_{1i}, ε_{2i} and ε_{3i}, $i = 1, 2, \ldots, n$, the overall closed-loop system (10.13) is stable with H_∞ performance $\|G_{z\bar{d}}\|_\infty < \gamma$, if the following optimization problem is feasible:*

$$\min \ \beta_1 \sum_{i=1}^{n} \bar{\gamma}_i + \beta_2 \sum_{i=1}^{n} \bar{\alpha}_i \tag{10.14}$$

$$\text{s.t.} \begin{bmatrix} \Pi_{11} & \Pi_{12} & \Pi_{13} & 0 & \Pi_{15} & 0 & \Pi_{17} & 0 & 0 \\ \star & \Pi_{22} & 0 & \Pi_{24} & 0 & \Pi_{26} & 0 & \Pi_{28} & \Pi_{29} \\ \star & \star & -\varepsilon_3^{-1} Z & 0 & 0 & 0 & 0 & 0 & 0 \\ \star & \star & \star & -\varepsilon_3 Z & 0 & 0 & 0 & 0 & 0 \\ \star & \star & \star & \star & -\epsilon I & 0 & 0 & 0 & 0 \\ \star & \star & \star & \star & \star & -\varepsilon_1 I & 0 & 0 & 0 \\ \star & \star & \star & \star & \star & 0 & -I & 0 & 0 \\ \star & \star & \star & \star & \star & \star & \star & -I & 0 \\ \star & \star & \star & \star & \star & \star & \star & \star & -\bar{\gamma} \end{bmatrix} \prec 0 \tag{10.15}$$

$$Z_i = Z_i^\top \succ 0, \ Q_i = Q_i^\top \succ 0, \ \bar{\alpha}_i > 0, \ \bar{\gamma}_i > 0, \ i = 1, \ldots, n \tag{10.16}$$

where

$$Z = \mathrm{diag}(Z_1, \ldots, Z_n), \ \varepsilon_1 = \mathrm{diag}(\varepsilon_{1i}, \ldots, \varepsilon_{1n}), \ \varepsilon_3 = \mathrm{diag}(\varepsilon_{3i}, \ldots, \varepsilon_{3n}),$$
$$\epsilon = \mathrm{diag}(\epsilon_1, \ldots, \epsilon_n), \ \epsilon_i = \bar{\alpha}_i(\varepsilon_{1i} + \varepsilon_{2i})^{-1}, \ \bar{\gamma} = \mathrm{diag}(\bar{\gamma}_1, \ldots, \bar{\gamma}_n),$$
$$\Pi_{11} = \mathrm{diag}\left(\Pi_{111}, \ldots, \Pi_{11n}\right), \ \Pi_{11i} = \mathrm{He}(A_i Z_i - B_i M_{1i}) + \varepsilon_{2i}^{-1} I,$$
$$\Pi_{12} = \mathrm{diag}\left([0 \ F_1], \ldots, [0 \ F_n]\right), \ \Pi_{13} = \mathrm{diag}(B_1 M_{1i}, \ldots, B_n M_{1n}),$$
$$\Pi_{15} = Z[H_{01}^\top \ \ldots \ H_{0n}^\top], \ \Pi_{17} = \mathrm{diag}(Z_1 C_{x1}^\top, \ldots, Z_n C_{xn}^\top),$$
$$\Pi_{22} = \mathrm{diag}(\Pi_{221}, \ldots, \Pi_{22n}), \ \Pi_{22i} = \mathrm{He}(Q_i \bar{A}_i - M_{3i}\bar{C}_i \bar{A}_i - M_{2i}\bar{C}_i),$$
$$\Pi_{24} = \mathrm{diag}(\Lambda, \ldots, \Lambda), \ \Pi_{26} = \mathrm{diag}(\Pi_{261}, \ldots, \Pi_{26n}), \ \Pi_{26i} = Q_i - M_{3i}\bar{C}_i,$$
$$\Pi_{28} = \mathrm{diag}(C_{e1}^\top, \ldots, C_{en}^\top), \ \Pi_{29} = \mathrm{diag}(\Pi_{291}, \ldots, \Pi_{29n}), \ \Pi_{29i} = (Q_i - M_{3i}\bar{C}_i)\bar{D}_i.$$

The gains are obtained as $\gamma = \sqrt{\bar{\gamma}}$, $K_{xi}{=}M_{1i}Z_i^{-1}$, $L_{i1}{=}Q_{i1}^{-1}M_{2i}$, $H_i = Q_{i1}^{-1}M_{3i}$.

Proof Consider the Lyapunov function $V_e = e^\top Qe$ with a s.p.d. matrix Q. According to Assumption 10.2, it holds that for some positive scalar $\varepsilon_1 = \text{diag}(\varepsilon_{11}, \ldots, \varepsilon_{1n})$,

$$\begin{aligned}\text{He}(e^\top Q\Xi\bar{h}(x,t)) &= -\left[\sqrt{\varepsilon_1}^{-1}\Xi^\top Qe - \sqrt{\varepsilon_1}\bar{h}(x,t)\right]^\top \times \left[\sqrt{\varepsilon_1}^{-1}\Xi^\top Qe - \sqrt{\varepsilon_1}\bar{h}(x,t)\right]\\ &\quad +\varepsilon_1^{-1}e^\top Q\Xi\Xi^\top Qe + \varepsilon_1\bar{h}^\top(x,t)\bar{h}(x,t)\\ &\leq \varepsilon_1^{-1}e^\top Q\Xi\Xi^\top Qe + \varepsilon_1 x^\top H_0^\top H_0 x.\end{aligned}$$

The time derivative of V_e along (10.9) is derived as

$$\begin{aligned}\dot{V}_e &= e^\top\text{He}(Q\tilde{A}_e)e + \text{He}(e^\top Q\Xi\bar{h}(x,t))\\ &\leq e^\top[\text{He}(Q\tilde{A}_e) + \varepsilon_1^{-1}Q\Xi\Xi^\top Q]e + \varepsilon_1 x^\top H_0^\top H_0 x.\end{aligned} \tag{10.17}$$

Consider another Lyapunov function $V_x = x^\top Px$. It holds that, for some positive scalar $\varepsilon_2 = \text{diag}(\varepsilon_{21}, \ldots, \varepsilon_{2n})$,

$$\begin{aligned}\text{He}(x^\top Ph(x,t)) &= -\left[\sqrt{\varepsilon_2}^{-1}Px - \sqrt{\varepsilon_2}h(x,t)\right]^\top \times \left[\sqrt{\varepsilon_2}^{-1}Px - \sqrt{\varepsilon_2}h(x,t)\right]\\ &\quad +\varepsilon_2^{-1}x^\top PPx + \varepsilon_2 h^\top(x,t)h(x,t)\\ &\leq \varepsilon_2^{-1}x^\top PPx + \varepsilon_2 x^\top H_0^\top H_0 x.\end{aligned}$$

The time derivative of V_x along (10.11) is

$$\begin{aligned}\dot{V}_x &= x^\top\text{He}(P\tilde{A})x - \text{He}(x^\top P\tilde{F}e) + \text{He}(x^\top Ph(x,t))\\ &\leq x^\top\left[\text{He}(P\tilde{A}) + \varepsilon_2^{-1}PP\right]x + \varepsilon_2 x^\top H_0^\top H_0 x - \text{He}(x^\top P\tilde{F}e).\end{aligned} \tag{10.18}$$

Let $\xi = [x^\top\ e^\top]^\top$, then the H_∞ performance $\|G_{z\bar{d}}\|_\infty < \gamma$ can be represented as

$$J = \int_0^\infty \left(\xi^\top(t)\xi(t) - \gamma^2\bar{d}(t)^\top\bar{d}(t)\right)\mathrm{d}t < 0. \tag{10.19}$$

Under zero initial conditions, it can be derived that

$$\begin{aligned}J &= \int_0^\infty \left(\xi(t)^\top\xi(t) - \gamma^2\bar{d}(t)^\top\bar{d}(t) + \dot{V}_x(t) + \dot{V}_e(t)\right)\mathrm{d}t - \int_0^\infty \left(\dot{V}_x(t) + \dot{V}_e(t)\right)\mathrm{d}t\\ &= \int_0^\infty \left(\xi(t)^\top\xi(t) - \gamma^2\bar{d}(t)^\top\bar{d}(t) + \dot{V}_x(t) + \dot{V}_e(t)\right)\mathrm{d}t\\ &\quad - (V_x(\infty) + V_e(\infty)) + (V_x(0) + V_e(0))\\ &\leq \int_0^\infty \left(\xi(t)^\top\xi(t) - \gamma^2\bar{d}(t)^\top\bar{d}(t) + \dot{V}_x(t) + \dot{V}_e(t)\right)\mathrm{d}t.\end{aligned}$$

A sufficient condition for (10.19) is

$$\xi^\top \xi - \gamma^2 \bar{d}^\top \bar{d} + \dot{V}_x + \dot{V}_e < 0. \tag{10.20}$$

Substituting (10.17) and (10.18) into (10.20) yields

$$\begin{bmatrix} x \\ e \\ \bar{d} \end{bmatrix}^\top \begin{bmatrix} J_{11} & J_{12} & 0 \\ \star & J_{22} & J_{23} \\ \star & \star & -\gamma^2 I \end{bmatrix} \begin{bmatrix} x \\ e \\ \bar{d} \end{bmatrix} < 0 \tag{10.21}$$

and equivalently,

$$\begin{bmatrix} J_{11} & J_{12} & 0 \\ \star & J_{22} & J_{23} \\ \star & \star & -\gamma^2 I \end{bmatrix} \prec 0 \tag{10.22}$$

where

$$J_{11} = \mathrm{He}(P\tilde{A}) + (\varepsilon_1 + \varepsilon_2) H_0^\top H_0 + \varepsilon_2^{-1} P P + C_x^\top C_x, \ J_{12} = P\tilde{F},$$
$$J_{22} = \mathrm{He}(Q\tilde{A}_e) + \varepsilon_1^{-1} Q \Xi \Xi^\top Q + C_e^\top C_e, \ J_{23} = Q\Xi\bar{D}.$$

Define $Z = P^{-1}$. Pre- and post-multiplying both sides of (10.22) with diag (Z, I, I) and its transpose, respectively, then (10.22) is equivalently converted into

$$\begin{bmatrix} J_{11} & J_{12} & 0 \\ \star & J_{22} & J_{23} \\ \star & \star & -\gamma^2 I \end{bmatrix} \prec 0 \tag{10.23}$$

where

$$J_{11} = \mathrm{He}(\tilde{A}Z) + (\varepsilon_1 + \varepsilon_2) Z H_0^\top H_0 Z + \varepsilon_2^{-1} I + Z C_x^\top C_x Z, \ J_{12} = \tilde{F},$$
$$J_{22} = \mathrm{He}(Q\tilde{A}_e) + \varepsilon_1^{-1} Q \Xi \Xi^\top Q + C_e^\top C_e, \ J_{23} = Q\Xi\bar{D}.$$

Define $Z = \mathrm{diag}(Z_1, \ldots, Z_n)$. Notice that

$$J_{12} = \mathrm{diag}\left([0 \ F_1], \ldots, [0 \ F_n]\right) + \mathrm{diag}\left([B_1 K_{x1} \ 0], \ldots, [B_n K_{xn} \ 0]\right).$$

By using the Young inequality (see Sect. 1.7.1), then there exist some positive scalars $\varepsilon_3 = \mathrm{diag}(\varepsilon_{31}, \ldots, \varepsilon_{3n})$ such that

$$\begin{bmatrix} 0 & 0 & 0 \\ 0 & B_i K_{xi} & 0 \\ 0 & 0 & 0 \end{bmatrix} \preceq \varepsilon_{3i}^{-1} \begin{bmatrix} 0 \\ \Lambda \\ 0 \end{bmatrix} Z_i^{-1} \begin{bmatrix} 0 \\ \Lambda \\ 0 \end{bmatrix}^\top + \varepsilon_{3i} \begin{bmatrix} 0 \\ B_i K_{xi} Z_i \\ 0 \end{bmatrix} Z_i^{-1} \begin{bmatrix} 0 \\ B_i K_{xi} Z_i \\ 0 \end{bmatrix}^\top,$$

where $\Lambda = [I_n \ 0]^\top$.

Define $Q = \text{diag}(Q_1, \ldots, Q_n)$, then partition $\Pi_{22} = \text{He}(Q\tilde{A}_e)$ into the diagonal form $\Pi_{22} = \text{diag}(\Pi_{221}, \ldots, \Pi_{22n})$ with $\Pi_{22i} = \text{He}(Q_i(\Xi_i \bar{A}_i - L_{i1}\bar{C}_i))$. Define $\bar{\alpha}_i = \alpha_i^{-1}$, $\epsilon_i = \bar{\alpha}_i(\varepsilon_{1i} + \varepsilon_{2i})^{-1}$ and $\bar{\gamma} = \gamma^2$, then $\bar{\gamma}$ can be partitioned into $\bar{\gamma} = \text{diag}(\bar{\gamma}_1, \ldots, \bar{\gamma}_n)$.

Furthermore, define $M_{1i} = K_{xi}Z_i$, $M_{2i} = Q_i L_{i1}$ and $M_{3i} = Q_i H_i$. Using the Schur Complement (see Sect. 1.7.1) repeatedly, (10.23) can be finally formulated as the optimization problem (10.14). □

It is worth noting that the optimization problem (10.14) seeks to comprise the robustness against the fault modelling errors $\bar{d}_i$ (i.e. $\dot{f}_i$) and against the uncertain system interactions $h_i(x, t)$, by tuning the weights β_1 and β_2. This is a trade-off between robust FE and FTC performances. Another purpose of using these weights is to scale the cost function to avoid numerical problems that lead to infeasibility.

10.4 Decentralized FE and FTC with Sensor Fault

This section presents the robust integration of decentralized FE and FTC for the large-scale systems in the presence of sensor faults. In this case, the ith ($i = 1, 2, \ldots, n$) subsystem is represented by

$$\begin{aligned} \dot{x}_i &= A_i x_i + B_i u_i + h_i(x, t) \\ y_i &= C_i x_i + F_{si} f_{si} \end{aligned} \tag{10.24}$$

where $f_{si} \in \mathbb{R}^{q_i}$ denotes the sensor fault with the known constant distribution matrix $F_{si} \in \mathbb{R}^{p_i \times q_i}$. The other variables and matrices are the same as in (10.1). The system (10.24) satisfies Assumption 10.2 and the following assumptions:

Assumption 10.5 The pair (A_i, B_i) is controllable, and the triple (A_i, C_i, F_{si}) has no invariant zero in the closed right-half complex plane.

Assumption 10.6 The sensor fault f_{si} and its first-order derivative are norm-bounded.

This section aims to design an integrated decentralized FE and FTC for the large-scale system (10.24) following the structure in Fig. 10.1.

10.4.1 Decentralized FE Observer Design

By defining f_{si} as auxiliary state, the ith subsystem (10.24) can be augmented into

$$\begin{aligned} \dot{\bar{x}}_i &= \bar{A}_i \bar{x}_i + \bar{h}_i(x, t) + \bar{B}_i u_i + \bar{D}_i \bar{d}_i \\ y_i &= \bar{C}_i \bar{x}_i \end{aligned} \tag{10.25}$$

where

$$\bar{x}_i = \begin{bmatrix} x_i \\ f_{si} \end{bmatrix},\ \bar{A}_i = \begin{bmatrix} A_i & 0 \\ 0 & 0 \end{bmatrix},\ \bar{h}_i(x,t) = \begin{bmatrix} h_i(x,t) \\ 0 \end{bmatrix},\ \bar{B}_i = \begin{bmatrix} B_i \\ 0 \end{bmatrix},\ \bar{D}_i = \begin{bmatrix} 0 \\ I_{q_i} \end{bmatrix},$$
$$\bar{d}_i = \dot{f}_{si},\ \bar{C}_i = [C_i\ F_{si}].$$

It can be seen that (10.25) is in a similar form of (10.3), thus the following ASUIO can be designed to estimate the augmented state $\bar{x}_i$:

$$\begin{aligned} \dot{z}_i &= M_i z_i + G_i u_i + L_i y_i \\ \hat{\bar{x}}_i &= z_i + H_i y_i \end{aligned} \tag{10.26}$$

where $z_i \in \mathbb{R}^{n_i+q_i}$ is the observer state and $\hat{\bar{x}}_i \in \mathbb{R}^{n_i+q_i}$ is the estimate of $\bar{x}_i$. The matrices M_i, G_i, L_i and H_i are to be designed.

Following a similar design procedure as in Sect. 10.3.1 and using the matrix equations (10.6)–(10.8), then the estimation error system is obtained as

$$\dot{e}_i = (\Xi_i \bar{A}_i - L_{i1}\bar{C}_i)e_i + \Xi_i \bar{h}_i(x,t). \tag{10.27}$$

A difference between the error systems (10.27) and (10.5) is that the disturbance term $\Xi_i \bar{D}_i \bar{d}_i$ ($\Xi_i = I_{n_i+q_i} - H_i\bar{C}_i$) is removed. This is possible because $\text{rank}(\bar{C}_i\bar{D}_i) = \text{rank}(\bar{D}_i) = q_i$ and thus the matching condition (see Sect. 1.7.3) is satisfied. In this case, there always exists a matrix $H_i = \bar{D}_i(\bar{C}_i\bar{D}_i)^\top(\bar{C}_i\bar{D}_i(\bar{C}_i\bar{D}_i)^\top)^\dagger$ such that $\Xi_i\bar{D}_i = 0$. Hence, in the error system (10.27), only the matrix L_{i1} needs to be determined, which reduces the FE design complexity.

10.4.2 Decentralized FTC Controller Design

An FTC controller for the ith subsystem (10.24) is designed as

$$u_i = -K_i\hat{\bar{x}}_i \tag{10.28}$$

where $K_i = [K_{xi}\ 0]$ and $K_{xi} \in \mathbb{R}^{m_i \times n_i}$ is the baseline control gain.

Substituting (10.28) into (10.24) yields the FTC closed-loop system

$$\begin{aligned} \dot{x}_i &= (A_i - B_i K_{xi})x_i + B_i K_i e_i + h_i(x,t) \\ y_{ci} &= y_i - F_{si}\hat{f}_{si} \end{aligned} \tag{10.29}$$

where y_{ci} is the system output with sensor fault compensation and $\hat{f}_{si} = [0\ I_{q_i}]\hat{\bar{x}}_i$ is the sensor fault estimate.

10.4.3 Robust Integration of FE and FTC

Combining (10.27) and (10.29) gives the composite closed-loop system for the ith subsystem

$$\begin{aligned} \dot{x}_i &= (A_i - B_i K_{xi})x_i + B_i K_i e_i + h_i(x,t) \\ \dot{e}_i &= (\Xi_i \bar{A}_i - L_{i1}\bar{C}_i)e_i + \Xi_i \bar{h}_i(x,t) \\ y_{ci} &= y_i - F_{si}\hat{f}_{si}. \end{aligned} \tag{10.30}$$

Hence, the composite closed-loop system for the large-scale system is obtained as

$$\begin{aligned} \dot{x} &= \tilde{A}x + \tilde{F}e + h(x,t) \\ \dot{e} &= \tilde{A}_e e + \Xi\bar{h}(x,t) \\ y_c &= y - F_s\hat{f}_s \end{aligned} \tag{10.31}$$

where

$$e = \begin{bmatrix} e_1 \\ \vdots \\ e_n \end{bmatrix},\ y_c = \begin{bmatrix} y_{c1} \\ \vdots \\ y_{cn} \end{bmatrix},\ y = \begin{bmatrix} y_1 \\ \vdots \\ y_n \end{bmatrix},\ \hat{f}_s = \begin{bmatrix} \hat{f}_{s1} \\ \vdots \\ \hat{f}_{sn} \end{bmatrix},\ \bar{h}(x,t) = \begin{bmatrix} \bar{h}_1(x,t) \\ \vdots \\ \bar{h}_n(x,t) \end{bmatrix},$$

$$\begin{aligned} &\tilde{A} = \mathrm{diag}(A_1 + B_1K_{x1}, \ldots, A_n + B_nK_{xn}),\ \tilde{F} = \mathrm{diag}(B_1K_{x1}, \ldots, B_nK_{xn}), \\ &\Xi = \mathrm{diag}(\Xi_1, \ldots, \Xi_n),\ F_s = \mathrm{diag}(F_{s1}, \ldots, F_{sn}), \\ &\tilde{A}_e = \mathrm{diag}\left(\Xi_1\bar{A}_1 - L_{11}\bar{C}_1, \ldots, \Xi_n\bar{A}_n - L_{n1}\bar{C}_n\right). \end{aligned}$$

Now the integrated design problem is reformulated as the problem of designing the controller gains K_{xi}, $i = 1, \ldots, n$, and observer gains L_{1i}, $i = 1, \ldots, n$, to ensure the robust stability of the overall closed-loop system (10.31). This design problem is solved using Theorem 10.2 with a single-step LMI formulation.

Theorem 10.2 *Under Assumptions 10.2, 10.5 and 10.6, given positive scalars β, ε_{1i}, ε_{2i} and ε_{3i}, $i = 1, 2, \ldots, n$, the overall closed-loop system (10.31) is stable if the following optimization problem is feasible:*

$$\min\ \beta \sum_{i=1}^{n} \bar{\alpha}_i \tag{10.32}$$

$$\text{s.t.} \begin{bmatrix} \Pi_{11} & 0 & \Pi_{13} & 0 & \Pi_{15} & 0 \\ \star & \Pi_{22} & 0 & \Pi_{24} & 0 & \Pi_{26} \\ \star & \star & -\varepsilon_3^{-1}Z & 0 & 0 & 0 \\ \star & \star & \star & -\varepsilon_3 Z & 0 & 0 \\ \star & \star & \star & \star & -\epsilon I & 0 \\ \star & \star & \star & \star & \star & -\varepsilon_1 I \end{bmatrix} \prec 0 \tag{10.33}$$

$$Z_i = Z_i^\top \succ 0,\ Q_i = Q_i^\top \succ 0,\ \bar{\alpha}_i > 0,\ i = 1, \ldots, n \tag{10.34}$$

where

$$\begin{aligned}
&Z = \text{diag}(Z_1, \ldots, Z_n),\ \varepsilon_1 = \text{diag}(\varepsilon_{1i}, \ldots, \varepsilon_{1n}),\ \epsilon = \text{diag}(\epsilon_1, \ldots, \epsilon_n),\\
&\epsilon_i = \bar{\alpha}_i(\varepsilon_{1i} + \varepsilon_{2i})^{-1},\ \varepsilon_3 = \text{diag}(\varepsilon_{3i}, \ldots, \varepsilon_{3n}),\ \Pi_{11} = \text{diag}\left(\Pi_{111}, \ldots, \Pi_{11n}\right),\\
&\Pi_{11i} = \text{He}(A_i Z_i - B_i M_{1i}) + \varepsilon_{2i}^{-1} I,\ \Pi_{13} = \text{diag}(B_1 M_{1i}, \ldots, B_n M_{1n}),\\
&\Pi_{15} = Z[H_{01}^\top\ \ldots\ H_{0n}^\top],\ \Pi_{22} = \text{diag}(\Pi_{221}, \ldots, \Pi_{22n}),\\
&\Pi_{22i} = \text{He}(Q_i \Xi_i \bar{A}_i - M_{2i}\bar{C}_i),\ \Pi_{24} = \text{diag}(\Lambda, \ldots, \Lambda),\ \Lambda = [I_n\ 0]^\top,\\
&\Pi_{26} = \text{diag}(\Pi_{261}, \ldots, \Pi_{26n}),\ \Pi_{26i} = \text{He}(Q_i \Xi_i).
\end{aligned}$$

Then the gains are obtained as $K_{xi} = M_{1i} Z_i^{-1}$ *and* $L_{i1} = Q_{i1}^{-1} M_{2i}$.

Proof Consider the Lyapunov function $V_e = e^\top Q e$ with a s.p.d. matrix Q. According to Assumption 10.2, it holds that for some positive scalar $\varepsilon_1 = \text{diag}(\varepsilon_{11}, \ldots, \varepsilon_{1n})$,

$$\begin{aligned}
\text{He}(e^\top Q \Xi \bar{h}(x,t)) &= -\left[\sqrt{\varepsilon_1}^{-1} \Xi^\top Q e - \sqrt{\varepsilon_1}\bar{h}(x,t)\right]^\top \times \left[\sqrt{\varepsilon_1}^{-1} \Xi^\top Q e - \sqrt{\varepsilon_1}\bar{h}(x,t)\right]\\
&\quad + \varepsilon_1^{-1} e^\top Q \Xi \Xi^\top Q e + \varepsilon_1 \bar{h}^\top(x,t)\bar{h}(x,t)\\
&\le \varepsilon_1^{-1} e^\top Q \Xi \Xi^\top Q e + \varepsilon_1 x^\top H_0^\top H_0 x.
\end{aligned}$$

The time derivative of V_e along the error subsystem in (10.31) is derived as

$$\begin{aligned}
\dot{V}_e &= e^\top \text{He}(Q\tilde{A}_e)e + \text{He}(e^\top Q \Xi \bar{h}(x,t))\\
&\le e^\top[\text{He}(Q\tilde{A}_e) + \varepsilon_1^{-1} Q \Xi \Xi^\top Q]e + \varepsilon_1 x^\top H_0^\top H_0 x.
\end{aligned} \tag{10.35}$$

Consider another Lyapunov function $V_x = x^\top P x$. It holds that, for some positive scalar $\varepsilon_2 = \text{diag}(\varepsilon_{21}, \ldots, \varepsilon_{2n})$,

$$\begin{aligned}
\text{He}(x^\top P h(x,t)) &= -\left[\sqrt{\varepsilon_2}^{-1} P x - \sqrt{\varepsilon_2} h(x,t)\right]^\top \times \left[\sqrt{\varepsilon_2}^{-1} P x - \sqrt{\varepsilon_2} h(x,t)\right]\\
&\quad + \varepsilon_2^{-1} x^\top P P x + \varepsilon_2 h^\top(x,t) h(x,t)\\
&\le \varepsilon_2^{-1} x^\top P P x + \varepsilon_2 x^\top H_0^\top H_0 x.
\end{aligned}$$

Then the time derivative of V_x along the control subsystem in (10.31) is

$$\begin{aligned}
\dot{V}_x &= x^\top \text{He}(P\tilde{A})x + \text{He}(x^\top P\tilde{F}e) + \text{He}(x^\top P h(x,t))\\
&\le x^\top\left[\text{He}(P\tilde{A}) + \varepsilon_2^{-1} P P\right]x + \varepsilon_2 x^\top H_0^\top H_0 x + \text{He}(x^\top P\tilde{F}e).
\end{aligned} \tag{10.36}$$

The composite system (10.31) is stable if

$$\dot{V}_x + \dot{V}_e < 0. \tag{10.37}$$

Substituting (10.35) and (10.36) into (10.22) yields

$$\begin{bmatrix} x \\ e \end{bmatrix}^{\top} \begin{bmatrix} J_{11} & J_{12} \\ \star & J_{22} \end{bmatrix} \begin{bmatrix} x \\ e \end{bmatrix} < 0 \tag{10.38}$$

and equivalently,

$$\begin{bmatrix} J_{11} & J_{12} \\ \star & J_{22} \end{bmatrix} \prec 0 \tag{10.39}$$

where

$$J_{11} = \text{He}(P\tilde{A}) + (\varepsilon_1 + \varepsilon_2)H_0^{\top} H_0 + \varepsilon_2^{-1} PP, \; J_{12} = P\tilde{F},$$
$$J_{22} = \text{He}(Q\tilde{A}_e) + \varepsilon_1^{-1} Q\Xi\Xi^{\top} Q.$$

Define $Z = P^{-1}$. Pre- and post-multiplying both sides of (10.39) with diag(Z, I) and its transpose, respectively, then (10.39) is equivalently converted into

$$\begin{bmatrix} \hat{J}_{11} & \hat{J}_{12} \\ \star & \hat{J}_{22} \end{bmatrix} \prec 0 \tag{10.40}$$

where

$$\hat{J}_{11} = \text{He}(\tilde{A}Z) + (\varepsilon_1 + \varepsilon_2)ZH_0^{\top} H_0 Z + \varepsilon_2^{-1} I, \; \hat{J}_{12} = \tilde{F},$$
$$\hat{J}_{22} = \text{He}(Q\tilde{A}_e) + \varepsilon_1^{-1} Q\Xi\Xi^{\top} Q.$$

Define Z=diag$(Z_1, \ldots, Z_n)$. Notice that $\hat{J}_{12}$=diag $([B_1 K_{x1}\ 0], \ldots, [B_n K_{xn}\ 0])$. By using the Young inequality, then there exist some positive scalars $\varepsilon_3 =$ diag$(\varepsilon_{31}, \ldots, \varepsilon_{3n})$ such that

$$\begin{bmatrix} 0 & 0 & 0 \\ 0 & B_i K_{xi} & 0 \\ 0 & 0 & 0 \end{bmatrix} \leq \varepsilon_{3i}^{-1} \begin{bmatrix} 0 \\ \Lambda \\ 0 \end{bmatrix} Z_i^{-1} \begin{bmatrix} 0 \\ \Lambda \\ 0 \end{bmatrix}^{\top} + \varepsilon_{3i} \begin{bmatrix} 0 \\ B_i K_{xi} Z_i \\ 0 \end{bmatrix} Z_i^{-1} \begin{bmatrix} 0 \\ B_i K_{xi} Z_i \\ 0 \end{bmatrix}^{\top},$$

where $\Lambda = [I_n\ 0]^{\top}$

Define $Q = \text{diag}(Q_1, \ldots, Q_n)$ and partition $\Pi_{22} = \text{He}(Q\tilde{A}_e)$ into $\Pi_{22} = \text{diag}(\Pi_{221}, \ldots, \Pi_{22n})$ with $\Pi_{22i} = \text{He}(Q_i(\Xi_i \bar{A}_i - L_{i1}\bar{C}_i))$. Further define $\bar{\alpha}_i = \alpha_i^{-1}$, $\epsilon_i = \bar{\alpha}_i(\varepsilon_{1i} + \varepsilon_{2i})^{-1}$, $M_{1i} = K_{xi} Z_i$ and $M_{2i} = Q_i L_{i1}$. By using the Schur Complement repeatedly, (10.40) can be formulated as the optimization problem (10.32).

□

10.5 Application to 3-Machine Power System

This section presents application of the proposed decentralized FE-based FTC to a 3-machine power system with the structure sketched in Fig. 10.2.

The system dynamics of the ith machine ($i = 1, 2, 3$) are represented by

$$\begin{aligned} \dot{x}_i &= A_i x_i + B_i u_i + h_i(x,t) \\ y_i &= C_i x_i \end{aligned} \tag{10.41}$$

where $x_i = [\Delta\sigma_i(t) \;\; \omega_i(t) \;\; \Delta P_{\mathrm{m}i}(t) \;\; \Delta X_{\mathrm{e}i}(t)]^\top$, $h_i(x,t) = \sum_{j=1,j\neq i}^{n} p_{ij} G_{ij} g_{ij}(x_i, x_j)$, $g_{ij}(x_i, x_j) = \sin(\sigma_i - \sigma_j) - \sin(\sigma_{i0} - \sigma_{j0})$, and

$$A_i = \begin{bmatrix} 0 & 1 & 0 & 0 \\ 0 & \frac{-D_i}{2H_i} & \frac{\omega_0}{2H_i} & 0 \\ 0 & 0 & \frac{-1}{T_{\mathrm{m}i}} & \frac{K_{\mathrm{m}i}}{T_{\mathrm{m}i}} \\ 0 & \frac{K_{\mathrm{e}i}}{T_{\mathrm{e}i} R_i \omega_0} & 0 & \frac{-1}{T_{\mathrm{e}i}} \end{bmatrix}, \; B_i = \begin{bmatrix} 0 \\ 0 \\ 0 \\ \frac{1}{T_{\mathrm{e}i}} \end{bmatrix}, \; C_i = I_4, \; G_{ij} = \begin{bmatrix} 0 \\ -\frac{\omega_0 E'_{qi} E'_{qj} B_{ij}}{2H_i} \\ 0 \\ 0 \end{bmatrix}.$$

The system parameters are defined in Table 10.1, with the values listed in Table 10.2. Other detailed specifications of the system can be found in Kalsi et al. (2009), Guo et al. (2000).

Following the method in Kalsi et al. (2009), it can be verified that the interactions satisfy Assumption 10.2 with the matrices H_{0i} given as

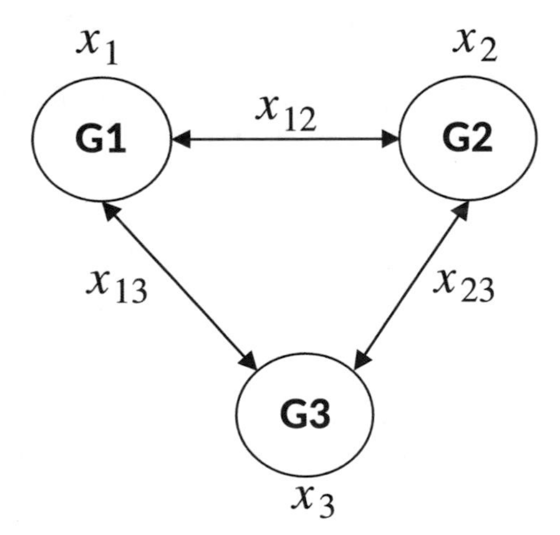

Fig. 10.2 Structure of the 3-machine power system

Table 10.1 Definition of the physical parameters

Parameter	Physical meaning
ΔX_{ei}	Control vector u_i
σ_i	Rotor angle
ω_i	Relative speed
P_{mi}	Per unit (*pu*) mechanical power
X_{ei}	*pu* steam valve aperture
p_{ij}	Index of connection of the ith and jth machines (0: disconnected; 1: connected)
H_i	Inertia constant
D_i	*pu* damping coefficient
T_{mi}	Time constant for the turbine
K_{mi}	The gain of the turbine
T_{ei}	Time constant for the speed governor
K_{ei}	The gain of the speed governor
R_i	*pu* regulation constant
B_{ij}	*pu* Nodal susceptance between the ith and jth machines
ω_0	Synchronous machine speed
σ_{i0}, P_{mi0}, X_{ei0}	Nominal values of σ_i, P_{mi}, X_{ei}
$\Delta\sigma_i$	Deviation of the rotor angle ($\sigma_i - \sigma_{i0}$)
ΔP_{mi}	Deviation of the mechanical power ($P_{mi} - P_{mi0}$)
ΔX_{ei}	Deviation of the steam valve aperture ($X_{ei} - X_{ei0}$)

Table 10.2 Machine parameters

Parameter	Machine 1	Machine 2	Machine 3
H_i	4	5.1	5.1
D_i	5	3	3
T_{mi}	0.35	0.35	0.35
K_{mi}	1	1	1
T_{ei}	0.1	0.1	0.1
K_{ei}	1	1	1
R_i	0.05	0.05	0.05
ω_0	314.159	314.159	314.159
σ_{i0}	67.6 deg	67.6 deg	67.6 deg

$$H_{01} = \begin{bmatrix} 1\,0\,0\,0\,-1\,0\,0\,0\;\;0\;\;0\,0\,0 \\ 1\,0\,0\,0\,-1\,0\,0\,0\,-1\,0\,0\,0 \end{bmatrix}^{\top},$$

$$H_{02} = \begin{bmatrix} -1\,0\,0\,0\,1\,0\,0\,0\;\;0\;\;0\,0\,0 \\ 0\;\;0\,0\,0\,1\,0\,0\,0\,-1\,0\,0\,0 \end{bmatrix}^{\top},$$

$$H_{03} = \begin{bmatrix} -1\,0\,0\,0\;\;0\;\;0\,0\,0\,1\,0\,0\,0 \\ 0\;\;0\,0\,0\,-1\,0\,0\,0\,1\,0\,0\,0 \end{bmatrix}^{\top}.$$

The following two simulation cases are considered: (1) The 3-machine power system has actuator faults acting on each machine. (2) The 3-machine power system suffers from sensor faults on the measured outputs of each machine.

10.5.1 Actuator Fault Case

Consider a serious situation when all the three machines suffer from actuator faults that are characterized by

$$F_1 = B_1,\ f_1(t) = \begin{cases} 0, & 0\text{ s} \le t \le 0.1\text{ s} \\ \sin(3t - 0.3), & t > 0.1\text{ s} \end{cases},$$

$$F_2 = B_2,\ f_2(t) = \begin{cases} 0, & 0\text{ s} \le t \le 0.5\text{ s} \\ 1, & 0.5\text{ s} < t \le 1\text{ s} \\ 2, & 1\text{ s} < t \le 1.5\text{ s} \\ 0, & 1.5\text{ s} < t \le 3\text{ s} \\ -1, & t > 3\text{ s} \end{cases},$$

$$F_3 = B_3,\ f_3(t) = \begin{cases} 0, & 0\text{ s} \le t \le 0.5\text{ s} \\ -0.5, & 0.5\text{ s} < t \le 1.5\text{ s} \\ 0.5, & 1.5\text{ s} < t \le 2\text{ s} \\ \cos(5t), & t > 2\text{ s} \end{cases}.$$

It is verified that Assumptions 10.2, 10.3 and 10.4 are satisfied for this 3-machine system in the presence of the actuator faults studied here. Given $\beta_1 = 10^{-9}$, $\beta_2 = 10^{-7}$, $\varepsilon_{11} = \varepsilon_{12} = \varepsilon_{13} = 0.1$, $\varepsilon_{21} = \varepsilon_{22} = \varepsilon_{23} = 100$ and $\varepsilon_{31} = \varepsilon_{32} = \varepsilon_{33} = 0.1$, solving the optimization problem in Theorem 10.1 gives

$$\bar{\alpha}_1 = 509.0925,\ \bar{\alpha}_2 = 354.65,\ \bar{\alpha}_3 = 354.65,$$
$$\bar{\gamma}_1 = 22.7195,\ \bar{\gamma}_2 = 22.6872,\ \bar{\gamma}_3 = 22.6872,$$

$$K_{x1} = \begin{bmatrix} 1.0009 \\ 0.7527 \\ 4.913 \\ 0.9987 \end{bmatrix}^{\top},\ M_1 = \begin{bmatrix} -1.0157 & -0.1784 & -0.8021 & -0.12 & 0 \\ 0 & -0.5759 & -0.4606 & -0.0775 & 0 \\ 0 & -0.053 & -2.4964 & -0.5054 & 0 \\ 0 & 0.0007 & 0.0079 & -0.6231 & 0 \\ 0 & 0 & 0 & 0 & -451.1381 \end{bmatrix},$$

$$L_1 = \begin{bmatrix} 0 & 0 & 0 & 0 \\ 0 & 0 & 0 & 0 \\ 0 & 0 & 0 & 0 \\ 0 & 0 & 0 & 0 \\ 0 & 29 & 0 & -19901 \end{bmatrix}, \; H_1 = \begin{bmatrix} 1 & 0 & 0 & 0 \\ 0 & 1 & 0 & 0 \\ 0 & 0 & 1 & 0 \\ 0 & 0 & 0 & 1 \\ 0 & 0 & 0 & 45.1138 \end{bmatrix}, \; G1 = \begin{bmatrix} 0 \\ 0 \\ 0 \\ 0 \\ -451.1381 \end{bmatrix},$$

$$K_{x2} = \begin{bmatrix} 1.4813 \\ 0.984 \\ 4.8667 \\ 0.9878 \end{bmatrix}^{\top}, \; M_2 = \begin{bmatrix} -1.297 & -0.3055 & -1.0288 & -0.1532 & 0 \\ 0 & -0.6014 & -0.5103 & -0.0826 & 0 \\ 0 & -0.0689 & -2.1721 & -0.4337 & 0 \\ 0 & -0.0092 & -0.0006 & -0.6089 & 0 \\ 0 & 0 & 0 & 0 & -103.379 \end{bmatrix},$$

$$L_2 = \begin{bmatrix} 0 & 0 & 0 & 0 \\ 0 & 0 & 0 & 0 \\ 0 & 0 & -0.0001 & 0 \\ 0 & 0 & 0 & 0 \\ 0 & 6.5813 & 0 & -965.3422 \end{bmatrix}, \; H_2 = \begin{bmatrix} 1 & 0 & 0 & 0 \\ 0 & 1 & 0 & 0 \\ 0 & 0 & 1 & 0 \\ 0 & 0 & 0 & 1 \\ 0 & 0 & 0 & 10.3379 \end{bmatrix},$$

$$K_{x3} = \begin{bmatrix} 1.4813 \\ 0.984 \\ 4.8667 \\ 0.9878 \end{bmatrix}^{\top}, \; G2 = \begin{bmatrix} 0 \\ 0 \\ 0 \\ 0 \\ -103.379 \end{bmatrix},$$

$$M_3 = \begin{bmatrix} 1.297 & -0.3055 & -1.0287 & -0.1531 & 0 \\ 0 & -0.6014 & -0.8234 & -0.0915 & 0 \\ -0.0001 & 0.2442 & -2.1721 & -0.4310 & 0 \\ 0 & -0.0002 & -0.0033 & -0.6089 & 0 \\ 0 & 0 & 0 & 0 & -103.379 \end{bmatrix},$$

$$L_3 = \begin{bmatrix} 0 & 0 & 0 & 0 \\ 0 & 0 & 0 & 0 \\ 0 & 0 & 0 & 0 \\ 0 & 0 & 0 & 0 \\ 0 & 6.5813 & 0 & -965.3422 \end{bmatrix}, \; H_3 = \begin{bmatrix} 1 & 0 & 0 & 0 \\ 0 & 1 & 0 & 0 \\ 0 & 0 & 1 & 0 \\ 0 & 0 & 0 & 1 \\ 0 & 0 & 0 & 10.3379 \end{bmatrix}, \; G3 = \begin{bmatrix} 0 \\ 0 \\ 0 \\ 0 \\ -103.379 \end{bmatrix}.$$

The simulations are performed under initial conditions: $x_1(0) = [0\ 1\ 0.5\ -1]^{\top}$, $x_2(0) = [0.1\ 0.5\ -0.1\ 0.5]^{\top}$, $x_3(0) = [1\ 0.2\ 0.5\ 0]^{\top}$, $z_1(0) = z_2(0) = z_3(0) = 0_{5\times 1}$.

The results in Figs. 10.3, 10.4, 10.5 and 10.6 show that the proposed FE-based FTC design achieves accurate estimation of the actuator faults, compensation of the fault in each subsystem and robust stability of the 3-machine system, in the presence of faults and uncertain nonlinear interactions.

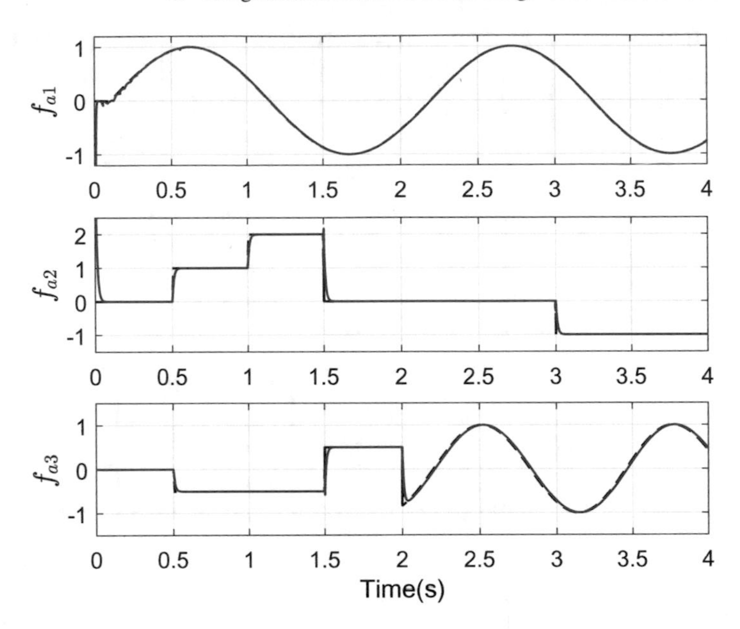

Fig. 10.3 Actuator faults (black dash) and their estimates (red solid)

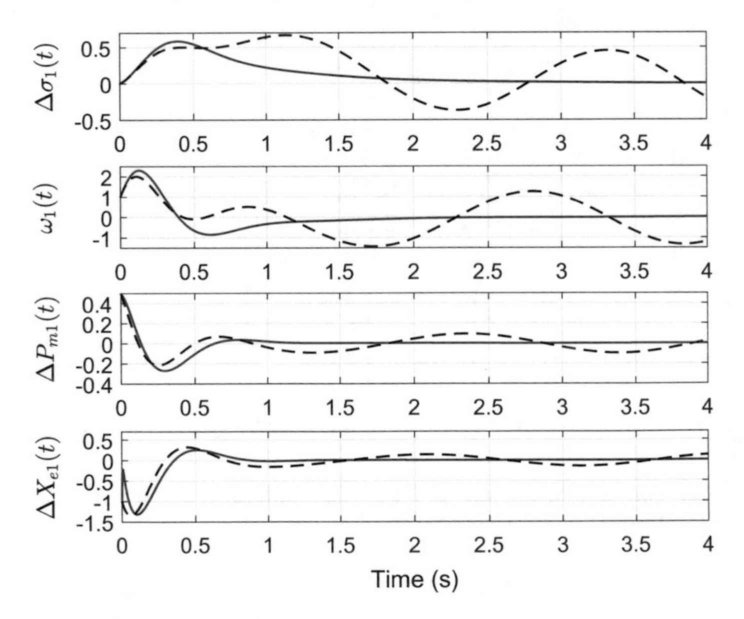

Fig. 10.4 State of machine 1 with (red solid) or without FTCs (black dash)

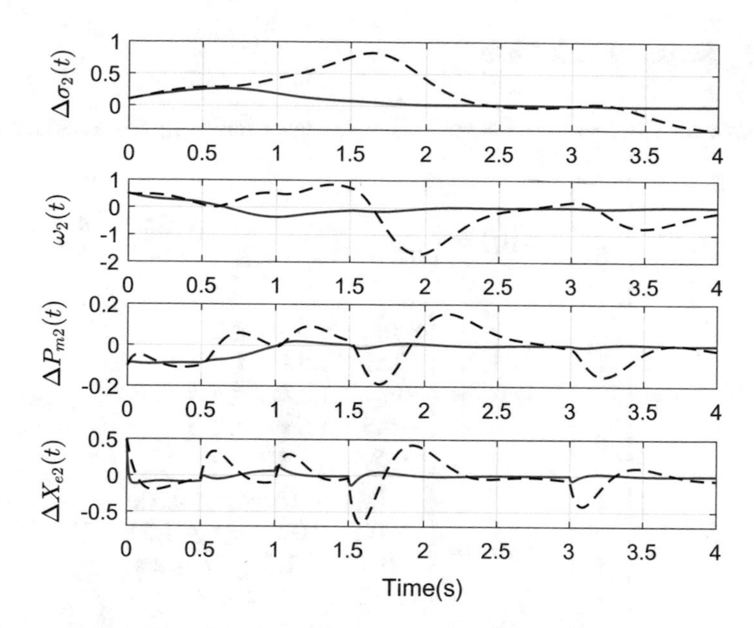

Fig. 10.5 State of machine 2 with (red solid) or without FTCs (black dash)

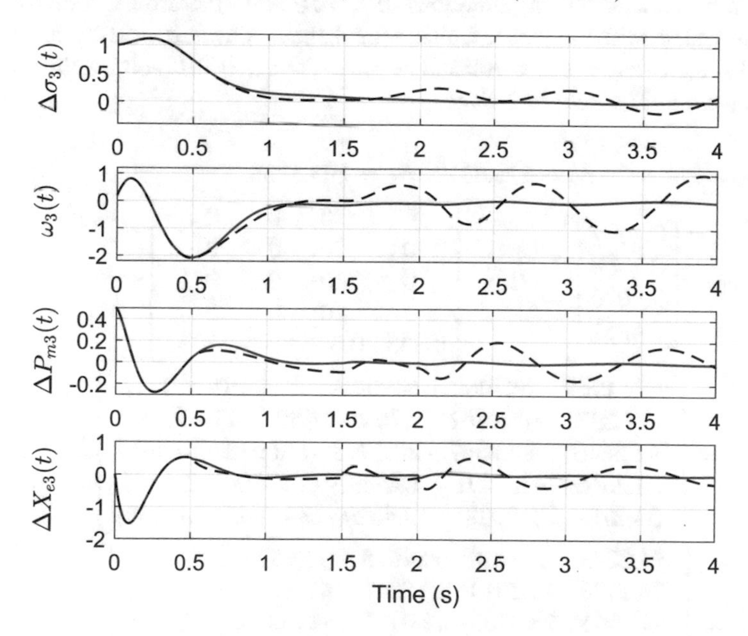

Fig. 10.6 State of machine 3 with (red solid) or without FTCs (black dash)

10.5.2 Sensor Fault Case

In this case the ith, $i = 1, 2, 3$, machines have sensor faults f_{si} characterized by

$$F_{s1} = \begin{bmatrix} 1 \\ 1 \\ 0 \\ -1 \end{bmatrix}, \; f_{s1}(t) = \begin{cases} 0, & 0\,\text{s} \le t \le 1\,\text{s} \\ 0.5\sin(5(t-1)), & t > 1\,\text{s} \end{cases},$$

$$F_{s2} = \begin{bmatrix} -1 \\ 1 \\ -1 \\ 0 \end{bmatrix}, \; f_{s2}(t) = \begin{cases} 0, & 0\,\text{s} \le t \le 0.5\,\text{s} \\ 0.5, & 0.5\,\text{s} < t \le 1\,\text{s} \\ 1, & 1\,\text{s} < t \le 1.5\,\text{s} \\ -0.5, & 1.5\,\text{s} < t \le 3\,\text{s} \\ 0.5, & t > 3\,\text{s} \end{cases},$$

$$F_{s3} = \begin{bmatrix} 1 \\ -1 \\ 1 \\ -1 \end{bmatrix}, \; f_{s3}(t) = \begin{cases} 0, & 0\,\text{s} \le t \le 0.5\,\text{s} \\ -0.5, & 0.5\,\text{s} < t \le 1.5\,\text{s} \\ 0.5, & 1.5\,\text{s} < t \le 2\,\text{s} \\ \cos(2t), & t > 2\,\text{s} \end{cases}.$$

It can be verified that Assumptions 10.2, 10.5 and 10.6 are satisfied for the 3-machine system with the sensor faults defined above. Given $\beta = 10^{-8}$, $\varepsilon_{11} = \varepsilon_{12} = \varepsilon_{13} = 0.1$, $\varepsilon_{21} = \varepsilon_{22} = \varepsilon_{23} = 100$ and $\varepsilon_{31} = \varepsilon_{32} = \varepsilon_{33} = 0.1$, solving the optimization problem in Theorem 10.2 gives

$$\bar{\alpha}_1 = 268.847, \; \bar{\alpha}_2 = 195.4312, \; \bar{\alpha}_3 = 193.3623,$$

$$K_{x1} = \begin{bmatrix} 2.1404 \\ 0.7968 \\ 3.6070 \\ 0.5851 \end{bmatrix}^{\top}, \; H_1 = \begin{bmatrix} 0 & 0 & 0 & 0 \\ 0 & 0 & 0 & 0 \\ 0 & 0 & 0 & 0 \\ 0 & 0 & 0 & 0 \\ 0.3333 & 0.3333 & 0 & -0.3333 \end{bmatrix},$$

$$M_1 = \begin{bmatrix} -52.1969 & 1.4706 & -60.4825 & -57.8210 & 6.0947 \\ -27.2971 & -16.9462 & 5.7658 & -40.2605 & -3.3578 \\ 41.5640 & -22.8547 & -20.8763 & 11.0048 & 10.5617 \\ 25.1365 & -6.7491 & 5.2523 & 6.5773 & 2.4467 \\ 34.2788 & 2.9028 & 19.4329 & 34.7945 & -0.6089 \end{bmatrix},$$

$$L_1 = \begin{bmatrix} 54.2285 & 1.5610 & 60.4825 & 55.7895 \\ 26.1778 & 15.2019 & 33.5041 & 41.3797 \\ -38.0434 & 26.3753 & 18.0191 & -11.6682 \\ -24.3209 & 6.9280 & -5.2523 & -17.3929 \\ -34.4818 & -3.4430 & -32.5229 & -37.9248 \end{bmatrix}, \; G1 = \begin{bmatrix} 0 \\ 0 \\ 0 \\ 10 \\ 3.3333 \end{bmatrix},$$

$$K_{x2} = \begin{bmatrix} 2.8225 \\ 1.1 \\ 4.3513 \\ 0.7146 \end{bmatrix}^{\top}, \; H_2 = \begin{bmatrix} 0 & 0 & 0 & 0 \\ 0 & 0 & 0 & 0 \\ 0 & 0 & 0 & 0 \\ 0 & 0 & 0 & 0 \\ -0.3333 & 0.3333 & -0.3333 & 0 \end{bmatrix},$$

$$M_2 = \begin{bmatrix} -55.9559 & -4.1253 & 58.6381 & -6.0251 & -7.8075 \\ -23.3241 & -18.9636 & 30.7629 & 0.6151 & 4.6916 \\ -79.8124 & -72.3103 & -13.3611 & -0.4051 & 18.0061 \\ 1.4839 & -2.3961 & -3.5802 & -19.7074 & 0.3369 \\ -36.4242 & -18.9985 & 4.5116 & -2.3037 & 1.2637 \end{bmatrix},$$

$$L_2 = \begin{bmatrix} 58.5584 & 2.5228 & -56.0356 & 6.0251 \\ 21.7603 & 20.2334 & -1.5269 & -0.6151 \\ 73.8104 & 78.3123 & 4.5019 & 3.2623 \\ -1.5962 & 1.8717 & 3.4679 & 9.7074 \\ 36.0030 & 19.8511 & -16.1519 & 3.2560 \end{bmatrix}, \; G2 = \begin{bmatrix} 0 \\ 0 \\ 0 \\ 10 \\ 0 \end{bmatrix},$$

$$K_{x3} = \begin{bmatrix} 2.7859 \\ 1.0715 \\ 4.3256 \\ 0.72 \end{bmatrix}^{\top}, \; H_3 = \begin{bmatrix} 0 & 0 & 0 & 0 \\ 0 & 0 & 0 & 0 \\ 0 & 0 & 0 & 0 \\ 0 & 0 & 0 & 0 \\ 0.25 & -0.25 & 0.25 & -0.25 \end{bmatrix},$$

$$M_3 = \begin{bmatrix} -43.6434 & 7.2952 & 58.2204 & -3.2805 & 11.5623 \\ -25.4191 & -20.3271 & 33.8291 & 1.3949 & -3.7518 \\ -70.7169 & -58.3854 & -17.9290 & -13.2977 & -11.2486 \\ 4.745 & 6.262 & -3.3595 & -16.9073 & 1.3942 \\ 22.4544 & 9.1346 & -2.2369 & 0.182 & -1.2103 \end{bmatrix},$$

$$L_3 = \begin{bmatrix} 46.534 & -9.1858 & -55.3299 & 0.39 \\ 24.4812 & 20.9709 & -3.9672 & -0.4569 \\ 67.9047 & 61.1975 & 12.2597 & 18.9669 \\ -4.3965 & -7.2471 & 3.7081 & 6.5587 \\ -22.757 & -9.3147 & 10.3485 & -3.0937 \end{bmatrix}, \; G3 = \begin{bmatrix} 0 \\ 0 \\ 0 \\ 10 \\ 2.5 \end{bmatrix}.$$

Simulations are performed under initial conditions: $x_1(0) = [0\ 1\ 0.5\ -1]^{\top}$, $x_2(0) = [0.1\ 0.5\ -0.1\ 0.5]^{\top}$, $x_3(0) = [1\ 0.2\ 0.5\ 0]^{\top}$, $z_1(0) = z_2(0) = z_3(0) = 0_{5\times 1}$.

The results in Figs. 10.7, 10.8, 10.9 and 10.10 show that the proposed FTC design achieves good fault estimation and ensures robust stability of the 3-machine system, with the sensor faults effect on the system outputs well compensated.

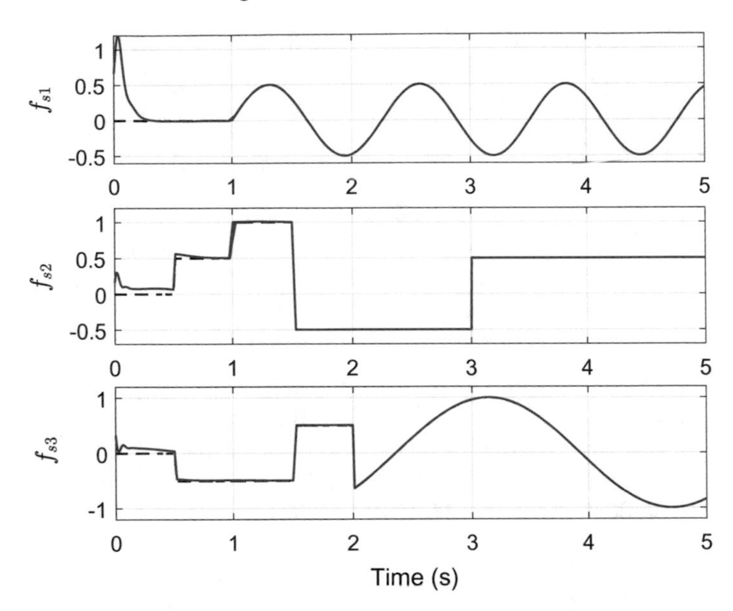

Fig. 10.7 Sensor faults estimation performance

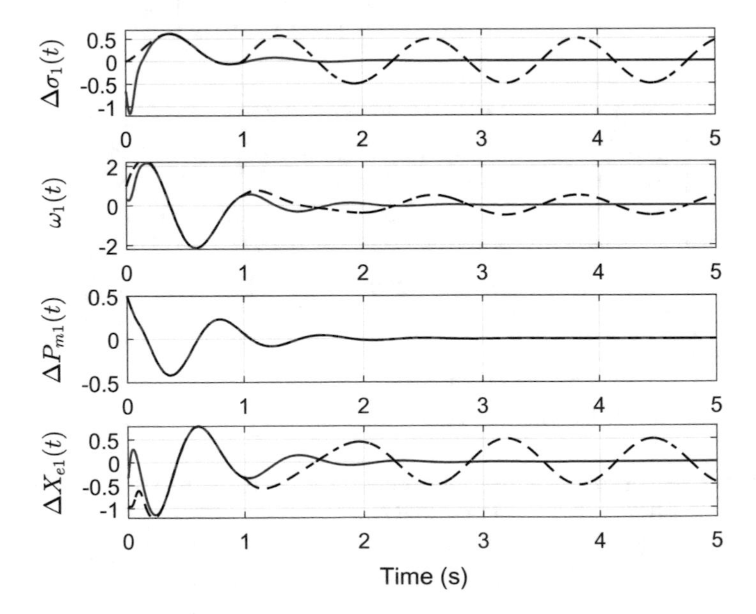

Fig. 10.8 Outputs of machine 1 with (red solid) or without FTCs (black dash)

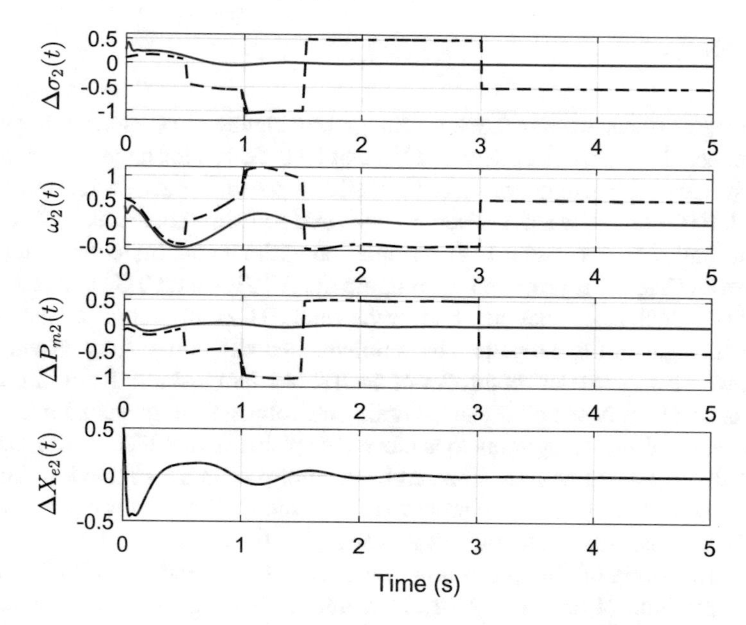

Fig. 10.9 Outputs of machine 2 with (red solid) or without FTCs (black dash)

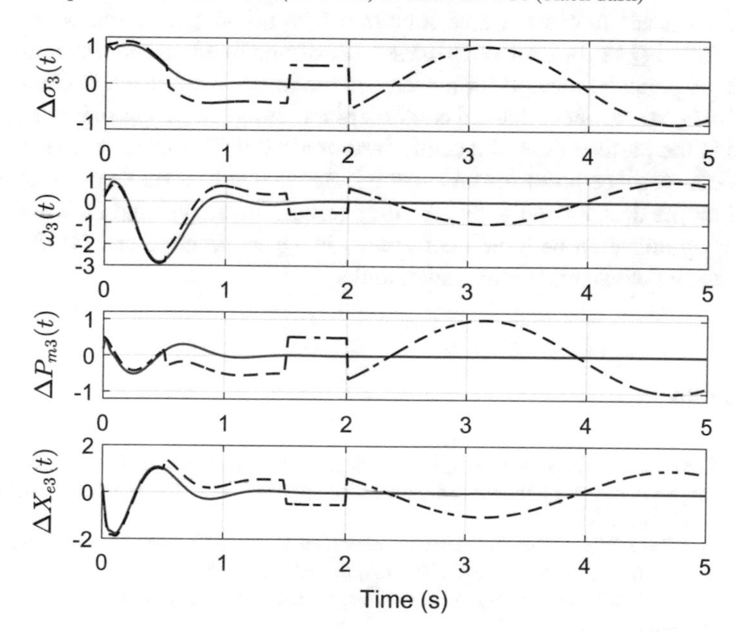

Fig. 10.10 Outputs of machine 3 with (red solid) or without FTCs (black dash)

10.6 Notes

This chapter extends the simultaneous integration strategy in Chap. 5 to large-scale interconnected systems. An integrated FE and FTC design for large-scale interconnected systems with actuator or sensor faults is proposed, using a decentralized state feedback FTC controller and a decentralized ASUIO for simultaneous estimation of state and faults. A single-step LMI formulation without equality constraint (which is needed in Chap. 5) is proposed to compute the ASUIO and FTC controller gains. It should be emphasized that the FE observers and FTC controllers of all the subsystems are designed offline together, but implemented online in a fully decentralized fashion where observer and controller of each subsystem just use the local information. The local FE-based FTC can estimate and compensate the actuator or sensor faults in each of the subsystems to achieve acceptable robust FTC performance of the overall large-scale systems. This has been confirmed in the simulation studies of a 3-machine power system. The proposed FE-based FTC design is easily modified to handle the case when actuator and sensor faults co-exist.

The formulation of the optimization problems (10.14) and (10.32) relies on the diagonal partition of the s.p.d. matrices Z and Q. This enforces the obtained FE observer and FTC controller to be fully decentralized. However, it reduces the design freedom and leads to conservative solutions. The non-diagonal forms of decision variables Z and Q can be explored to have more design freedom and less conservative solutions. A potential method for this can be the iterative approach in Geromel and Peres (1985), where decentralized controllers are designed to approximate as much as possible the performance of a centralized controller. However, there is a lack of rigorous convergence proof for this method. Alternatively, a rank constraint can be imposed on the decision variables, see for example Jovanović and Dhingra (2016). However, optimization with rank constraints is a hard problem and is still difficult to solve using the existing optimization tools.

References

Amani AM, Poorjandaghi SS, Afshar A, Menhaj MB (2014) Fault tolerant control of large-scale systems subject to actuator fault using a cooperative approach. Proc Inst Mech Eng Part I: J Syst Control Eng 228(2):63–77

Bakule L, Rodellar J (1996) Decentralised control design of uncertain nominally linear symmetric composite systems. IEE Proc-Control Theory Appl 143(6):530–536

Bakule L, Rossel JM (2008) Overlapping controllers for uncertain delay continuous-time systems. Kybernetika 44(1):17–34

Benigni A, D'Antona G, Ghisla U, Monti A, Ponci F (2010) A decentralized observer for ship power system applications: implementation and experimental validation. IEEE Trans Instrum Meas 59(2):440–449

Chen F, Zhang K, Jiang B, Wen C (2016) Adaptive sliding mode observer-based robust fault reconstruction for a helicopter with actuator fault. Asian J Control 18(4):1558–1565

Geromel J, Peres P (1985) Decentralised load-frequency control. In: IEE Proc D (Control Theory Appl) (IET) 132:225–230

Guo Y, Hill DJ, Wang Y (2000) Nonlinear decentralized control of large-scale power systems. Automatica 36(9):1275–1289

Hashemi M, Askari J, Ghaisari J (2016) Adaptive decentralised dynamic surface control for nonlinear large-scale systems against actuator failures. IET Control Theory Appl 10(1):44–57

Huang Z, Patton RJ (2015) Output feedback sliding mode FTC for a class of nonlinear interconnected systems. IFAC-PapersOnLine 48(21):1140–1145

Ikeda M (1989) Decentralized control of large scale systems. In: Three decades of mathematical system theory, vol 135, pp 219–242

Jin XZ, Yang GH (2009) Distributed adaptive robust tracking and model matching control with actuator faults and interconnection failures. Int J Control Autom Syst 7(5):702–710

Jovanović MR, Dhingra NK (2016) Controller architectures: tradeoffs between performance and structure. Eur J Control 30:76–91

Kalsi K, Lian J, Zak SH (2009) Decentralized control of multimachine power systems. In: Proceedings of the American control conference, IEEE, Missouri, United States, pp 2122–2127

Kalsi K, Lian J, Zak SH (2010) Decentralized dynamic output feedback control of nonlinear interconnected systems. IEEE Trans Autom Control 55(8):1964–1970

Khalili M, Zhang X, Polycarpou M, Parisini T, Cao Y (2015) Distributed adaptive fault-tolerant control of uncertain multi-agent systems. IFAC-PapersOnLine 48(21):66–71

Naghavi SV, Safavi A, Kazerooni M (2014) Decentralized fault tolerant model predictive control of discrete-time interconnected nonlinear systems. J Frankl Inst 351(3):1644–1656

Pagilla PR, Zhu Y (2004) A decentralized output feedback controller for a class of large-scale interconnected nonlinear systems. In: Proceedings of the American control conference, IEEE, Boston, United States, vol 4, pp 3711–3716

Panagi P, Polycarpou MM (2011) Distributed fault accommodation for a class of interconnected nonlinear systems with partial communication. IEEE Trans Autom Control 56(12):2962–2967

Patton RJ, Klinkhieo S (2009) A two-level approach to fault-tolerant control of distributed systems based on the sliding mode. In: Proceedings of the IFAC symposium on fault detection, supervision and safety of technical processes, pp 1043–1048

Patton RJ, Kambhampati C, Casavola A, Zhang P, Ding S, Sauter D (2007) A generic strategy for fault-tolerance in control systems distributed over a network. Eur J Control 13(2):280–296

Sauter D, Boukhobza T, Hamelin F (2006) Decentralized and autonomous design for FDI/FTC of networked control systems. IFAC Proc Vol 39(13):138–143

Shafai B, Ghadami R, Saif M (2011) Robust decentralized PI observer for linear interconnected systems. In: Proceedings of the IEEE international symposium on computer-aided control system design, IEEE, Denver, CO, United States, pp 650–655

Šiljak DD (1991) Decentralized control of complex systems. Academic, Boston

Šiljak DD, Zečević AI (2005) Control of large-scale systems: beyond decentralized feedback. Annu Rev Control 29(2):169–179

Tlili AS, Braiek NB (2009) Decentralized observer based guaranteed cost control for nonlinear interconnected systems. Int J Control Autom 2(2):19–34

Yang H, Jiang B, Staroswiecki M, Zhang Y (2015) Fault recoverability and fault tolerant control for a class of interconnected nonlinear systems. Automatica 54:49–55

Chapter 11
Summary and Future Perspectives

11.1 Summary of the Book

This book addresses the problem of robust integration of model-based FE and FTC, aiming to establish effective closed-loop fault-tolerant dynamic systems in the presence of system uncertainties, disturbances and faults.

The first part (Chaps. 1–2) provides necessary background of fault diagnosis and FTC, summarizes the current literature and illustrates the importance and challenges of robust integration. Both theoretic analysis and motivating example are used for illustration. Several important concepts are discussed in this part, including *unidirectional robustness interaction*, *bidirectional robustness interactions* and robust integration.

The second part (Chaps. 3–7) describes five different robust integration strategies for linear systems: sequential strategy, iterative strategy, simultaneous strategy, robust decoupling strategy and adaptive decoupling strategy. Each strategy comes with one or more tutorial examples to demonstrate its effectiveness. Discussions of the cons and pros of each strategy are also provided at the end of each chapter. For ease of reading, the main points of the discussions are summarized below:

- The sequential strategy (see Chap. 3) uses a two-step manner to reduce the design complexity. This strategy is applicable to a wide range of dynamic systems. Compared to the separated strategy in the literature, the sequential strategy is advantageous because the effects of system uncertainty on FE are addressed. However, it covers only the *unidirectional robustness interaction* because the effects of FE uncertainty on the FTC system are ignored. Moreover, this strategy can only obtain a suboptimal solution to the robust integration problem.
- The iterative strategy (see Chap. 4) builds on the sequential strategy but can address the *bidirectional robustness interactions*. The FTC controller and FE observer are designed iteratively until the robust performance index reaches its prescribed accuracy. Although converging in finite iterations, the iterative procedure normally

© The Author(s), under exclusive license to Springer Nature Switzerland AG 2021

J. Lan and R. J. Patton, *Robust Integration of Model-Based Fault Estimation and Fault-Tolerant Control*, Advances in Industrial Control,
https://doi.org/10.1007/978-3-030-58760-4_11

needs more computational power, making it possibly less preferable for online implementation. Moreover, similar to the sequential strategy, the iterative strategy can only generate suboptimal solutions to the robust integration problem.
- The simultaneous strategy (see Chap. 5) has the key advantage of determining the optimal FE observer and FTC controller in one shot. However, the formulated optimization problem is in a BMI form that is difficult to solve using the off-the-shelf solvers. Although the BMI can be linearized as LMI using the equality constraint or Young inequality, it comes with design conservativeness. Moreover, the one-shot design reduces the design freedom. This can be a drawback in some applications where the users need more freedom to tune the observer and controller gains, while optimal solutions are less concerned.
- The robust decoupling strategy (see Chap. 6) can avoid the BMI issue and enjoy more design freedom, by approximately recovering the Separation Principle in the spirit of Small Gain Theorem. Even though the FE observer and FTC controller are designed separately, the *bidirectional robustness interactions* are addressed. This is advantageous when compared with the sequential strategy. Under the robust decoupling scheme, an iterative algorithm is further adopted to improve attenuation of the coupling effects. The iteration also gives a chance to balance the robustness against external disturbance and the coupling.
- The adaptive decoupling strategy (see Chap. 7) also approximately recovers the Separation Principle allowing the FE and FTC designs separated. This offers great design freedom. Moreover, the perturbations are estimated and compensated using the proposed controller, which improves the control robustness. Compared to all the other robust integration strategies, this strategy can handle larger classes of faults, including differentiable, non-differentiable, matched and unmatched faults. However, in order to use the adaptive decoupling strategy, all the system uncertainties and disturbances need to be lumped into a single term. Moreover, distribution matrices of the lumped perturbation signals have to satisfy certain rank assumptions to ensure observability.

The third part (Chaps. 8–10) shows extensions and further developments of the above strategies for nonlinear and large-scale systems. The strategies are applied to wind turbine pitch control system, nonlinear 3-DOF helicopter system and multi-area power systems.

11.2 Future Perspectives

This book has provided a first study of the important subject "robust integration of FE and FTC". However, there are still many open questions, which include but are not limited to the following:

- *Robust integration of FE and FTC for large-scale systems*. The principal examples of large-scale systems are critical infrastructures, such as electric power systems, water distribution networks, transportation systems, as well as systems for emergency management and response (Kyriakides and Polycarpou 2014). Such systems are essential for economic growth and social well-being. However, they are normally large scale, complex in operation, time-varying, uncertain, heterogeneous and safety critical. Moreover, they are prone to various sources of faults due to the increased complexity. Therefore, it is important and challenging to develop integrated FE and FTC strategies for large-scale systems. Furthermore, the strategies must be quick, effective and economical to prevent serious system failure and economical loss.

- *Robust integration of FE and FTC for systems with co-existence of faults and cyber attacks*. Modern controlled systems use information flow and communication extensively, making them exposed to potential malicious cyber attacks. Typical attacks are denial of service attacks, replay attacks, false data injection attacks, zero dynamics attacks and covert attacks. Therefore, it is important to design algorithms to detect and mitigate the attacks, especially for safety-critical systems. Many studies have been conducted on the detection of cyber attacks from a control perspective, details of which can be found in the recent survey paper (Sánchez et al. 2019). However, the situation becomes more challenging when there is a co-existence of faults and cyber attacks. Some attacks may influence the dynamic systems in a way similar to faults, e.g. the false data injection attacks are similar to the additive actuator or sensor faults. The detection of certain attacks is difficult when the attackers have enough knowledge of the system dynamics, e.g. the zero dynamics attacks use the system invariant zeros to make themselves undetectable, and the covert attacks hide the state attacks from the output measurement. Therefore, it is important and challenging to design effective FE and FTC strategies that are able to estimate and mitigate both faults and attacks for dynamic systems, especially safety-critical systems.
- *Robust integration of data-driven FE and FTC*. The robust integration strategies developed in this book lay the basis on the availability of mathematical system models. In practice, it may be difficult or economically expensive to obtain the system models through identification or physical modelling. Hence, there is a line of research aiming to overcome the requirement of accurate mathematical modelling, by using data-driven designs. Data-driven approaches can also take full advantage of the abundant data available from past observations and those collected online in real time to improve the monitoring and control performances. Several data-driven fault diagnosis and FTC methods have been published, see for example, the monographs (Ding 2014; Jain et al. 2018) and the survey papers (MacGregor and Cinar 2012; Jiang et al. 2019, 2018). However, the robust integration of data-driven FE and FTC remains open. A similar line of research is learning-based FE and FTC. In the past decade, machine learning has been the spotlight of academia and industry. There have been significant developments in this area and the power of machine learning in system modelling and control has been continuously released. Therefore, it is also interesting to investigate learning-based FE and FTC. This can help to make the FE and FTC algorithms more intelligent and adaptive to changes in the system and operating environment.

References

Ding SX (2014) Data-driven design of fault diagnosis and fault-tolerant control systems. Springer, Berlin

Jain T, Yamé JJ, Sauter D (2018) Active fault-tolerant control systems

Jiang Q, Yan X, Huang B (2019) Review and perspectives of data-driven distributed monitoring for industrial plant-wide processes. Ind & Eng Chem Res 58(29):12899–12912

Jiang Y, Yin S, Kaynak O (2018) Data-driven monitoring and safety control of industrial cyber-physical systems: basics and beyond. IEEE Access 6:47374–47384

Kyriakides E, Polycarpou M (2014) Intelligent monitoring, control, and security of critical infrastructure systems, vol 565. Springer, Berlin

MacGregor J, Cinar A (2012) Monitoring, fault diagnosis, fault-tolerant control and optimization: data driven methods. Comput & Chem Eng 47:111–120

Sánchez HS, Rotondo D, Escobet T, Puig V, Quevedo J (2019) Bibliographical review on cyber attacks from a control oriented perspective. Ann Rev Control 48:103–128. Elsevier

Index

© The Editor(s) (if applicable) and The Author(s), under exclusive license to Springer Nature Switzerland AG 2021

J. Lan and R. J. Patton, *Robust Integration of Model-Based Fault Estimation and Fault-Tolerant Control*, Advances in Industrial Control,
https://doi.org/10.1007/978-3-030-58760-4

Printed in the United States
by Baker & Taylor Publisher Services